Proceedings of the

Eighth International Conference *on* Difference Equations *and* Applications

Proceedings of the

Eighth International Conference *on* Difference Equations *and* Applications

EDITED BY

Saber Elaydi
Gerasimos Ladas
Bernd Aulbach
Ondrej Dosly

CRC Press
Taylor & Francis Group
Boca Raton London New York

CRC Press is an imprint of the
Taylor & Francis Group, an **informa** business
A CHAPMAN & HALL BOOK

CRC Press
Taylor & Francis Group
6000 Broken Sound Parkway NW, Suite 300
Boca Raton, FL 33487-2742

First issued in paperback 2019

ISBN-13: 978-1-58488-536-8 (hbk)
ISBN-13: 978-0-367-39293-2 (pbk)

Library of Congress Card Number 2005045766

Library of Congress Cataloging-in-Publication Data

International Conference on Difference Equations and Applications (8th : 2003 : Brno, Czech Republic)
Proceedings of the Eighth International Conference on Difference Equations and Applications / edited by Saber Elaydi ... [et al.].
p. cm.
Includes bibliographical references and index.
ISBN 1-58488-536-X
1. Difference equations--Congresses. 2. Differential equations--Congresses. I. Elaydi, Saber, 1943- . II. Title.

QA431.I15145 2003
515'.625--dc22 2005045766

Visit the Taylor & Francis Web site at
http://www.taylorandfrancis.com

and the CRC Press Web site at
http://www.crcpress.com

In Memory of Bernd Aulbach (1947-2005)

On January 14, 2005, Professor Bernd Aulbach suddenly and unexpectedly passed away at the age of 57. The Institute for Mathematics lost a valued colleague, a respected scientist, and a popular university teacher. We are all mourning his death.

Bernd Aulbach was born in Aschaffenburg on December 23, 1947, where he also went to school and graduated with the German high school diploma (Abitur) in 1967. Subsequently, he studied mathematics with a minor in physics in Würzburg. Early on, he discovered his interest in differential equations and consequently wrote his Master's thesis with Professor H.W. Knobloch on "The Domain of Attractivity of an Asymptotically Stable Solution for Nonautonomous Periodic Differential Equations". He graduated from Würzburg with the Master's degree (Diplom) in 1973 and remained in Würzburg as a scientific assistant to the chair, Professor Knobloch, with whom he completed the Ph.D. degree in 1976 by writing a thesis also on domains of attractivity of stable periodic solutions. He spent the academic year 1978/79 as a Visiting Assistant Professor at State University of New York in Albany. From 1983 until 1986 he had a fellowship from the "Volkswagenwerk Foundation" in order to study the project "Qualitative Analysis of Nonlinear Dynamic Systems by Means of Invariant Manifolds". In the context of this project, his State Doctoral thesis (Habilitationsschrift) "Continuous and Discrete Dynamics near Manifolds of Equilibria" emerged, which also appeared as Lecture Notes published by Springer in 1984. In August 1984, he became a lecturer (Privatdozent) at the University of Würzburg. Subsequently, he was awarded a highly competitive Heisenberg Scholarship from the DFG (German Research Society), which he used to finance a longer stay at the University of California in Berkeley in 1986/87. In 1987, he finally accepted a position at the University of Augsburg. In 1970, still as a student, he married Gudrun Nöll with whom he had one daughter and two sons; the children were born in the years 1971, 1976, and 1980.

As can be inferred from the quoted topic of his first research project, his interest was devoted to the qualitative theory of dynamical systems. And though he remained faithful to this area, it was natural that the emphasis of his work shifted. His main legacy is without doubt his steady effort to regard continuous and discrete dynamics from a common point of view. This begins 1984 with his Habilitationsschrift and culminates 2001 in his presidency of the "International Society of Difference Equations". He was thereby not only concerned with a new unification of different approaches, but also with bringing together scientists who work and perform research to a large extent separately in the areas of "differential equations" and "difference equations". This effort succeeded with the "Sixth International Conference on Difference Equations", which he organized in Augsburg in 2001. Until his death, he was considerably involved with the continuation of this conference series. Hand in hand with this, one must also recognize his activity as an editor of the "Journal of Difference Equations and Applications", to which, however, his editorial activity was not limited: The journals "Differential Equations and Dynamical Systems" and "Nonlinear Dynamics and Systems Theory" are to be mentioned as well. A crucial tool to build the bridge between continuous and discrete dynamics is the "measure chain calculus", and he was probably the first one to recognize its importance: He had already put the decisive basic principles together with his Ph.D. student S. Hilger in the late Eighties. A further emphasis of his research was nonautonomous systems, which do not produce dynamical systems in the classical sense. Here as well, he emphasized the importance of only measurable time dependence, i.e., he was concerned with a unified qualitative theory for both nonautonomous differential equations and nonautonomous difference equations. At all times he was particularly open for applications.

Together with Professor Colonius he also directed the working group "Dynamics and Control of Ordinary Differential Equations" within the scope of Augsburg's Graduate School (Graduiertenkolleg) in "Nonlinear Problems in Analysis, Geometry, and Physics". This Graduate School was granted to Augsburg in 1996 by the DFG, and from the very beginning, Bernd Aulbach was its speaker. It was also mainly due to his dedicated input that the Graduate School was extended twice and thus the maximum support duration was granted. Unfortunately he was unable to see his favorite project (as he admitted) through to its completion in 2005.

As a scientist, Bernd Aulbach was always active and successful, which is documented by his over 60 scientific publications. His text book on "Ordinary Differential Equations", which just appeared in its second edition, is much in demand.

Professor Aulbach was a highly gifted teacher and consequently was very popular among his students. This also resulted in a higher-than-average number of graduate students. His student S. Siegmund received a renowned Emmy Noether Scholarship. Like his lectures, his presentations were characterized by extreme clarity. For this reason, he was invited to many national and international conferences, workshops, and colloquia. In order to participate, he did not hesitate to even take the longest journeys, e.g., to Vietnam, China, or India. As a colleague, Bernd Aulbach was extremely cooperative, and he never refused to serve in committees and advisory commissions. His balanced personality was esteemed, and his ideas advanced many a committee.

The Institute for Mathematics mourns the loss of a universally qualified colleague, a popular teacher, and a friend. We will remember him forever.

Hansjörg Kielhöfer, Augsburg
(translated by Martin Bohner, Rolla)

PREFACE

The Eighth International Conference on Difference Equations and Applications (ICDEA 2003) was held at Masaryk University in Brno, Czech Republic, July 28–August 1, 2003. The idea for this conference was born in 1994 during the First International Conference on Difference Equations, which was organized at Trinity University, San Antonio, TX, May 25–28, 1994. The next three ICDEAs were held at Veszprém (Hungary), Taipei (Republic of China) and Poznań (Poland). In 1999 it was at the University of Frontera-Temuco (Chile), where the 5th ICDEA was held, while the 6th ICDEA was organized by the University of Augsburg (Germany). Finally, ICDEA 2002 was held at Changsha (Republic of China).

At ICDEA 2003, more than 80 participants represented 21 countries. The majority of them presented talks that discussed the theory of difference equations, open problems and conjectures, as well as a wide spectrum of applications. The diversity of theoretical and applied topics presented during this conference is reflected in the proceedings. As with the previous proceedings, all papers in this book have been reviewed and accepted for publication.

The editors would like to thank the local organizing committee for their help with the preparations for the meeting.

Preparation of this volume was partially supported by the Grants and Research Projects 201/01/0079, J07/98/141300001, and MSM0022162409.

CONTRIBUTORS

Adamec, Ladislav, Department of Mathematical Analysis, Masaryk University, Janáčkovo nám. 2a, Brno CZ-66295, Czech Republic
adamec@math.muni.cz

Alves, J.F., Departamento de Matemática, Instituto Superior Técnico, Av. Rovisco Pais, 1049-001, Lisbon, Portugal
jalves@math.ist.utl.pt

Andres, Jan, Palacký University, Dept. of Math. Analysis, Tomkova 40, 779 00 Olomouc, Czech Republic
andres@rics.upol.cz

Aulbach, Bernd, Department of Mathematics, University of Augsburg, D-86135 Augsburg, Germany
aulbach@math.uni-augsburg.de

Baštinec, Jaromír, UMAT FEKT VUT, Technická 8, CZ-616 00 Brno, Czech Republic
bastinec@feec.vutbr.cz

Boichuk, Alexander, Department of Applied Mathematic, University of Žilina, J.M. Hurbana 15, Žilina, Slovak Republic
boichuk@fpv.utc.sk, boichuk@imath.kiev.ua

Camouzis, Elias, The American College of Greece, Dept. of Mathematics, Gravias 6, Aghia Paraskevi 15342, Athens, Greece
ecamouzis@aegean.gr., camouzis@otenet.gr

Cecchi, Mariella, Department of Electr. and Telecom., University of Florence, Via S. Marta 3, 50139 Florence, Italy
cecchi@det.unifi.it

Čermák, Jan, Institute of Mathematics, Brno University of Technology, Technická 2, CZ 61669 Brno, Czech Republic
cermakh@um.fme.vutbr.cz

Cheung, Wing-Sum, Department of Mathematics, University of Hong Kong, Pokfulam Road, Hong Kong
wscheung@hku.hk

Dannan, Fozi, Department of Mathematics, Faculty of Science, Qatar University, P.O.Box 2713 Doha, Qatar
fmdannan@qu.edu.qa

Diblík, Josef, Department of Mathematics and Descriptive Geometry, Faculty of Civil Engineering, Žižkova 17, CZ-602 00 Brno, Czech Republic
diblik.j@fce.vutbr.cz

Došlá, Zuzana, Department of Mathematics, Masaryk University, Janáčkovo nám. 2a, CZ 66295 Brno, Czech Republic
dosla@math.muni.cz

Elaydi, Saber, Department of Mathematics, Trinity University, San Antonio, Texas 78209, USA
selaydi@trinity.edu

Elyseeva, Julia, Moscow State Univ. of Technology, Dept. of Mathematics, Vadkovskii per. 3a, 101 472 Moscow, Russia
elyseeva@mtu-net.ru

Fachada, Jose-Luis, Instituto Superior Tecnico, Dep. Matemática, Av. Rovisco Pais, 1, 1049-001 Lisbon, Portugal
fachada@math.ist.utl.pt

Fernandes, Sara, R. Monte Redondo e Torrão, N.6, 7000 - 736 Évora, Portugal
saf@uevora.pt

Fišer, Jiří, Palacký University, Dept. of Math. Analysis, Tomkova 40, 779 00 Olomouc, Czech Republic
fiser@aix.upol.cz

Hamaya, Yoshihiro, Okayama Univ. of Sci., Dept. of Information Sci., 1-1 Ridai chyo, Okayama 700-0005, Japan
hamaya@icity.or.jp

Hilger, Stefan, Catholic University in Eichstätt, D-85071 Eichstätt, Germany
stefan.hilger@ku-eichstaett.de

Hommel, Angela, Bauahus-University Weimar, Inst. of Math. and Physics, Coudraystrasse 13A, Room 117, D 99423 Weimar, Germany
angela.hommel@bauing.uni-weimar.de

Janglajew, Klara, University of Bialystok, Institute of Mathematics, ul. Akademicka 2, 15-267 Bialystok, Poland
jang@math.uwb.edu.pl

Kobza, Aleš, Department of Mathematics, Masaryk University, Janáčkovo nám. 2a, CZ 66295 Brno, Czech Republic
akob@math.muni.cz

Krause, Ulrich, Department of Mathematics, University of Bremen, Bibliothekstr. MZH, 28334 Bremen, Germany
krause@math.uni-bremen.de

Lukš, Antonín, Palacký University, Laboratory of Quantum Optics, Tř. Svobody 26, 771 46 Olomouc, Czech Republic
luks@optnw.upol.cz

Marini, Mauro, Department of Electr. and Telecom., University of Florence, Via S. Marta 3, 50139 Florence, Italy
marini@ing.unifi.it

Migda, Małgorzata, Poznan University of Technology, Institute of Mathematics, ul. Piotrowo 3a, 60-965 Poznan, Poland
mmigda@math.put.poznan.pl

Musielak, Anna, Poznan University of Technology, Institute of Mathematics, ul. Piotrowo 3a, 60-965 Poznan, Poland
musielak@math.put.poznan.pl

Okazaki, Hiromitsu, Department of Mathematics, Faculty of Education, Kumamoto University, Kumamoto 860, Japan
okazaki@educ.kumamoto-u.ac.jp

Oliveira, Henrique, Instituto Superior Técnico, Av. Rovisco Pais, 1, 1049 001 Lisbon, Portugal
holiv@math.ist.utl.pt

Papaschinopoulos, Garyfalos, Democritus Univ. of Thrace, Department of Electr. and Comp. Engin., 67 100 Xanthi, Greece
gpapas@ee.duth.gr

Peřinová, Vlasta, Palacký University, Laboratory of Quantum Optics, Tř. Svobody 26, 771 46 Olomouc, Czech Republic
perinova@optnw.upol.cz

Pötzsche, Christian, Department of Mathematics, University of Augsburg, D-86135 Augsburg, Germany
christian.poetzsche@math.uni-augsburg.de

Ramos, J. Sousa, Instituto Superior Tecnico, Dep. Matemática, Av. Rovisco Pais, 1, 1049-001 Lisbon, Portugal
sramos@math.ist.utl.pt

Rasmussen, Martin, Department of Mathematics, University of Augsburg, D-86135 Augsburg, Germany
rasmussen@math.uni-augsburg.de

Růžičková, Miroslava, Department of Applied Mathematic, University of Žilina, J.M. Hurbana 15, Žilina, Slovak Republic
ruzickova@fpv.utc.sk

Sacker, Robert J., Department of Mathematics, University of Southern California, 1042 Downey Way, Los Angeles, CA 90089-1113, USA
rsacker@math.usc.edu

Schinas, Christos, Democritus University of Thrace, School of Engineering, 67 100 Xanthi, Greece
cshinas@pme.duth.gr

Schmeidel, Ewa, Poznan Univ. of Technology, Institute of Mathematics, ul. Piotrowo 3a, 60-965 Poznan, Poland
eschmeid@math.put.poznan.pl

Siegmund, Stefan, Department of Mathematics, University of Frankfurt, D-60325 Frankfurt, Germany
siegmund@math.uni-frankfurt.de

Stefanidou, Gesthimani, Democritus Univ. of Thrace, Dept. of Electrical and Comp. Engin., 671 00 Xanthi, Greece
tfele@yahoo.gr

Takano, Katsuo, Dept. Math., Ibaraki University, Faculty of Science, Mito, Ibaraki 310, Japan
ktaka@mx.ibaraki.ac.jp

Valeev, Kim, Department of Mathematics, Kiev National University of E-conomics, 252057 Kiev, Ukraine

Vinagre, Sandra, University of Évora, Rua Romão Ramalho 59, 7000 Évora, Portugal
smv@uevora.pt

Contents

Simple Floquet Theory on Time Scales

LADISLAV ADAMEC[1]
Department of Mathematics, Masaryk University Brno
BRNO, Czech Republic
and
Mathematical Institute, Czech Academy of Sciences
BRNO, Czech Republic.

Abstract In this paper we introduce a possible representation of solutions of a simple first order dynamic equation on some time scales that are not more complicated than, e.g., the classical middle-third Cantor set. By using this representation we investigate properties of solutions of some periodic linear dynamic equations on such time scales.

Keywords Floquet theory, time scales, linear equations, unification of difference, and differential calculus

AMS Subject Classification 39A10

1 Introduction

In this note we will intentionally limit ourselves to scalar equations. Most results presented here are valid if the scalar valued function $a(t)$ in (1) is replaced by an $n \times n$ matrix valued function $A(t)$ with only minor changes in proofs. The only exception is Equation (6), where the change from the single valued equation to the system of n^2 unknowns leads to substantial complications and will appear in another paper.

We shall study the linear equation

$$x^{\Delta} = a(t)x, \qquad x \in \mathbb{R},\ t \in \mathbb{T}, \tag{1}$$

where $t \mapsto a(t)$ is an rd-continuous, regressive, and T-periodic function (we tacitly suppose that $T > 0$) such that (1) is uniquely solvable and $\mathbb{T}$ is a suitable time scale. This equation is considered to be well known and the unique solution of (1) fulfilling the initial condition $x(t_0) = 1$ is denoted by $e_a(\cdot, t_0)$ and called the *(generalized) exponential function.* Many properties of $e_a(\cdot, t_0)$ are known (e.g., [4], [5]), still there is much to answer, even in the case when $a(t) \equiv a(= Const)$. One such question is concerned with an inverse mapping to $e_a(\cdot, t_0)$ called a *(generalized) logarithm.*

[1]Supported by the Czech Grant Agency under grant 201/01/0079.

2 The Floquet theory

To see what we are looking for, let us recall the following classical result for ordinary differential equations.

Theorem 2.1. *Let $t \mapsto a(t)$ be a T-periodic function defined on $\mathbb{R}$. Then every nontrivial solution $y(t)$ of the differential equation $y' = a(t)y$ has the form*

$$y(t) = p(t)e^{bt}, \qquad b > 0, \tag{2}$$

where $p(t + 2T) \equiv p(t)$ for all $t \in \mathbb{R}$.

Formula (2) is called *the Floquet normal form* of the solution $y(t)$. Theorem 2.1 is a well-known and, in our setting, rather elementary assertion of the geometric theory of ODEs (e.g. [8]). A similar assertion for difference equations is, in the scalar case, again an easy result of the geometric theory of DEs (e.g., [6]).

However for dynamical systems on time scales the situation is not entirely clear. Recently C.D. Ahlbrandt and J. Ridenhour, in their paper [3] published a version of Floquet theorem on time scales. They proposed the (complex) Floquet representation of the fundamental matrix in the form $p(t)e^{bt}$, which is, in the case $\mathbb{T} = \mathbb{N}$, different from the form $p(n)b^n$ proposed in [6] (for maps). We believe, that the solution in the spirit of the time scales theory could be something like $p(t)e_b(t, t_0)$. Now the questions are

1. Is such representation possible?

2. Does such b exist?

To (particularly) solve this question we shall start with the development of Floquet theory on time scales. If we want to formulate the Floquet theorem on time scales, we must introduce a new assumption concerning "the time space" $\mathbb{T}$.

Definition 2.2. *A time scale $\mathbb{T}$ is called* periodic with a period T (T-periodic) *if it is invariant with respect to T-translations ($t \in \mathbb{T} \Longrightarrow t + T \in \mathbb{T}$).*

Clearly both time scales $\mathbb{R}$ and $\mathbb{Z}$ are periodic, the first one with any period T, the second one with the smallest period $T = 1$. Not every time scale is periodic, so we will henceforth suppose that the time scale $\mathbb{T}$ is periodic with the period $T > 0$. Clearly we *cannot* suppose that T is the smallest period of $\mathbb{T}$. If $\phi(t)$ is an arbitrary solution of (1), then the function $\psi(t) := \phi(t+T)$ is again a solution of (1) as

$$\psi^{\Delta}(t) = \phi^{\Delta}(t+T) = a(t+T)\phi(t+T) = a(t)\psi(t),$$

so there is a constant $c \neq 0$ such that

$$\psi(t) \equiv \phi(t)c,$$

for all $t \in \mathbb{T}$. We can write $c := \phi^{-1}(t_0)\psi(t_0) = \phi^{-1}(t_0)\phi(t_0+T)$. Therefore

$$\phi(t_0+2T) = \phi((t_0+T)+T) = \phi(t_0+T)c = \phi(t_0)c^2.$$

From the well-known proof of the Floquet theorem for ODEs and DEs it is clear that we would like to prove that a constant $b > 0$ exists such that

$$c^2 = e_b(t_0+2T, t_0). \tag{3}$$

If this is true, we can proceed as follows. Let us write

$$p(t) := \phi(t)\frac{1}{e_b(t,t_0)} = \phi(t)e_b(t_0,t),$$

then

$$\begin{aligned} p(t+2T) &= \phi(t+2T)e_b(t_0,t+2T) \\ &= \phi(t)c^2 e_b(t_0,t+2T) \\ &= \phi(t)e_b(t_0+2T,t_0)e_b(t_0,t+2T) \\ &= \phi(t)e_b(t_0+2T,t+2T) \\ &= \phi(t)e_b(t_0,t) \\ &= p(t); \end{aligned}$$

therefore there is a $2T$-periodic function $p(t)$ such that

$$\phi(t) = p(t)e_b(t,t_0),$$

which is exactly the proof of the (real) Floquet theorem. It remains to prove the existence of the positive constant b.

3 The real generalized logarithm

Of course (3) is nothing but an equation for b. The usual approach in ODEs or DEs is to explicitly construct b, which is often called "the real (matrix) logarithm of c." This is the most complicated part of Floquet theory. But any known usage of Floquet theory is nonconstructive, so in fact it is sufficient to prove the existence of such b. To do this, let us recall that $e_b(t,t_0)$ is the unique solution of the initial problem

$$x^{\Delta}(t) = bx(t), \qquad x(t_0) = 1, \quad t, t_0 \in \mathbb{T}, \tag{4}$$

whence we have to find a representation of the solution $x(t)$ of (4).

To this end we may replace (4) by an ordinary differential equation on $\mathbb{R}$ as in [1] or [7] (clear but lengthy) or to use a formal approach as in [2]. We shall follow here the second route.

Any time scale $\mathbb{T}$ is a closed set, so for any $t \geq t_0$ there exist uniquely determined *complementary intervals* (a_i, b_i), $i = 0, 1, \ldots, \omega_1(t), (0 \leq \omega_1(t) \leq \infty)$ that are pairwise disjoint, and

$$[t_0, t] \setminus \mathbb{T} = \bigcup_{i=1}^{\omega_1(t)} (a_i, b_i). \tag{5}$$

If $\omega_1(t) = 0$, the right-hand side of (5) is the empty set. We may suppose that the sequence $(b_i - a_i)_{i=1}^{\omega_1(t)}$ is nonincreasing. Therefore

$$[t_0, t] \setminus \bigcup_{i=1}^{\omega_1(t)} (a_i, b_i) = \bigcup_{i=1}^{\omega_2(t)} [c_i, d_i] \cup P,$$

where all intervals $[c_i, d_i]$ are nondegenerated and pairwise disjoint. Again we may suppose that the sequence $(d_i - c_i)_{i=1}^{\omega_2(t)}$ is nonincreasing.

The set P is a nowhere dense set that could be divided to two disjoint sets $P = P_1 \cup P_2$. All points of P_1 are (with respect to $\mathbb{T}$) either isolated or left-dense and right-scattered or left-scattered and right-dense. All such points are boundary points of suitable complementary intervals, in particular P_1 is at most countable.
The set P_2 is the set of all those points of P that are (with respect to $\mathbb{T}$) left- and right-dense, so they are not among the boundary points of complementary intervals.

Hence we can in a unique way (ignoring any ordering) write

$$\mathbb{T} \cap [t_0, t] = \bigcup_{i=1}^{\omega_1(t)} (a_i, b_i) \cup \bigcup_{i=1}^{\omega_2(t)} [c_i, d_i] \cup P_1 \cup P_2,$$

where any of the four sets on the right-hand side $(\bigcup(a_i, b_i), \ldots, P_2)$ could be the empty set.

Let us consider the middle-third Cantor set $\mathcal{C}$. It is well known that $\mathcal{C} \subseteq [0, 1]$ is perfect, totally disconnected, and uncountable and at the same time has Lebesgue measure $m(\mathcal{C}) = 0$. The set $\mathcal{C}$ is an example of a time scale, which suggests that P_2 could be uncountable. Clearly P_2 could be a quite complicated object. To mollify this we will suppose the hypothesis

H $m(P_2) = 0$.

For example, the hypothesis **H** is satisfied for $\mathbb{T} = \mathbb{R}$ or $\mathbb{T} = \mathbb{Z}$; similarly it is satisfied for the periodic extension of the middle-third Cantor set on $\mathbb{R}$, $\mathbb{T} = i + \mathcal{C}$, $i \in \mathbb{Z}$.

It is clear that, if $t \in P_2$, then there exist sequences $(\alpha_i)_{i \geq 1}$ and $(\beta_i)_{i \geq 1}$– subsets of the sets of boundary points of complementary intervals $\{a_i, b_i\}$ such that $\alpha_i \uparrow t$ and $\beta_i \downarrow t$ for $i \to \infty$.

It is possible to prove [2] that at any $t \in M := \bigcup_{i=1}^{\omega(\infty)}[c_i, d_i] \cup P_1$, the function

$$\tilde{x}(t) := \prod_{i=1}^{\omega_1(t)} [1 + (b_i - a_i)b] \exp\left(b \sum_{i=1}^{\omega_2(t)} (d_i - c_i) \right),$$

fulfills the equation $\tilde{x}^{\Delta}(t) = bx(t)$; hence its continuous extension $\tilde{x}(t)$ must be the solution of the initial problem $\tilde{x}^{\Delta} = bx,\ x(t_0) = 1$.

Now, if we consider the function $F : \mathbb{R} \to \mathbb{R}$,

$$F(b) := c^2 - \prod_{i=1}^{\omega_1(2T)} [1 + (b_i - a_i)b] \exp\left(b \sum_{i=1}^{\omega_2(2T)} (d_i - c_i) \right),$$

then any solution of the equation

$$F(b) = 0, \tag{6}$$

is the hoped-for constant b. If any of the index sets

$$\{1, \ldots, \omega_1\}, \quad \text{or} \quad \{1, \ldots, \omega_2\}$$

is empty, then the product, respectively the sum is replaced by 1 respectively by 0. In both cases the existence of such b is guaranteed by the Floquet theory for ODEs or DEs.

The only interesting case is when both index sets are nonempty. But then $F(b)$ is a real analytic (hence continuous) function such that $F((a_1 - b_1)^{-1}) = c^2 > 0$ and $\lim_{b\to\infty} F(b) = -\infty$, so there is a $b \in ((a_1 - b_1), \infty)$ such that $F(b) = 0$. In fact the equation $F(b) = 0$ could have in this case more than one or even infinitely many solutions; hence there is no usual inverse to $e_b(\cdot, t_0)$, which is a rich source of difficulties of the Floquet theory on time scales.

In any case, we have proved the following variant of the (alas one-dimensional) Floquet theorem.

Theorem 3.1. *Suppose that a T-periodic time scale $\mathbb{T}$ fulfills the hypothesis* **H**, *that $a : \mathbb{T} \to \mathbb{R}$ is a T-periodic function that is rd-continuous and regressive. Then any nontrivial solution $\phi(t)$ of (1) has the form $\phi(t) = p(t)e_b(t, t_0)$, where $b > 0$ and $p(t + 2T) = p(t)$ for all $t \in \mathbb{T}$.*

References

[1] L. Adamec, A remark on matrix equation $x^{\Delta} = A(t)x$ on small time scales., *J. Difference Equ. Appl.*, **10** (2004), 1107–1117

[2] L. Adamec, A note on the matrix solution of the problem $X^{\Delta} = AX$, $X(t_0) = I$., *Comput. Math. Appl.* (in press).

[3] C.D. Ahlbrandt, J. Ridenhour, Floquet theory for time scales and Putzer representations of matrix logarithms. *J. Difference Equ. Appl.*, **9** (2003), 77–92.

[4] M. Bohner, A. Peterson, *Dynamic Equations on Time Scales.* Birkhäuser, Boston-Basel-Berlin, 2001.

[5] M. Bohner, A. Peterson (ed.), *Advances in Dynamic Equations on Time Scales.* Birkhäuser, Boston-Basel-Berlin, 2003.

[6] S. Elaydi, *An Introduction to Difference Equations.* Springer-Verlag New York, Inc., 2nd ed., 1999.

[7] B.M. Garay, S. Hilger, Embeddability of some time scale dynamics in ode dynamics. *Nonlinear Analysis*, **47** (2001), 1357–1371.

[8] P. Hartman, *Ordinary Differential Equations.* John Wiley & Sons Inc., New York-London-Sydney, 1964.

A Condition for Transitivity of Lorenz Maps

J.F. ALVES, J.L. FACHADA, and J. SOUSA RAMOS[1]

Departamento de Matemática, Instituto Superior Técnico Lisbon, Portugal

Abstract We introduce in this article a topological invariant, $t(f) \geq 0$, that allows us to study the topological transitivity of a Lorenz map, f, with positive topological entropy. Since it is possible to calculate $t(f)$ with any intended precision, this result supplies an algorithm to detect the topological transitivity for a relevant class of these maps.

Keywords Symbolic dynamics, one-dimensional maps, transitivity, Lorenz maps, kneading theory, topological invariants

AMS Subject Classification 37B10, 37E05, 37A35, 37B20

1 Introduction

Lorenz maps have been studied by several authors (see [5] and [3]) because of the important role they play in the study of the Lorenz attractor (see [6] and [4]). In this paper we study conditions under which we can say the Lorenz maps are topologically transitive. We will answer this question in terms of positivity of a topological invariant $t(f)$, to be defined.

First some basic definitions:

Definition 1.1. *A map $f : I \backslash \{c\} \to I$, with $I = [0,1]$ and $c \in \,]0,1[$, will be called a Lorenz map if: i) f is continuous and strictly increasing on $[0,c[$ and $]c,1]$; ii) $\lim_{x\uparrow c} f(x) = 1$ and $\lim_{x\downarrow c} f(x) = 0$.*

Definition 1.2. *A Lorenz map $f : I \backslash \{c\} \to I$ is topologically transitive on I if for any two open sets $U, V \subset I$ there exists $n \geq 0$ such that $f^n(U) \cap V \neq \emptyset$.*

The usual definition of topological entropy for continuous maps, using (n,ϵ)-separated sets, can be used to define the entropy for piecewise continuous maps. In what follows we shall use the notation $h(f)$ to denote the topological entropy of a Lorenz map f.

[1]Supported by FCT (Portugal) through program POCTI.

We will now introduce the topological invariant $t(f)$. Let $f : I \backslash \{c\} \to I$ be a Lorenz map with positive topological entropy $h(f) = \log(s)$. Define the numbers

$$A = \sum_{n \in S_0} s^{-n} \text{ and } B = \sum_{n \in S_1} s^{-n},$$

where $S_i \subseteq \mathbb{N}$ is defined by

$$S_i = \left\{n \in \mathbb{N} : f^j(i) \neq c, \text{ for all } j < n\right\}, \text{ for } i \in \{0, 1\}. \tag{1}$$

Since $s > 1$, we can define maps $a, b : [0, 1] \to \mathbb{R}$ by

$$a(x) = \sum_{n \in S_0} a_n(x) \text{ and } b(x) = \sum_{n \in S_1} b_n(x),$$

where $a_n, b_n : [0, 1] \to \mathbb{R}$ are the step functions defined by

$$a_n(x) = \begin{cases} s^{-n} & \text{if } x < f^n(0) \\ (1/2)\, s^{-n} & \text{if } x = f^n(0) \\ 0 & \text{if } x > f^n(0) \end{cases}$$

and

$$b_n(x) = \begin{cases} s^{-n} & \text{if } x < f^n(1) \\ (1/2)\, s^{-n} & \text{if } x = f^n(1) \\ 0 & \text{if } x > f^n(1) \end{cases}.$$

Finally we define the topological invariant $t(f) \geq 0$ by setting

$$t(f) = \inf \{t_f(x) : x \in [0, 1]\},$$

where $t_f : [0, 1] \to \mathbb{R}$ is the map defined by

$$t_f(x) = |A\ b(x) - B\ a(x)|.$$

In order to understand the relationship between $t(f)$ and topological transitivity it is convenient to introduce more notation. By an expansive s-Lorenz map we mean a Lorenz map that is piecewise linear with slope $\pm s$ everywhere and $s > 1$. Recall that Parry [8], and later Milnor and Thurston [7], showed that every continuous piecewise monotone map with positive topological entropy $\log(s)$ is semiconjugate to a continuous piecewise linear map with slope $\pm s$ everywhere. With a similar argument (see [9] or [10]), it can be proven that every Lorenz map with positive topological entropy $\log(s)$ is semiconjugate to an expansive s-Lorenz map. Furthermore, if f is topologically transitive, then the mentioned semiconjugation is in fact a conjugation (see [9]). So from now on we shall restrict the discussion to the class of expansive s-Lorenz maps.

We are now in position to state the main result, which shows that, in a certain sense, the topological transitivity of f can be measured by the topological invariant $t(f)$.

Theorem 1.3. *Let f be an expansive s-Lorenz map. If $t(f) > 0$ then f is topologically transitive.*

Before proving Theorem 1.3 we want to show its scope in the following example.

Example 1.4. *The Baker's map $f_s : [0,1] \to [0,1]$, defined by*

$$f_s(x) = \begin{cases} s\left(x - \frac{1}{2}\right) + 1 & 0 \le x < \frac{1}{2} \\ s\left(x - \frac{1}{2}\right) & \frac{1}{2} < x \le 1 \end{cases}, \text{ with } s \in \left]1, 2\right],$$

is discussed in almost all introductory texts on dynamical systems. It is known that f_s is topologically transitive on $[0,1]$ if and only if $s \in \left[\sqrt{2}, 2\right]$.
Now, the Baker´s map is a particular member of the family of maps $f_{s,c}$: $[0,1] \to [0,1]$, defined by

$$f_{s;c}(x) = \begin{cases} s\,(x - c) + 1 & 0 \le x < c \\ s\,(x - c) & c < x \le 1 \end{cases}, \text{ with } c \in \left]0,1\right[\text{ and } 1 < s < \frac{1}{1-c}.$$

Let us notice however that for $c \neq 1/2$ the values of s for which $f_{s,c}$ is topologically transitive are not necessarily the same as in the particular case of the Baker's map. As an example, for $c = 0.49$ there are values of $s < \sqrt{2}$, for which $f_{s,c}$ is topologically transitive. Indeed we can evoke Theorem 1.3 to prove that, for instance, $f_{1.35;0.49}$ is topologically transitive, since simple calculations show that $t\left(f_{1.35;0.49}\right) \ge 0.54 > 0$.

2 Proof of Theorem 1.3

The proof of Theorem 1.3 uses the kneading theory introduced by Milnor and Thurston in [7]. In particular we will use some results of [1], as well as the main result of [2].

In order to establish a connection between $t(f)$ and the topological transitivity of f, let us begin by defining the generating function $\Lambda\,(J;x;z)$. Let $f : I \backslash \{c\} \to I$ be a Lorenz map with $h(f) = \log(s)$. For each $x \in I$, and $J = [a,b] \subseteq I$, define the formal power series (in the indeterminate z)

$$\Lambda\,(J;x;z) = 1 - \sum_{n \ge 0} \gamma_n z^{n+1}$$

with

$$\gamma_n = \#\left\{y \in \left]a,b\right[: f^n(y) = x\right\} + \frac{1}{2}\#\left\{y \in \{a,b\} : f^n(y) = x\right\}$$

and the corresponding radius of convergence

$$r(J;x) = \frac{1}{\overline{\lim_n} \sqrt[n]{\gamma_n}}.$$

It is known (see, for instance, [3]) that

$$r(J;x) \geqslant s^{-1} \text{ and } r(I;c) = s^{-1}. \tag{2}$$

Notice that if $r(J;x) < \infty$, then $\Lambda(J;x;z) \neq 0$ and as an immediate consequence of the definitions we obtain the following:

Proposition 2.1. *Let $f : I\backslash\{c\} \to I$ be a Lorenz map such that $r(J;x) < \infty$, for all $J = [a,b] \subseteq I$ and $x \in I$. Then f is topologically transitive on I.*

So, using Proposition 2.1, Theorem 1.3 follows from the following result:

Proposition 2.2. *Let $f : I\backslash\{c\} \to I$ be an s-Lorenz map with $s > 1$, and $x \in I$. If $t_f(x) > 0$ then $r(J;x) = s^{-1}$, for all $J = [a,b] \subseteq I$.*

In order to prove Proposition 2.2, we begin by recalling the definition of the kneading determinant of a Lorenz map. Let $f : I\backslash\{c\} \to I$ be a Lorenz map, the keading matrix of f is the matrix, with entries in $\mathbb{Z}[[z]]$, defined by

$$M_f(z) = \begin{bmatrix} \sum\limits_{n\in S_0} \alpha(f^n(0)) z^n & \sum\limits_{n\in S_1} \alpha(f^n(1)) z^n \\ \sum\limits_{n\in S_0} \beta(f^n(0)) z^n & \sum\limits_{n\in S_1} \beta(f^n(1)) z^n \end{bmatrix},$$

where S_0 and S_1 are the sets defined in (1), and $\alpha, \beta : [0,1] \to \{-1, 0\}$ are step functions defined by

$$\alpha(y) = \begin{cases} -1 & y > c \\ 0 & y \leq c \end{cases} \text{ and } \beta(y) = \begin{cases} 0 & y \geq c \\ -1 & y < c \end{cases}.$$

The kneading determinant of f, $D_f(z)$, is the formal power series defined by

$$D_f(z) = \det\left(\begin{bmatrix} 1 & 0 \\ 0 & 1 \end{bmatrix} - zM_f(z)\right)$$

(see [9] or [3] for an alternative definition of kneading determinant). Recall that $D_f(z)$ is a holomorphic function in the unit disk $\{z \in \mathbb{C} : |z| < 1\}$. Furthermore if f has positive topological entropy $h(f) = \log(s)$, then s^{-1} is the smallest zero of $D_f(z)$ in $]0,1[$. Notice that the limit $\lim_{z\to s^{-1}} \Lambda(I;x;z)$ may not exist, because the radius of convergence of $\Lambda(I;x;z)$ can be equal to s^{-1}. Nevertheless, the limit

$$\lim_{z\to s^{-1}} D_f(z)\Lambda(I;x;z)$$

always exists. Indeed, this is a consequence of the main identity (see [ASR])

$$\Lambda(I;x;z) = \frac{\det\left(\begin{bmatrix} 1 & 0 & 0 \\ 0 & 1 & 0 \\ 0 & 0 & 1 \end{bmatrix} - zN_f(x;z)\right)}{D_f(z)},$$

where $N_f(x;z)$ denotes the matrix, with entries in $\mathbb{Q}[[z]]$, defined by

$$\begin{bmatrix} \sum\limits_{n\in S_0} \alpha\left(f^n(0)\right) z^n & \sum\limits_{n\in S_1} \alpha\left(f^n(1)\right) z^n & \sum\limits_{n\in S_1} \alpha\left(f^n(1)\right) z^n - \sum\limits_{n\in S_0} \alpha\left(f^n(0)\right) z^n \\ \sum\limits_{n\in S_0} \beta\left(f^n(0)\right) z^n & \sum\limits_{n\in S_1} \beta\left(f^n(1)\right) z^n & \sum\limits_{n\in S_1} \beta\left(f^n(1)\right) z^n - \sum\limits_{n\in S_0} \beta\left(f^n(0)\right) z^n \\ \sum\limits_{n\in S_0} \delta_x\left(f^n(0)\right) z^n & \sum\limits_{n\in S_1} \delta_x\left(f^n(1)\right) z^n & \sum\limits_{n\in S_1} \delta_x\left(f^n(1)\right) z^n - \sum\limits_{n\in S_0} \delta_x\left(f^n(0)\right) z^n \end{bmatrix},$$

and $\delta_x : [0,1] \to \mathbb{R}$ is the step function defined by

$$\delta_x(y) = \begin{cases} 1 & y > x \\ 1/2 & y = x \\ 0 & y < x \end{cases}.$$

Since the entries of $N_f(x;z)$ are convergent in $\{z \in \mathbb{C} : |z| < 1\}$, and $s > 1$, we obtain

$$\lim_{z\to s^{-1}} D_f(z)\Lambda\left(I;x;z\right) = \det\left(\begin{bmatrix} 1 & 0 & 0 \\ 0 & 1 & 0 \\ 0 & 0 & 1 \end{bmatrix} - s^{-1}N_f(x;s^{-1})\right). \tag{3}$$

We can now prove the following lemma:

Lemma 2.3. *Let $f : I\backslash\{c\} \to I$ be a Lorenz map with positive topological entropy $h(f) = \log(s)$, and $x \in I$. If $t_f(x) > 0$ then $r(I;x) = s^{-1}$.*

Proof. Simple computations show that $s\left|\det\left(Id - s^{-1}N_f(x;s^{-1})\right)\right| = t_f(x)$, and from (3) we obtain $\lim_{z\to s^{-1}} D_f(z)\Lambda\left(I;x;z\right) \neq 0$. Since $D_f(s^{-1}) = 0$, it follows that $\Lambda\left(I;x;z\right)$ does not converge at s^{-1}, therefore $r(I;x) \leqslant s^{-1}$, and from (2) it follows that $r(I;x) = s^{-1}$. □

Remark that for a general Lorenz map f , with $h(f) > 0$, we can have $r(I;x) = s^{-1}$ and $r(J;x) > s^{-1}$ for some interval $J \subseteq I$. We will see, however, that if f is an s-Lorenz map this cannot happen. Proposition 2.2 follows from Lemmas 2.3 and 2.4.

Lemma 2.4. *Let $f : I\backslash\{c\} \to I$ be an expansive s-Lorenz map, and $x \in I$. If $r(I;x) = s^{-1}$ then $r(J;x) = s^{-1}$, for all $J = [a,b] \subseteq I$.*

To prove Lemma 2.4 we have to recall the notion of semiconjugacy.

Definition 2.5. *Let $f : I\backslash\{c\} \to I$ be a Lorenz map with positive topological entropy $h(f) = \log(s)$. A continuous, onto and increasing (not necessarily strictly increasing) map $\lambda : I \to I$ will be called an s-semiconjugacy of f if there exists an s-Lorenz map $g : I\backslash\{d\} \to I$ such that*

$$\lambda \circ f = g \circ \lambda.$$

It is known that for any Lorenz map $f : I \setminus \{c\} \to I$ with positive topological entropy $\log(s)$ there exists at least an s-semiconjugacy. We will see next that for the proof of Lemma 2.4 it is important to discuss the *unicity* of such a semiconjugacy. Indeed we have the following:

Lemma 2.6. *Let $f : I \setminus \{c\} \to I$ be an expansive s-Lorenz map, and $x \in I$ such that: 1) $r(I;x) = s^{-1}$; 2) there exists an interval $J \subseteq I$ such that $r(J;x) > s^{-1}$. Then there exist at least two different s-semiconjugacies of f.*

Proof. A standard argument (see [9] or [7]) allows us to define a s-semiconjugacy of f by setting

$$\lambda(x) = \lim_{z \to s^{-1}} \frac{\Lambda([0,x];x;z)}{\Lambda(I;x;z)},$$

for all $x \in I$. Since $r(J;x) > s^{-1}$, the same arguments also show that λ is constant on J. Thus, since f is a s-linear Lorenz map, the identity map $i : I \to I$ and $\lambda : I \to I$ are different s-semiconjugacies of f. □

So the set of all s-semiconjugacies of f plays a relevant role in the study of topological transitivity of f. On the other hand it was shown in [2] that this set can be characterized in terms of the real matrix $M(f)$, defined by

$$M(f) = M_f\left(s^{-1}\right)^T = \begin{bmatrix} \sum\limits_{n\geq 0} \alpha\left(f^n(0)\right) s^{-n} & \sum\limits_{n\geq 0} \beta\left(f^n(0)\right) s^{-n} \\ \sum\limits_{n\geq 0} \alpha\left(f^n(1)\right) s^{-n} & \sum\limits_{n\geq 0} \beta\left(f^n(1)\right) s^{-n} \end{bmatrix}.$$

The following result (with the obvious modifications) can be found in [2].

Theorem 2.7. *Let $f : I \setminus \{c\} \to I$ be a Lorenz map with $h(f) = \log(s) > 0$. Then: 1) s is an eigenvalue of $M(f)$; 2) If the eigenspace $Ker(M(f) - sI)$ is one-dimensional then there exists one and only one s-semiconjugacy λ of f.*

Note that, under the conditions of the previous theorem, it is easy to prove that $\dim Ker(M(f) - sI) = 1$. In fact, since $M(f)$ is a 2×2 matrix, if we had $\dim Ker(M(f) - sI) = 2$, then we would have

$$M(f) = \begin{bmatrix} s & 0 \\ 0 & s \end{bmatrix},$$

but this is a contradiction because $\sum\limits_{n\geq 0} \alpha\left(f^n(1)\right) s^{-n} \neq 0$. So, we can conclude the following.

Corollary 2.8. *Let $f : I \setminus \{c\} \to I$ be a Lorenz map with $h(f) = \log(s) > 0$. Then there exists one and only one s-semiconjugacy of f.*

And this finishes the proof of the Theorem 1.3, since Lemma 2.4 follows from Lemma 2.6 and Corollary 2.8.

References

[1] Alves, J.F. and Sousa Ramos, J.: Kneading theory: a functorial approach. *Comm. Math. Physics*, **204** (1999), 89–114.

[2] Alves, J.F. and Sousa Ramos, J.: One-dimensional semiconjugacy revisited. *International Journal of Bifurcation and Chaos*, **13** (2003), 1657–1663.

[3] Glendinning, P. and Hall, T.: Zeros of the kneading invariant and topological entropy for Lorenz maps. *Nonlinearity* **9** (1996), 999-1014.

[4] Guckenheimer, J. and Williams, R. F.: Structural stability of Lorenz attractors, *Publ. Math. IHES*, **50** (1979), 59–72.

[5] Hubbard, J. and Sparrow, C.: The classification of topologically expansive Lorenz maps. *Communications on Pure and Applied Mathematics*, **XLIII** (1990), 431–443.

[6] Lorenz, E.: Deterministic non-periodic flow, *J. Atmos. Sci.*, **20** (1963), 130–141.

[7] Milnor, J. and Thurston, W.: On iterated maps of the interval. *In Dynamical Systems: Proceedings 1986–1987*, 465–563. Springer-Verlag, 1988. Lecture Notes in Mathematics 1342.

[8] Parry, W.: Symbolic dynamics and transformations of unit interval. *Trans. Amer. Math. Soc.* **122** (1964), 368–378.

[9] Preston, C.: What you need to knead. *Advances in Math.* **78** (1989), 192–252.

[10] Silva, Luis and Sousa Ramos, J.: Topological invariants and renormalization of Lorenz maps. *Phys. D*, **162** (2002), 233–243.

Iterated Multifunction Systems

JAN ANDRES[1] and JIŘÍ FIŠER[1,2]

Department of Mathematical Analysis
Faculty of Science, Palacký University
Olomouc-Hejčín, Czech Republic

Abstract Our recent results concerning multivalued fractals are presented. The existence, uniqueness and multiplicity (including topological structure) criteria are based on the application of various fixed-point theorems to certain induced operators in hyperspaces.

Keywords Multivalued fractals, metric fractals, topological fractals, existence criteria, multiplicity criteria, topological structure, fixed points, hyperspaces

AMS Subject Classification 28A80, 34A60, 39B12, 47H09, 54B20

1 Introduction

J. Hutchinson [Hu] and M. Barnsley [Ba] defined fractals as compact invariant subsets of a complete metric space w.r.t. the union of contractions.

Theorem 1.1 (Hutchinson, Barnsley). *Assume that (X, d) is a complete metric space and*

$$\{f_i : X \to X; i = 1, \ldots, n; n \in \mathbb{N}\} \tag{1}$$

is a system of contractions. Then there exists exactly one compact invariant subset $A^* \subset X$ of the Hutchinson-Barnsley map $\bigcup_{i=1}^n f_i(x)$, $x \in X$, called the attractor (fractal) of (1) or, equivalently, exactly one fixed-point $A^* \in \mathcal{K}(X) := \{A \subset X | A \text{ is nonempty and compact}\}$ of the induced Hutchinson-Barnsley operator $\bigcup_{x\in A} \bigcup_{i=1}^n f_i(x)$, $A \in \mathcal{K}(X)$, in the hyperspace $(\mathcal{K}(X), d_H)$.

Remark 1.2. *The proof is apparently based on the Banach contraction principle and since fractals can be obtained as limits (in the Hausdorff metric d_H) of the iterates of the Hutchinson-Barnsley operators, for any $A \in \mathcal{K}(X)$, one usually speaks about iterated function systems (IFS) (1). As observed in [AG1], Theorem 1.1 holds if we replace (1) by multivalued contractions with compact values, namely*

$$\{\varphi_i : X \to \mathcal{K}(X); i = 1, \ldots, n; n \in \mathbb{N}\}, \tag{2}$$

[1]Supported by the Council of Czech Government (J 14/98: 153 100 011).
[2]Delivered by the second author.

where $d_H(\varphi_i(x), \varphi_i(y)) \leq L_i d(x,y)$, *for all* $x, y \in X$, $L_i \in [0,1)$, $i = 1, \ldots, n$. *Thus, we can also speak about iterated multifunction systems (IMS) (2) (whence the title) and about multivalued fractals as compact invariant subsets (attractors) of the related Hutchinson-Barnsley maps.*

2 Metric multivalued fractals

If, instead of the Banach contraction principle, another metric fixed-point theorem is applied, then we will speak about *metric (multivalued) fractals.* Hence, applying the fixed-point theorem of B.E. Rhoades [Rh] for weak contractions, Theorem 1.1 as well as its multivalued analogy in [AG1] can be generalized as follows.

Theorem 2.1 ([AF1]). *Assume that* (X, d) *is a complete metric space and let (2) be a system of weak contractions, i.e.*

$$d_H(\varphi_i(x), \varphi_i(y)) \leq d(x,y) - h(d(x,y)), \text{ for all } x, y \in X, i = 1, \ldots, n,$$

where $h : [0, \infty) \to [0, \infty)$ *is a continuous and nondecreasing function such that*

(i) $h(t) = 0 \iff t = 0$ *(i.e.* $h(t) > 0$, *for* $t \in (0, \infty)$*),*

(ii) $\lim_{t\to\infty} h(t) = \infty$.

Then there exists a unique (metric) multivalued fractal for (2). In other words, the Hutchinson-Barnsley operator $\bigcup_{x\in X} \bigcup_{i=1}^n \varphi_i(x)$, $A \in \mathcal{K}(X)$, *of a weakly contractive system (2) has exactly one fixed-point* $A^* \in (\mathcal{K}(X), d_H)$.

Another possibility consists in approximation of multivalued maps in (2) by (multivalued) contractions. The Hutchinson-Barnsley operator for such maps should, however, contain a certain amount of compactness.

Definition 2.2. *We say that (2) satisfies the Palais-Smale condition if each map* φ_i, $i = 1, \ldots, n$, *sends compact sets onto compact sets and the Hutchinson-Barnsley operator*

$$F^* := \overline{\bigcup_{x\in A} \bigcup_{i=1}^n \varphi_i(x)}, \ A \in \mathcal{K}(X) \tag{3}$$

satisfies:
if $\{A_k\}$, $A_k \subset \mathcal{K}(X)$, *is a sequence with* $d_H(A_k, F^*(A_k)) \to 0$, *then it possesses a subsequence convergent w.r.t.* d_H.

Remark 2.3. *Although the closure in (3) can be omitted in Theorems 1.1 and 2.1 (even more generally, for metric and topological fractals, in the present paper), we need it for condensing IMS below.*

Remark 2.4. *If φ_i, $i = 1, \ldots, n$, are compact upper-semicontinuous maps (with compact values) and (X, d) is a complete metric space, then (2) satisfies (see [AFGL]) the Palais-Smale condition.*

Theorem 2.5 (([AFGL]). *Assume that X is a complete absolute retract and let maps φ_i, $i = 1, \ldots, n$, in (2) be approximated by multivalued contractions (with constants L_k) $\varphi_{i,k} : X \to \mathcal{K}(X)$, $i = 1, \ldots, n$, $k \geq 1$, as follows:*

$$d_H\left(\varphi_{i,k}(x), \varphi_i(x)\right) \leq \min\left\{\frac{1}{k}, c(1 - L_k)\right\},$$

for all $x, y \in X$, $L_k \in [0, 1)$, $i = 1, \ldots, n$, $k \geq 1$, for some constant $c \geq 0$. If (2) satisfies the Palais-Smale condition (see Definition 2.2), then the Hutchinson-Barnsley operator F^* in (3) admits a fixed-point $A^* \in (\mathcal{K}(X), d_H)$ (multivalued fractal). Moreover,

$$\operatorname{Fix} F^* := \{A^* \in \mathcal{K}(X) | A^* \in F^*(A^*)\} \tag{4}$$

is an R_δ-set, i.e. in particular, nonempty, compact and connected.

Let us recall (cf. [AG2])

Definition 2.6. *A metric space X is called an absolute retract (respectively an absolute neighborhood retract) if each homeomorphic embedding $h : X \hookrightarrow Y$ into a metrizable space Y such that $h(X) \subset Y$ is closed is a retract (respectively a neighborhood retract) of Y.*

If there exists a decreasing sequence $\{X_n\}$ of compact absolute retracts X_n such that $X = \bigcap\{X_n | n = 1, 2, \ldots\}$, then X is called an R_δ-set.

Remark 2.7. *In Theorem 2.5, multivalued maps φ_i, $i = 1, \ldots, n$, in (2), satisfying the Palais-Smale condition, can be, e.g., "only" nonexpansive. Theorem 2.5 deals for the first time (there is no analogy in the literature) with the topological structure of the set of fractals.*

3 Topological multivalued fractals

If, instead of metric fixed-point theorems, a topological fixed-point theorem is applied, then we will speak about *topological (multivalued) fractals.* Hence, applying the (Lefschetz-type) fixed-point theorem of A. Granas [Gr] for compact continuous maps, we can formulate a *continuation principle for fractals,* provided (X, d) is a locally connected metric space, and subsequently (see [Cu]) $(\mathcal{K}(X), d_H)$ is an absolute neighborhood retract (see Definition 2.6).

Let

$$F_\lambda(x) := \bigcup_{i=1}^{n} \varphi_i(x, \lambda), x \in X, \lambda \in [0, 1],$$

be a one-parameter family of Hutchinson-Barnsley maps, where $\varphi_i : X \times [0, 1] \to \mathcal{K}(X)$, $i = 1, \ldots, n$, are Hausdorff-continuous compact maps, i.e., continuous w.r.t. the metric d in X and the Hausdorff metric d_H in $\mathcal{K}(X)$. Then we have the induced Hutchinson-Barnsley operators

$$F_\lambda^*(A) := \overline{\bigcup_{x \in A} F_\lambda(x)}, A \in \mathcal{K}(X), \lambda \in [0,1],$$

so that $F_\lambda^* : \mathcal{K}(X) \times [0,1] \to \mathcal{K}(X)$, where F_λ^* becomes a compact (continuous) homotopy (see [AF1] or [AFGL]).

We can, therefore, associate with F_λ^* the generalized Lefschetz number $\Lambda(F_\lambda^*) \in \mathbb{Z}$, and if $\Lambda(F_\lambda^*) \neq 0$, for some $\lambda \in [0,1]$, then the Granas fixed-point theorem (see [Gr, GD]) implies the existence of a fixed-point A^* of F_λ^* (i.e., $F_\lambda^*(A^*) = A^*$), for such a $\lambda \in [0,1]$.

Definition 3.1. *If A^* belongs to some essential Nielsen class (observe that, in view of $\Lambda(F_\lambda^*) \neq 0$, at least one of the Nielsen classes is essential; for more details, see [AG2]), then the fixed-point A^* is called a (homotopically)* essential multivalued fractal *of the IMS $\{\varphi_i(x,\lambda), i = 1, \ldots, n\}$, for a given $\lambda \in [0,1]$.*

We are ready to formulate the following continuation principle for (multivalued) fractals.

Theorem 3.2 ([AFGL]). *Let X be a locally connected metric space (e.g., an absolute neighborhood retract; see Definition 2.6) and $\{\varphi_i : X \times [0,1] \to \mathcal{K}(X), i = 1, \ldots, n\}$ be a system of Hausdorff-continuous compact maps. Then an essential fractal exists for the system $\{\varphi_i(\,.\,,0) : X \to \mathcal{K}(X), i = 1, \ldots, n\}$ if and only if the same is true for $\{\varphi_i(\,.\,,1) : X \to \mathcal{K}(X), i = 1, \ldots, n\}$.*

Remark 3.3. *If X is still connected, then $\mathcal{K}(X)$ becomes an absolute retract (see [Cu]), and subsequently (cf. [AG2]) $\Lambda(F_\lambda^*) = 1$, for every $\lambda \in [0,1]$, by which (cf. (4))* $\mathrm{Fix}\, F_\lambda^* \neq \emptyset$, *for every $\lambda \in [0,1]$.*

Remark 3.4. *Under the assumptions of Theorem 3.2, we can even associate with F_λ^* another homotopic invariant, namely the Nielsen number $N(F_\lambda^*)$, defined as the number of essential Nielsen classes, allowing us to make a lower estimate of the number of fractals for the system $\{\varphi_i : X \times [0,1] \to \mathcal{K}(X), i = 1, \ldots, n\}$; for more details, see [AFGL].*

Remark 3.5. *Theorem 3.2 can be still improved by means of the Conley-type (integer-valued) index defined as a fixed-point index (or other topological invariants) in the induced hyperspaces; for more details, see [AFGL].*

4 Multivalued fractals for condensing IMS

The most general existence criterion can be, however, obtained by means of the Knaster-Tarski fixed-point theorem (see, e.g., [GD, p. 25]). Before we state it, let us recall that a bounded mapping $F : X \to \mathcal{K}(X)$ is *condensing* if $\mu[F(B)] < \mu[B]$, whenever $B \subset X$ is a bounded subset and $\mu[B] > 0$, where μ denotes the (Kuratowski or Hausdorff) *measure of noncompactness*. Compact self-maps or contractions with compact values (in linear spaces, also their sum) are well-known to be condensing. In [AFGL], we have shown that weak contractions with compact values are also condensing. For more details, see, e.g., [AG2].

Theorem 4.1 ([AFGL]). *Assume that (X, d) is a complete metric space and let (2) be an IMS of condensing maps such that $\varphi_i(X)$ is bounded, for every $i = 1, \ldots, n$. Then there exists a (minimal) nonempty, compact, invariant set $A^* \subset X$ w.r.t. the Hutchinson-Barnsley map $\bigcup_{i=1}^{n} \varphi_i(x), x \in X$, i.e., a (minimal) fractal $A^* \in \mathcal{K}(X)$ with $F^*(A^*) = A^*$, where the Hutchinson-Barnsley operator F^* is defined in (3).*

Remark 4.2. *One can readily check that Theorem 4.1 significantly improves the existence part of Theorem 3.2. One can also check (see [AFGL]) that the same is true w.r.t. the existence part of Theorem 2.1. On the other hand, the fractal A^* in Theorem 4.1 need not be essential in the sense of Definition 3.1. Moreover, a lower estimate of the number of fractals need not be available (see Remark 3.4).*

Multivalued fractals can also be generated implicitly by means of differential equations or, more generally, inclusions, just on the basis of Theorem 4.1. IMS (2) is determined here by the related Poincaré translation operators along the trajectories of given systems. These (multivalued) operators are well known, under natural restrictions, to be admissible in the sense of L. Górniewicz and compact (in particular, condensing) on any bounded subdomain of the phase-space (see, e.g., [AG2]).

Hence, consider the not necessarily coupled system of ordinary differential equations (with or without uniqueness) [or, more generally, inclusions]

$$X' = F_i(t, X) \quad [X' \in F_i(t, X)], \; i = 1, \ldots, n, \tag{5}$$

where $F_i : [t_0, \infty) \times \mathbb{R}^n \to \mathbb{R}^n$ are Carathéodory maps [$F_i : [t_0, \infty) \times \mathbb{R}^n \to \mathcal{K}(\mathbb{R}^n)$ are upper-Carathéodory maps with convex (compact) values].

Theorem 4.3 ([AF2, AFGL]). *Assume the existence of locally Lipschitzian (in all variables) Liapunov functions ${}_iV(t, X)$, $i = 1, \ldots, n$, defined for $t \geq t_0$, $|X| \geq R$, where R can be large, such that the conditions*

(i) ${}_iW_1(|X|) \leq {}_iV(t, X) \leq {}_iW_2(|X|)$,

(ii) ${}_iV'_{(5)}(t, X) := \limsup_{h \to 0^+} \dfrac{{}_iV(t + h, X + hF_i(t, X)) - {}_iV(t, X)}{h} \leq -{}_iW_3(|X|)$,

are satisfied for (5), where the wedges ${}_iW_1(r)$, ${}_iW_2(r)$ are continuous increasing functions such that ${}_iW_1(r) \to \infty$, whenever $r \to \infty$, and ${}_iW_3(r)$ are positive continuous functions, $i = 1, \ldots, n$. Then there exists $\omega_0 > t_0$ such

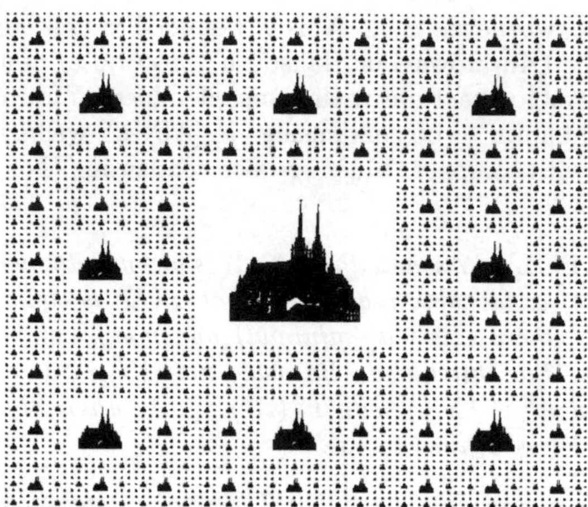

Figure 1: Fractal panorama of Brno

that, for every $\omega \geq \omega_0$*, the Hutchinson-Barnsley operator* F^* *in (3), where* $\varphi_i(x) := \{X_i(\omega, t_0, x) | X_i(t, t_0, x)$ *is a solution of (5) with* $X_i(t_0, t_0, x) = x\}$, $i = 1, \ldots, n$*, admits a fixed-point* $A^* \in (\mathcal{K}(\mathbb{R}^n), d_H)$ *called a (multivalued) fractal generated by (5).*

5 Concluding remarks

For those who are interested in the constructive aspects of fractals (cf. Remark 1.2), we recommend the papers [Ki, LM]. The numerical analysis is performed in [AFGL, Fi]. We restrict ourselves here only to plotting some symbolic fractal picture (see Fig. 1). For more details concerning multivalued fractals, see [An, AF1, AFGL, AG1, AG2, Fi, L1, L2, LM, PR].

References

[An] Andres, J., Some standard fixed-point theorems revisited, *Atti Sem. Mat. Fis. Univ. Modena* **49** (2001), 455–471.

[AF1] Andres, J. and Fišer, J., Metric and topological multivalued fractals, *Int. J. Bifurc. Chaos* **14** (2004), 1277–1289.

[AF2] Andres, J. and Fišer, J., Fractals generated by differential equations, *Dynam. Syst. Appl.* **11** (2002), 471–480.

[AFGL] Andres, J., Fišer, J., Gabor, G. and Leśniak, K., Multivalued fractals, submitted.

[AG1] Andres, J. and Górniewicz, L., On the Banach contraction principle for multivalued mappings, In: *Approximation, Optimization and Mathematical Economics. Proceedings of the 5th International Conference on Approximation and Optimization in the Caribbean, Guadeloupe, French West Indies, March 29–April 2, 1999* (M. Lassonde, ed.), Heidelberg: Physica-Verlag. 1–23, 2001.

[AG2] Andres, J. and Górniewicz, L., *Topological Fixed Point Principles for Boundary Value Problems*, Kluwer, Dordrecht, 2003.

[Ba] Barnsley, M.F., *Fractals Everywhere*, Academic Press, New York, 1988.

[Cu] Curtis, D.W., Hyperspaces of noncompact metric spaces, *Compositio Math.* **40** (1980), 139–152.

[GD] Granas, A. and Dugundji, J., *Fixed Point Theory*, Springer, Berlin, 2003.

[Fi] Fišer, J., Iterated function and multifunction systems, attractors and their basins of attraction, Ph.D. thesis, Palacký University, Olomouc, 2002, in Czech.

[Gr] Granas, A., Generalizing the Hopf-Lefschetz fixed point theorem for non-compact ANR-s, *Sympos. Infinite Dim. Topology, Baton Rouge 1967, Ann. Math. Studies* **69**, 119–130, 1972.

[Hu] Hutchinson, J.E., Fractals and self similarity, *Indiana Univ. Math. J.* **30** (1981), 713–747.

[Ki] Kieninger, B., Iterated function systems on compact Hausdorff spaces, Ph.D. thesis, Institut für Mathematik der naturwissenschaftlichen Fakultät der Universität Augsburg, Germany, June 2002.

[LM] Lasota, A. and Myjak, J., Attractors of multifunctions, *Bull. Pol. Ac. Math.* **48** (2000), 319–334.

[L1] Leśniak, K., Extremal sets as fractals, *Nonlin. Anal. Forum.* **7** (2002), 199–208.

[L2] Leśniak, K., Stability and invariance of multivalued iterated function systems, *Math. Slovaca* **53** (2003), 393–405.

[PR] Petruşel, A. and Rus, I.A., Dynamics on $(P_{cp}(X), H_d)$ generated by a finite family of multi-valued operators on (X, d), *Math. Moravica* **5** (2001), 103–110.

[Rh] Rhoades, B.E., Some theorems on weakly contractive maps, *Nonlinear Analysis* **47** (2001), 2683–2693.

Invariant Manifolds as Pullback Attractors of Nonautonomous Difference Equations

BERND AULBACH and MARTIN RASMUSSEN[1]
Department of Mathematics, University of Augsburg
Augsburg, Germany,
and
STEFAN SIEGMUND[2]
Department of Mathematics, University of Frankfurt
Frankfurt, Germany

Abstract We consider nonautonomous difference equations that give rise to a hierarchy of invariant fiber bundles, the so-called nonautonomous manifolds. It is our aim to point out the relationship between nonautonomous manifolds and pullback attractors, which is also important for the numerical approximation of these manifolds. In a first step we show that the unstable manifold is the pullback attractor of the system. Under the assumption of invertibility, the stable manifold is then related to the pullback attractor of inverted systems. Finally, using spectral transformations, our main result yields that every nonhyperbolic manifold is a pullback attractor of a related system.

Keywords Nonautonomous difference equation, noninvertible difference equation, pullback attractor, invariant manifold

AMS Subject Classification 37B55, 37C70, 37D10, 39A10

1 Introduction

In the theory of discrete dynamical systems (i.e., autonomous difference equations) the stable and unstable manifold theorem – first proved by POINCARÉ [8] and HADAMARD [6] – plays an important role in the study of the flow near a hyperbolic fixed point x_0 of a diffeomorphism f: The stable manifold is the set of points of the phase space that converge to x_0 in forward time; the unstable manifold is the stable manifold of f^{-1}, i.e., it consists of all points

[1]Supported by the Deutsche Forschungsgemeinschaft, grant GK 283.
[2]Supported by the Deutsche Forschungsgemeinschaft, grant Si801/1-3.

that converge to x_0 in backward time. However, the assumption of hyperbolicity of the fixed point is in some situations too restrictive. For nonhyperbolic fixed points, which admit a spectral separation of the linearization, one can show analogous results in which the manifolds are characterized by the growth behavior of the solutions contained in it. In this paper we consider a further generalization of the classical theorem by considering *nonautonomous* difference equations. In this context, the invariant manifolds are no longer subsets of the phase space but of the extended phase space (see AULBACH [1]).

Although their existence is clear, the analytical computation of the invariant manifolds is only possible in rare cases. Therefore it is reasonable to develop numerical tools for the approximation of these sets. For stable and unstable manifolds of autonomous difference equations this topic is well examined. In DELLNITZ, HOHMANN [4, 5] a subdivision and a continuation algorithm is introduced to approximate the global attractor of the system. Since the global attractor contains all unstable manifolds, these algorithms seem convenient to obtain information about the unstable manifolds. We want to address the questions concerning the approximation of nonhyperbolic autonomous manifolds as well as nonautonomous manifolds. We will use a suitable notion of a pullback attractor and nonautonomous versions of the subdivision and continuation algorithm for general nonautonomous dynamical systems (see [2]).

In this article we complete our considerations with respect to nonautonomous difference equations by showing that there exist strong connections between the manifolds and pullback attractors of transformed systems. In a forthcoming paper we will discuss this topic in the context of nonautonomous differential equations. Although there are many similarities, the case of difference equations seems to be more subtle since we do not assume invertibility. This general setting has no consequences for the existence of the manifolds, but our approach for approximating pseudo-stable manifolds is based on time inversion and therefore not applicable to noninvertible systems.

2 Preliminaries

As usual we denote by $\mathbb{Z}$ and $\mathbb{R}$ the sets containing all integers and reals, respectively, and we set $\mathbb{Z}_\kappa^+ := \{k \in \mathbb{Z} : k \geq \kappa\}$ for all $\kappa \in \mathbb{Z}$. With

$$\|(x_1, \ldots, x_N)\| := \sum_{i=1}^{N} |x_i| \quad \text{for all } (x_1, \ldots, x_N) \in \mathbb{R}^N$$

the $\mathbb{R}^N$ is a normed vector space, $\mathbb{R}^{N \times N}$ is the set of all real $N \times N$ matrices. We write $U_\epsilon(x_0) = \{x \in \mathbb{R}^N : \|x - x_0\| < \epsilon\}$ for the ϵ-neighborhood of a point $x_0 \in \mathbb{R}^N$. For arbitrary nonempty sets $A, B \subset \mathbb{R}^N$ and $x \in \mathbb{R}^N$ let $d(x, A) := \inf\{\|x - y\| : y \in A\}$ be the *distance* of x to A and $d(A|B) := \sup\{d(x, B) : x \in A\}$ be the *Hausdorff semidistance* of A and B. A function

g from a set of integers to $\mathbb{R}^N$ is called γ^+-*quasibounded* if

$$\|g\|_{\kappa,\gamma}^+ := \sup\{\|g(k)\|\gamma^{-k} : k \geq \kappa\} < \infty$$

for some $\kappa \in \mathbb{Z}$. Accordingly, the function g is called γ^--*quasibounded* if

$$\|g\|_{\kappa,\gamma}^- := \sup\{\|g(k)\|\gamma^{-k} : k \leq \kappa\} < \infty$$

for some $\kappa \in \mathbb{Z}$.

In this article we are concerned with difference equations

$$x' = f(k, x), \tag{1}$$

where $f : \mathbb{Z} \times \mathbb{R}^N \to \mathbb{R}^N$ is a continuous function. This equation gives rise to a *general solution* $\lambda(k, \kappa, \xi)$, which is the solution satisfying the initial condition $x(\kappa) = \xi$ with κ and ξ treated as additional parameters. Note that in general $\lambda(k, \kappa, \xi)$ is defined only for $k \in \mathbb{Z}_\kappa^+$. However, if the mappings $f(\cdot, k) : \mathbb{R}^N \to \mathbb{R}^N$ are invertible for all $k \in \mathbb{Z}$, then $\lambda(k, \kappa, \xi)$ is also defined for $k < \kappa$. In this case we say that the difference equation (1) is *invertible*.

A subset A of $\mathbb{Z} \times \mathbb{R}^N$ is called a *nonautonomous set*, if for all $k \in \mathbb{Z}$ the so-called *k-fibers* $A(k) := \{x \in \mathbb{R}^N : (k, x) \in A\}$ are nonempty. We call A *closed* or *compact*, if all k-fibers are closed or compact, respectively. Finally, a nonautonomous set A is called *invariant*, if $\lambda(k, \kappa, A(\kappa)) = A(k)$ for all $\kappa \in \mathbb{Z}$ and $k \in \mathbb{Z}_\kappa^+$.

In the literature pullback attractors are usually defined to be compact (see, e.g., KLOEDEN [7]), but the (global) nonautonomous manifolds under our consideration are always noncompact. If we want to establish connections between these two objects, we need a more general notion of a pullback attractor that is prepared by the following definition. An invariant nonautonomous set A is said to be *compactly generated* if there exists a compact set $K \subset \mathbb{R}^N$, a so-called generator of A, with the following property: For any compact set $C \subset \mathbb{R}^N$ there exists a number $\mathcal{T}(K, C) > 0$ such that for any $\kappa \in \mathbb{Z}$ we have

$$\lambda(\kappa, \kappa - k, A(\kappa - k) \cap K) \supset A(\kappa) \cap C \quad \text{for all } k > \mathcal{T}(K, C).$$

An invariant, closed and compactly generated nonautonomous set A is called a *global pullback attractor*, if for any compact set $C \subset X$ we have

$$\lim_{k \to -\infty} d(\lambda(\kappa, k, C) | A(\kappa)) = 0 \quad \text{for all } \kappa \in \mathbb{Z}.$$

A global pullback attractor is always unique. We have proved this result in [2], where we have also introduced algorithms to approximate pullback attractors.

3 Theory of nonautonomous manifolds

In the sequel we consider nonautonomous difference equations of the form

$$\begin{aligned} x_1' &= B_1(k)x_1 + F_1(k, x_1, x_2) \\ x_2' &= B_2(k)x_2 + F_2(k, x_1, x_2), \end{aligned} \tag{2}$$

where $B_1 : \mathbb{Z} \to \mathbb{R}^{N\times N}$, $B_2 : \mathbb{Z} \to \mathbb{R}^{M\times M}$, $F_1 : \mathbb{Z} \times \mathbb{R}^N \times \mathbb{R}^M \to \mathbb{R}^N$ and $F_2 : \mathbb{Z} \times \mathbb{R}^N \times \mathbb{R}^M \to \mathbb{R}^M$ are functions satisfying $F_1(k,0,0) = 0$ and $F_2(k,0,0) = 0$ for all $k \in \mathbb{Z}$. We suppose furthermore that the matrix $B_2(k)$ is invertible for all $k \in \mathbb{Z}$. Moreover we assume:

(H1) Hypotheses on linear part: The evolution operators Φ_1 and Φ_2 of the linear equations $x_1' = B_1(k)x_1$ and $x_2' = B_2(k)x_2$, respectively, satisfy the estimates

$$\begin{aligned} \|\Phi_1(l,m)\| &\leq K\alpha^{l-m} \quad \text{for all } l \geq m, \\ \|\Phi_2(l,m)\| &\leq K\beta^{l-m} \quad \text{for all } l \leq m \end{aligned}$$

with real constants $K \geq 1$ and $0 < \alpha < \beta$.

(H2) Hypotheses on perturbation: For all (x_1, x_2), $(\bar{x}_1, \bar{x}_2) \in \mathbb{R}^N \times \mathbb{R}^M$ and $k \in \mathbb{Z}$ we have

$$\begin{aligned} \|F_1(k,x_1,x_2) - F_1(k,\bar{x}_1,\bar{x}_2)\| &\leq L\|x_1 - \bar{x}_1\| + L\|x_2 - \bar{x}_2\|, \\ \|F_2(k,x_1,x_2) - F_2(k,\bar{x}_1,\bar{x}_2)\| &\leq L\|x_1 - \bar{x}_1\| + L\|x_2 - \bar{x}_2\|, \end{aligned}$$

where the constant L satisfies the estimate $0 \leq L < \frac{\beta-\alpha}{4K}$.

We denote the general solution of this system by

$$\lambda(k,\kappa,\xi,\eta) = (\lambda_1(k,\kappa,\xi,\eta), \lambda_2(k,\kappa,\xi,\eta)) \in \mathbb{R}^N \times \mathbb{R}^M$$

and choose an arbitrary constant $\delta \in (2KL, \frac{\beta-\alpha}{2}]$.

The following theorems are slight modifications of the results obtained in Aulbach [1] and Aulbach & Wanner [3]. First we state the fundamental existence theorem on nonautonomous manifolds, which says that system (2) gives rise to two nonautonomous manifolds, the *pseudo-stable manifold* S_0 and the *pseudo-unstable manifold* R_0. If system (2) is hyperbolic, i.e., $\alpha < 1 < \beta$, then S_0 and R_0 are called *stable manifold* and *unstable manifold*, respectively.

Theorem 3.1. *There exists a uniquely determined continuous mapping $s_0 : \mathbb{Z} \times \mathbb{R}^N \to \mathbb{R}^M$ whose graph*

$$S_0 := \{(\kappa, \xi, s_0(\kappa,\xi)) : \kappa \in \mathbb{Z}, \xi \in \mathbb{R}^N\}$$

allows the representation

$$S_0 = \{(\kappa,\xi,\eta) : \lambda(\cdot,\kappa,\xi,\eta) \text{ is } \gamma^+\text{-quasibounded}\}$$

for every choice of $\gamma \in [\alpha + \delta, \beta - \delta]$. Moreover, there exists a uniquely determined continuous mapping $r_0 : \mathbb{Z} \times \mathbb{R}^M \to \mathbb{R}^N$ whose graph

$$R_0 := \{(\kappa, r_0(\kappa,\eta), \eta) : \kappa \in \mathbb{Z}, \eta \in \mathbb{R}^M\}$$

allows the representation

$$R_0 = \{(\kappa, \xi, \eta) : \text{There exists a solution } \lambda^*(\cdot, \kappa, \xi, \eta) : \mathbb{Z} \to \mathbb{R}^N \times \mathbb{R}^M \text{ of (2)} \\ \text{with } \lambda^*(\kappa, \kappa, \xi, \eta) = (\xi, \eta) \text{ and } \lambda^*(\cdot, \kappa, \xi, \eta) \text{ is } \gamma^-\text{-quasibounded}\}$$

for every choice of $\gamma \in [\alpha + \delta, \beta - \delta]$. *The nonautonomous sets* S_0 *and* R_0 *are invariant and their intersection is the trivial solution of (2).*

We say that S_0 and R_0 are the nonautonomous manifolds of the trivial solution. Not only the trivial solution but every solution of (2) admits nonautonomous manifolds. This is the central statement of the following theorem.

Theorem 3.2. *There exists a mapping* $s : \{(\kappa, \xi, \kappa_0, \xi_0, \eta_0) \in \mathbb{Z} \times \mathbb{R}^N \times \mathbb{Z} \times \mathbb{R}^N \times \mathbb{R}^M : \kappa \geq \kappa_0\} \to \mathbb{R}^M$ *such that for every* $(\kappa_0, \xi_0, \eta_0) \in \mathbb{Z} \times \mathbb{R}^N \times \mathbb{R}^M$ *the graph*

$$S_{\kappa_0, \xi_0, \eta_0} := \{(\kappa, \xi, s(\kappa, \xi, \kappa_0, \xi_0, \eta_0)) : \kappa \in \mathbb{Z}^+_{\kappa_0}, \xi \in \mathbb{R}^N\}$$

allows the representation

$$S_{\kappa_0, \xi_0, \eta_0} = \{(\kappa, \xi, \eta) : \lambda(\cdot, \kappa, \xi, \eta) - \lambda(\cdot, \kappa_0, \xi_0, \eta_0) \text{ is} \\ \gamma^+\text{-quasibounded}, \kappa \geq \kappa_0\}$$

for every choice of $\gamma \in [\alpha + \delta, \beta - \delta]$. *The function* s *is continuous with the following properties:*

(S1) For all $\kappa, \kappa_0 \in \mathbb{Z}$ *with* $\kappa \geq \kappa_0$, $\xi_0, \xi_1, \xi_2 \in \mathbb{R}^N$ *and* $\eta_0 \in \mathbb{R}^M$ *we have*

$$\|s(\kappa, \xi_1, \kappa_0, \xi_0, \eta_0) - s(\kappa, \xi_2, \kappa_0, \xi_0, \eta_0)\| \leq \frac{K^2 L(\delta - KL)}{\delta(\delta - 2KL)} \|\xi_1 - \xi_2\|.$$

(S2) For all $\kappa, \kappa_0 \in \mathbb{Z}$ *with* $\kappa \geq \kappa_0$, $\xi, \xi_0 \in \mathbb{R}^N$, $\eta_0 \in \mathbb{R}^M$ *and* $\gamma \in [\alpha + \delta, \beta - \delta]$ *we have*

$$\begin{aligned} &\|\lambda_1(\cdot, \kappa, \xi, s(\kappa, \xi, \kappa_0, \xi_0, \eta_0)) - \lambda_1(\cdot, \kappa_0, \xi_0, \eta_0)\|^+_{\kappa,\gamma} \\ &\leq \left(K + \frac{K^3 L^2}{\delta(\delta - 2KL)}\right) \|\xi - \lambda_1(\kappa, \kappa_0, \xi_0, \eta_0)\| \gamma^{-\kappa}, \\ &\|\lambda_2(\cdot, \kappa, \xi, s(\kappa, \xi, \kappa_0, \xi_0, \eta_0)) - \lambda_2(\cdot, \kappa_0, \xi_0, \eta_0)\|^+_{\kappa,\gamma} \\ &\leq \frac{K^2 L(\delta - KL)}{\delta(\delta - 2KL)} \|\xi - \lambda_1(\kappa, \kappa_0, \xi_0, \eta_0)\| \gamma^{-\kappa}. \end{aligned}$$

If the difference equation (2) is invertible, then there exists a mapping $r : \mathbb{Z} \times \mathbb{R}^M \times \mathbb{Z} \times \mathbb{R}^N \times \mathbb{R}^M \to \mathbb{R}^N$ *such that for every* $(\kappa_0, \xi_0, \eta_0) \in \mathbb{Z} \times \mathbb{R}^N \times \mathbb{R}^M$ *the graph*

$$R_{\kappa_0, \xi_0, \eta_0} := \{(\kappa, r(\kappa, \eta, \kappa_0, \xi_0, \eta_0), \eta) : \kappa \in \mathbb{Z}, \eta \in \mathbb{R}^M\}$$

allows the representation

$$R_{\kappa_0,\xi_0,\eta_0} = \{(\kappa,\xi,\eta) : \lambda(\cdot,\kappa,\xi,\eta) - \lambda(\cdot,\kappa_0,\xi_0,\eta_0) \textit{ is } \gamma^- \textit{-quasibounded}\}$$

for every choice of $\gamma \in [\alpha+\delta, \beta-\delta]$. *The function* r *is continuous with the following properties:*

(R1) For all $\kappa, \kappa_0 \in \mathbb{Z}$, $\xi_0 \in \mathbb{R}^N$ *and* $\eta_0, \eta_1, \eta_2 \in \mathbb{R}^M$ *we have*

$$\|r(\kappa,\eta_1,\kappa_0,\xi_0,\eta_0) - r(\kappa,\eta_2,\kappa_0,\xi_0,\eta_0)\| \le \frac{K^2L(\delta-KL)}{\delta(\delta-2KL)}\|\eta_1-\eta_2\|.$$

(R2) For all $\kappa, \kappa_0 \in \mathbb{Z}$, $\xi_0 \in \mathbb{R}^N$, $\eta, \eta_0 \in \mathbb{R}^M$ *and* $\gamma \in [\alpha+\delta, \beta-\delta]$ *we have*

$$\begin{aligned}
&\|\lambda_1(\cdot,\kappa,r(\kappa,\eta,\kappa_0,\xi_0,\eta_0),\eta) - \lambda_1(\cdot,\kappa_0,\xi_0,\eta_0)\|^-_{\kappa,\gamma} \\
&\le \frac{K^2L(\delta-KL)}{\delta(\delta-2KL)}\|\eta - \lambda_2(\kappa,\kappa_0,\xi_0,\eta_0)\|\gamma^{-\kappa}, \\
&\|\lambda_2(\cdot,\kappa,r(\kappa,\eta,\kappa_0,\xi_0,\eta_0),\eta) - \lambda_2(\cdot,\kappa_0,\xi_0,\eta_0)\|^+_{\kappa,\gamma} \\
&\le \left(K + \frac{K^3L^2}{\delta(\delta-2KL)}\right)\|\eta - \lambda_2(\kappa,\kappa_0,\xi_0,\eta_0)\|\gamma^{-\kappa}.
\end{aligned}$$

Remark 3.3. *The properties (R1) and (R2) also hold true for the functions* r_0 *and* λ^* *of Theorem 3.1, instead of* r *and* λ, *respectively, even in the case of noninvertibility. This fact is needed in the proof of Theorem 4.1.*

The nonautonomous set

$$F_H(\kappa_0,\eta_0) := S_{\kappa_0, r_0(\kappa_0,\eta_0),\eta_0}$$

is called the *horizontal fiber bundle* through (κ_0, η_0). If the difference equation (2) is invertible, then the *vertical fiber bundle* through (κ_0, ξ_0) is defined by

$$F_V(\kappa_0,\xi_0) := R_{\kappa_0,\xi_0,s_0(\kappa_0,\xi_0)}.$$

The next theorem says – provided the constant L is small enough – that every point of the extended phase space lies on exactly one horizontal and in case of invertibility on one vertical fiber bundle.

Theorem 3.4. *Suppose that for the system (2) the constant* L *satisfies*

$$0 \le L < \frac{\beta-\alpha}{4K^2}(2+K-\sqrt{4+K^2}).$$

Then there exists a continuous mapping $\mathcal{F}_1 : \mathbb{Z} \times \mathbb{R}^N \times \mathbb{R}^M \to \mathbb{R}^M$ *with the following property: Every point* $(\kappa_0, \xi_0, \eta_0) \in \mathbb{Z} \times \mathbb{R}^N \times \mathbb{R}^M$ *lies exactly on one horizontal fiber bundle, namely* $F_H(\kappa_0, \eta)$ *with* $\eta = \mathcal{F}_1(\kappa_0, \xi_0, \eta_0)$. *Furthermore, for all* $\kappa \in \mathbb{Z}, \xi \in \mathbb{R}^N$ *and* $\eta \in \mathbb{R}^M$ *we get the estimate*

$$\|\mathcal{F}_1(\kappa, \xi, \eta)\| \leq \frac{1}{1 - \mathcal{C}(L)^2}(\|\xi\| + \|\eta\|), \tag{3}$$

where the constant $\mathcal{C}(L)$ *is defined by*

$$\mathcal{C}(L) := \frac{K^2 L(\delta - KL)}{\delta(\delta - 2KL)}.$$

If the difference equation (2) is invertible, then there exists a continuous mapping $\mathcal{F}_2 : \mathbb{Z} \times \mathbb{R}^N \times \mathbb{R}^M \to \mathbb{R}^N$ *with the following property: Every point* $(\kappa_0, \xi_0, \eta_0) \in \mathbb{Z} \times \mathbb{R}^N \times \mathbb{R}^M$ *lies exactly on one vertical fiber bundle, namely* $F_V(\kappa_0, \xi)$ *with* $\xi = \mathcal{F}_2(\kappa_0, \xi_0, \eta_0)$. *Furthermore, for all* $\kappa \in \mathbb{Z}, \xi \in \mathbb{R}^N$ *and* $\eta \in \mathbb{R}^M$ *the estimate*

$$\|\mathcal{F}_2(\kappa, \xi, \eta)\| \leq \frac{1}{1 - \mathcal{C}(L)^2}(\|\xi\| + \|\eta\|)$$

holds.

4 Hyperbolic manifolds and pullback attractors

We suppose now that system (2) is hyperbolic. For simplicity we refer to (2) also by writing $x' = f(k, x)$. In the following theorem we provide sufficient conditions concerning the spectral gap, the constant L and the invertibility of the system in order to establish connections between the two hyperbolic manifolds and pullback attractors.

Theorem 4.1. *We suppose that*

$$1 \in (\alpha + 2KL, \beta - 2KL), \tag{4}$$

and that the constant L *satisfies*

$$0 \leq L < \frac{\beta - \alpha}{4K^2}(2 + K - \sqrt{4 + K^2}).$$

Then the unstable manifold R_0 *is the global pullback attractor of (2). If the system (2) is invertible then we have the following connection between the global pullback attractor* A *of the inverted system*

$$x' = f^{-1}(-k - 1, x)$$

and the stable manifold S_0 *of system (2):*

$$A(k) = S_0(-k) \quad \textit{for all } k \in \mathbb{Z}.$$

Proof. First of all we show that R_0 is the global pullback attractor of system (2). We state that R_0 is an invariant nonautonomous set and that R_0 is the graph of the continuous function r_0 and therefore a closed nonautonomous set. Let us prove now that R_0 is compactly generated with generator $\overline{U_1(0)}$. For this we choose an arbitrary compact set $C \subset \mathbb{R}^N$. Since C is bounded, we have an $M > 0$ with $\|x\| < M$ for all $x \in C$. Due to (4) there exists a number $\delta \in (2KL, \frac{\beta-\alpha}{2}]$ with $\gamma := \beta - \delta > 1$. Thus we get a $\mathcal{T}(\overline{U_1(0)}, C) > 0$ such that for all $k > \mathcal{T}(\overline{U_1(0)}, C)$ we have

$$\left(\frac{K^2L(\delta - KL)}{\delta(\delta - 2KL)} + K + \frac{K^3L^2}{\delta(\delta - 2KL)}\right) M\gamma^{-k} < 1. \tag{5}$$

Choose $k_0 \in \mathbb{Z}$ and $(\xi, \eta) = (r_0(k_0, \eta), \eta) \in C \cap R_0(k_0)$ arbitrarily. Then for all $k > \mathcal{T}(\overline{U_1(0)}, C)$ we have

$$\begin{aligned}
&\|\lambda^*(k_0 - k, k_0, r_0(k_0, \eta), \eta)\| \\
&= \|\lambda_1^*(k_0 - k, k_0, r_0(k_0, \eta), \eta)\| + \|\lambda_2^*(k_0 - k, k_0, r_0(k_0, \eta), \eta)\| \\
&\overset{(R2)}{\leq} \left(\frac{K^2L(\delta - KL)}{\delta(\delta - 2KL)} + K + \frac{K^3L^2}{\delta(\delta - 2KL)}\right) \underbrace{\|\eta\|}_{<M} \gamma^{-k} \overset{(5)}{<} 1.
\end{aligned}$$

Thus for all $k > \mathcal{T}(\overline{U_1(0)}, C)$

$$\begin{aligned}
\{(\xi, \eta)\} &= \lambda(k_0, k_0 - k, \{\lambda^*(k_0 - k, k_0, r_0(k_0, \eta), \eta)\}) \\
&\subset \lambda(k_0, k_0 - k, \overline{U_1(0)})
\end{aligned}$$

is fulfilled. Since $(\xi, \eta) \in C \cap R_0(k_0)$ has been chosen arbitrarily we have

$$C \cap R_0(k_0) \subset \lambda(k_0, k_0 - k, \overline{U_1(0)}) \quad \text{for all } k > \mathcal{T}(\overline{U_1(0)}, C),$$

which means that R_0 is compactly generated with generator $\overline{U_1(0)}$. Finally we prove that R_0 attracts every compact set in the sense of pullback attraction. Therefore choose $k_0 \in \mathbb{Z}$ and a compact set $C \subset \mathbb{R}^N$ arbitrarily. Due to (4) there exists a number $\delta \in (2KL, \frac{\beta-\alpha}{2}]$ with $\gamma := \alpha + \delta < 1$. We state that for all $(\xi, \eta) \in C$ and $k \in \mathbb{Z}$ we have

$$\begin{aligned}
&\|r_0(k, \mathcal{F}_1(k, \xi, \eta)) - \xi\| \leq \|r_0(k, \mathcal{F}_1(k, \xi, \eta))\| + \|\xi\| \\
&\overset{(R1)}{\leq} \frac{K^2L(\delta - KL)}{\delta(\delta - 2KL)} \|\mathcal{F}_1(k, \xi, \eta)\| + \|\xi\| \\
&\overset{(3)}{\leq} \frac{K^2L(\delta - KL)}{\delta(\delta - 2KL)} \frac{1}{1 - \mathcal{C}(L)^2} (\|\xi\| + \|\eta\|) + \|\xi\| \leq M_1
\end{aligned}$$

with $M_1 > 0$ since C is bounded. Moreover, for all $(\xi, \eta) \in C$ and $k \in \mathbb{Z}$ the relation

$$s(k, r_0(k, \mathcal{F}_1(k, \xi, \eta)), k, \xi, \eta) = \mathcal{F}_1(k, \xi, \eta)$$

if fulfilled since

$$(k, r_0(k, \mathcal{F}_1(k, \xi, \eta)), \mathcal{F}_1(k, \xi, \eta)) \in S_{k,\xi,\eta} = F_H(k, \mathcal{F}_1(k, \xi, \eta))$$

holds for all $(\xi, \eta) \in C$ and $k \in \mathbb{Z}$. Hence for all $(\xi, \eta) \in C$ and $k \leq k_0$

$$\begin{aligned}
& \|\lambda(k_0, k, r_0(k, \mathcal{F}_1(k, \xi, \eta)), \mathcal{F}_1(k, \xi, \eta)) - \lambda(k_0, k, \xi, \eta)\| \\
= \; & \|\lambda_1(k_0, k, r_0(k, \mathcal{F}_1(k, \xi, \eta)), s(k, r_0(k, \mathcal{F}_1(k, \xi, \eta)), k, \xi, \eta)) \\
& \quad -\lambda_1(k_0, k, \xi, \eta)\| \\
& \| + \lambda_2(k_0, k, r_0(k, \mathcal{F}_1(k, \xi, \eta)), s(k, r_0(k, \mathcal{F}_1(k, \xi, \eta)), k, \xi, \eta)) \\
& \quad -\lambda_2(k_0, k, \xi, \eta)\| \\
\overset{(S2)}{\leq} \; & \left(K + \frac{K^3 L^2}{\delta(\delta - 2KL)}\right) \|r_0(k, \mathcal{F}_1(k, \xi, \eta)) - \xi\| \gamma^{k_0 - k} \\
& + \frac{K^2 L(\delta - KL)}{\delta(\delta - 2KL)} \|r_0(k, \mathcal{F}_1(k, \xi, \eta)) - \xi\| \gamma^{k_0 - k} \\
\leq \; & \left(K + \frac{K^3 L^2}{\delta(\delta - 2KL)} + \frac{K^2 L(\delta - KL)}{\delta(\delta - 2KL)}\right) M_1 \gamma^{k_0 - k} =: M_2 \gamma^{k_0 - k}
\end{aligned}$$

with $M_2 > 0$. We choose an $\epsilon > 0$. Then there exists a $\tilde{k} > 0$ with $M_2 \gamma^{\tilde{k}} < \epsilon$. Thus for all $(\xi, \eta) \in C$ and $k \geq \tilde{k}$

$$\begin{aligned}
& d(\lambda(k_0, k_0 - k, \xi, \eta), R_0(k_0)) \\
\leq \; & \|\lambda(k_0, k_0 - k, r_0(k_0 - k, \mathcal{F}_1(k_0 - k, \xi, \eta)), \mathcal{F}_1(k_0 - k, \xi, \eta)) \\
& \quad -\lambda(k_0, k_0 - k, \xi, \eta)\| \\
\leq \; & M_2 \gamma^k < \epsilon.
\end{aligned}$$

This leads immediately to $d(\lambda(k_0, k_0 - k, C)|R_0(k_0)) < \epsilon$ for all $k \geq \tilde{k}$, and this finishes the proof that R_0 is the global pullback attractor of (2). The verification of the second assertion of the theorem can be done similarly by considering backward time instead of forward time and the vertical fiber bundle instead of the horizontal fiber bundle.

5 Nonhyperbolic manifolds and pullback attractors

In this section we generalize the results of the previous one by omitting the assumption of hyperbolicity. We use spectral transformations to attribute the situation to the hyperbolic case.

Theorem 5.1. *We suppose that there exists a* $c > 0$ *with*

$$1 \in (c(\alpha + 2KL), c(\beta - 2KL)),$$

and that the constant L satisfies

$$0 \le L < \frac{\beta - \alpha}{4K^2}(2 + K - \sqrt{4 + K^2}).$$

We use the spectral transformation $y = c^{k-k_0}x$ ($k_0 \in \mathbb{Z}$) to transform (2) into

$$y' = c^{k+1-k_0} f(k, c^{k_0-k}y). \tag{6}$$

Then the following connection between the global pullback attractor A_1 of system (6) and the pseudo-unstable manifold R_0 of system (2) holds:

$$A_1(k) = c^{k-k_0} R_0(k) \quad \textit{for all } k \in \mathbb{Z}.$$

If the system (2) is invertible, then we have the following connection between the global pullback attractor A_2 of the inverted version of (6),

$$y' = c^{-k-k_0-1} f^{-1}(-k-1, c^{k_0+k}y), \tag{7}$$

and the pseudo-stable manifold S_0 of system (2):

$$A_2(k) = c^{-k-k_0} S_0(-k) \quad \textit{for all } k \in \mathbb{Z}.$$

Proof. We can write system (6) in the following form:

$$\begin{aligned} x_1' &= cB_1(k)x_1 + c^{k+1-k_0} F_1(k, c^{k_0-k}x_1, c^{k_0-k}x_2) \\ x_2' &= cB_2(k)x_2 + c^{k+1-k_0} F_2(k, c^{k_0-k}x_1, c^{k_0-k}x_2). \end{aligned}$$

This system fulfills the assumptions of Section 3, since the evolution operators $\tilde{\Phi}_1$ and $\tilde{\Phi}_2$ of $x_1' = cB_1(k)x_1$ and $x_2' = cB_2(k)x_2$, respectively, satisfy the estimates

$$\begin{aligned} \|\tilde{\Phi}_1(l,m)\| &= \|c^{l-m}\Phi_1(l,m)\| \le K(c\alpha)^{l-m} \quad \text{for all } l \ge m, \\ \|\tilde{\Phi}_2(l,m)\| &= \|c^{l-m}\Phi_2(l,m)\| \le K(c\beta)^{l-m} \quad \text{for all } l \le m, \end{aligned}$$

and for all $k \in \mathbb{Z}$ and (x_1, x_2), $(\bar{x}_1, \bar{x}_2) \in \mathbb{R}^N \times \mathbb{R}^M$ we have

$$\begin{aligned} \|c^{k+1-k_0}(F_1(k, c^{k_0-k}x_1, c^{k_0-k}x_2) &- F_1(k, c^{k_0-k}\bar{x}_1, c^{k_0-k}\bar{x}_2))\| \\ &\le cL\|x_1 - \bar{x}_1\| + cL\|x_2 - \bar{x}_2\|, \\ \|c^{k+1-k_0}(F_2(k, c^{k_0-k}x_1, c^{k_0-k}x_2) &- F_2(k, c^{k_0-k}\bar{x}_1, c^{k_0-k}\bar{x}_2))\| \\ &\le cL\|x_1 - \bar{x}_1\| + cL\|x_2 - \bar{x}_2\|, \end{aligned}$$

where the constant cL satisfies the estimate $0 \le cL < \frac{c\beta - c\alpha}{4K}$. We can apply Theorem 4.1 since $1 \in (c\alpha + 2KcL, c\beta - 2KcL)$. The unstable manifold $\tilde{R}_0$ of system (6) is therefore identical with the global pullback attractor of system (6). To prove the first assertion of this theorem it is thus sufficient to show that

$$\tilde{R}_0(k) = c^{k-k_0} R_0(k) \quad \text{for all } k \in \mathbb{Z}$$

holds. An easy calculation yields that $I \ni k \mapsto \mu(k) \in \mathbb{R}^{N+M}$ is a solution of (2) if and only if $I \ni k \mapsto c^{k-k_0}\mu(k) \in \mathbb{R}^{N+M}$ is a solution of (6). Hence for every $k \in \mathbb{Z}$ we have

$$\begin{aligned}
x \in \tilde{R}_0(k) \quad &\Leftrightarrow \quad \text{There exists a } 1^{-}\text{-quasibounded solution } \tilde{\mu} : \mathbb{Z} \to \mathbb{R}^{N+M} \\
&\qquad \text{of (6) with } \tilde{\mu}(k) = x \\
&\Leftrightarrow \quad \text{There exists a solution } \tilde{\mu} : \mathbb{Z} \to \mathbb{R}^{N+M} \text{ of (6) with } \tilde{\mu}(k) = x, \\
&\qquad \text{and } c^{k_0 - \cdot}\tilde{\mu}(\cdot) \text{ is } \left(\tfrac{1}{c}\right)^{-}\text{-quasibounded} \\
&\Leftrightarrow \quad \text{There exists a } \left(\tfrac{1}{c}\right)^{-}\text{-quasibounded solution } \mu : \mathbb{Z} \to \mathbb{R}^{N+M} \\
&\qquad \text{of (2) with } \mu(k) = c^{k_0 - k}x \\
&\Leftrightarrow \quad c^{k_0 - k}x \in R_0(k) \\
&\Leftrightarrow \quad x \in c^{k-k_0}R_0(k).
\end{aligned}$$

Analogously one shows that for the stable manifold $\tilde{S}_0$ of system (6) the relation

$$\tilde{S}_0(k) = c^{k-k_0}S_0(k) \quad \text{for all } k \in \mathbb{Z}$$

if fulfilled. If the system (2) is invertible, Theorem 4.1 implies that the pull-back attractor A_2 of system (7) satisfies

$$A_2(k) = \tilde{S}_0(-k) = c^{-k-k_0}S_0(-k) \quad \text{for all } k \in \mathbb{Z},$$

if system (7) is the inversion of (6) as introduced in Theorem 4.1. An obvious calculation yields that this is indeed true, so all assertions of the theorem are proved.

Remark 5.2.

- *Specializing* $c = 1$ *we see that Theorem 5.1 is a generalization of Theorem 4.1.*

- *The systems (6) and (7) are only suitable for the approximation of the* k_0*-fiber of the manifolds* R_0 *and* S_0*, respectively, because easy examples show that an exponential transformation of the numerical covering of the pullback attractors* A_1 *or* A_2*, respectively, leads to unusable results (see* Rasmussen *[9]).*

- *Even for the approximation of pseudo-stable or pseudo-unstable manifolds of* autonomous *systems the* nonautonomous *theory is essential since the spectral transformation yields a nonautonomous system.*

6 Hierarchies of nonautonomous manifolds

In this section we generalize our situation by considering the system

$$\begin{aligned} x_1' &= B_1(k)x_1 + F_1(k, x_1, x_2, \dots, x_n) \\ x_2' &= B_2(k)x_2 + F_2(k, x_1, x_2, \dots, x_n) \\ &\vdots \\ x_n' &= B_n(k)x_n + F_n(k, x_1, x_2, \dots, x_n), \end{aligned} \tag{8}$$

where $B_i : \mathbb{Z} \to \mathbb{R}^{N_i \times N_i}$, $F_i : \mathbb{Z} \times \mathbb{R}^N \to \mathbb{R}^{N_i}$ ($i = 1, \dots, n$) are mappings with $F_i(k, 0, 0) = 0$ for all $k \in \mathbb{Z}$ ($N = N_1 + \dots + N_n$). We suppose furthermore that the matrix $B_i(k)$ is invertible for all $k \in \mathbb{Z}$ and $i \in \{2, \dots, n\}$. Moreover we assume:

($\overline{\text{H}}$1) Hypotheses on linear part: There exist real constants $K \geq 1$ and $\alpha_i < \beta_i$ ($i = 1, \dots, n-1$) with $\beta_i \leq \alpha_{i+1}$ ($i = 1, \dots, n-2$) such that the evolution operators Φ_i of the linear equations $x_i' = B_i(k)x_i$ ($i = 1, \dots, n$) satisfy the estimates

$$\begin{aligned} \|\Phi_i(l, m)\| &\leq K\alpha_i^{l-m} \quad \text{for all } l \geq m, \\ \|\Phi_{i+1}(l, m)\| &\leq K\beta_i^{l-m} \quad \text{for all } l \leq m \end{aligned}$$

for all $i = 1, \dots, n-1$.

($\overline{\text{H}}$2) Hypotheses on perturbation: For all $x = (x_1, \dots, x_n), \bar{x} = (\bar{x}_1, \dots, \bar{x}_n) \in \mathbb{R}^N$, $k \in \mathbb{Z}$ and $i \in \{1, \dots, n\}$ we have

$$\|F_i(k, x_1, \dots, x_n) - F_i(k, \bar{x}_1, \dots, \bar{x}_n)\| \leq L \sum_{j=1}^{n} \|x_j - \bar{x}_j\| = L\|x - \bar{x}\|,$$

where the constant L satisfies the estimate

$$0 \leq L < \min\left\{ \frac{\beta_i - \alpha_i}{4K(n-1)} : i \in \{1, \dots, n-1\} \right\}.$$

We choose constants $\delta_i \in (2KL(n-1), \frac{\beta_i - \alpha_i}{2}]$ ($i = 1, \dots, n-1$).

Theorem 6.1. *There exist nonautonomous manifolds $\mathcal{W}_{i,j}$ ($1 \leq i \leq j \leq n$), the so-called hierarchy of nonautonomous manifolds, with the following characterizations:*

- $\mathcal{W}_{1,n} = \mathbb{Z} \times \mathbb{R}^N$.
- *For all $i \in \{1, \dots, n-1\}$ and $\gamma \in [\alpha_i + \delta_i, \beta_i - \delta_i]$ we have*

$$\begin{aligned} \mathcal{W}_{1,i} &= \{(\kappa, \xi) : \lambda(\cdot, \kappa, \xi) \textit{ is } \gamma^+\textit{-quasibounded}\}, \\ \mathcal{W}_{i+1,n} &= \{(\kappa, \xi) : \textit{There exists a solution } \lambda^*(\cdot, \kappa, \xi) : \mathbb{Z} \to \mathbb{R}^N \textit{ of (8)} \\ &\qquad \textit{with } \lambda^*(\kappa, \kappa, \xi) = \xi \textit{ and } \lambda^*(\cdot, \kappa, \xi) \textit{ is } \gamma^-\textit{-quasibounded}\}. \end{aligned}$$

- *For all* $1 < i \leq j < n$ *we have* $\mathcal{W}_{i,j} = \mathcal{W}_{1,j} \cap \mathcal{W}_{i,n}$. *Hence for every choice of* $\gamma_1 \in [\alpha_j + \delta_j, \beta_j - \delta_j]$ *and* $\gamma_2 \in [\alpha_{i-1} + \delta_{i-1}, \beta_{i-1} - \delta_{i-1}]$

$$\begin{aligned}\mathcal{W}_{i,j} = \{(\kappa, \xi) : \; & \text{There exists a solution } \lambda^*(\cdot, \kappa, \xi) : \mathbb{Z} \to \mathbb{R}^N \text{ of (8)} \\ & \text{with } \lambda^*(\kappa, \kappa, \xi) = \xi \text{ and } \lambda^*(\cdot, \kappa, \xi) \text{ is } \gamma_1^+\text{- and} \\ & \gamma_2^-\text{-quasibounded}\}\end{aligned}$$

  is fulfilled.

Proof. For abbreviation we write for $1 \leq i \leq j \leq n$

$$B_{i,j}(k) := \begin{pmatrix} B_i(k) & & & 0 \\ & B_{i+1}(k) & & \\ & & \ddots & \\ 0 & & & B_j(k) \end{pmatrix}, \quad x_{i,j} := \begin{pmatrix} x_i \\ x_{i+1} \\ \vdots \\ x_j \end{pmatrix}$$

and

$$F_{i,j}(k,x) := \begin{pmatrix} F_i(k, x_1, x_2, \ldots, x_n) \\ F_{i+1}(k, x_1, x_2, \ldots, x_n) \\ \vdots \\ F_j(k, x_1, x_2, \ldots, x_n) \end{pmatrix}, \quad N_{i,j} := \sum_{k=i}^{j} N_k.$$

We fix a number $i \in \{1, \ldots, n-1\}$ and write system (8) in the form

$$\begin{aligned} x'_{1,i} &= B_{1,i}(k)x_{1,i} + F_{1,i}(k, x_{1,i}, x_{i+1,n}), \\ x'_{i+1,n} &= B_{i+1,n}(k)x_{i+1,n} + F_{i+1,n}(k, x_{1,i}, x_{i+1,n}). \end{aligned}$$

Now we can apply Theorem 3.1 with the constants K, α_i, β_i and $L(n-1)$, since the evolution operators $\hat{\Phi}_1$ and $\hat{\Phi}_2$ of the linear equations $x'_{1,i} = B_{1,i}(k)x_{1,i}$ and $x'_{i+1,n} = B_{i+1,n}(k)x_{i+1,n}$ satisfy the estimates

$$\begin{aligned} \|\hat{\Phi}_1(l,m)\| &= \max\{\|\Phi_j(l,m)\| : j \in \{1, \ldots, i\}\} \\ &\leq K\alpha_i^{l-m} \quad \text{for all } l \geq m, \\ \|\hat{\Phi}_2(l,m)\| &= \max\{\|\Phi_j(l,m)\| : j \in \{i+1, \ldots, n\}\} \\ &\leq K\beta_i^{l-m} \quad \text{for all } l \leq m \end{aligned}$$

with $K \geq 1$ and $\alpha_i < \beta_i$, and for all $x = (x_{1,i}, x_{i+1,n}), \bar{x} = (\bar{x}_{1,i}, \bar{x}_{i+1,n}) \in \mathbb{R}^{N_{1,i}} \times \mathbb{R}^{N_{i+1,n}}$ and $k \in \mathbb{Z}$ we have

$$\begin{aligned} & \|F_{1,i}(k,x) - F_{1,i}(k,\bar{x})\| \leq iL \sum_{j=1}^{n} \|x_j - \bar{x}_j\| \\ \leq \; & (n-1)L(\|x_{1,i} - \bar{x}_{1,i}\| + \|x_{i+1,n} - \bar{x}_{i+1,n}\|), \\ & \|F_{i+1,n}(k,x) - F_{i+1,n}(k,\bar{x})\| \leq (n-i)L \sum_{j=1}^{n} \|x_j - \bar{x}_j\| \\ \leq \; & (n-1)L(\|x_{1,i} - \bar{x}_{1,i}\| + \|x_{i+1,n} - \bar{x}_{i+1,n}\|). \end{aligned}$$

Therefore we get the nonautonomous manifolds $\mathcal{W}_{1,i} := S_0$ and $\mathcal{W}_{i+1,n} := R_0$ with the asserted properties. For $1 < i \leq j < n$ the set $\mathcal{W}_{i,j} := \mathcal{W}_{1,j} \cap \mathcal{W}_{i,n}$ is a nonautonomous manifold since it is an intersection of two nonautonomous manifolds.

The following diagram visualizes the relations between the manifolds $\mathcal{W}_{i,j}$ of the hierarchy.

$$\begin{array}{ccccccccc}
\mathcal{W}_{1,1} & \subset & \mathcal{W}_{1,2} & \subset & \cdots & \subset & \mathcal{W}_{1,n-1} & \subset & \mathcal{W}_{1,n} \\
 & & \cup & & & & \cup & & \cup \\
 & & \mathcal{W}_{2,2} & \subset & \cdots & \subset & \mathcal{W}_{2,n-1} & \subset & \mathcal{W}_{2,n} \\
 & & & & & & \cup & & \cup \\
 & & & & \ddots & & \vdots & & \vdots \\
 & & & & & & \mathcal{W}_{n-1,n-1} & \subset & \mathcal{W}_{n-1,n} \\
 & & & & & & & & \cup \\
 & & & & & & & & \mathcal{W}_{n,n}
\end{array}$$

We state now the version of Theorem 5.1, which applies to system (8). For simplicity we refer to (8) also by writing $x' = f(k, x)$.

Theorem 6.2. *We suppose that there exists an $i \in \{1, \ldots, n-1\}$ and a $c > 0$ with*

$$1 \in \big(c(\alpha_i + 2KL(n-1)), c(\beta_i - 2KL(n-1))\big),$$

and that the constant L satisfies

$$0 \leq L < \frac{\beta_i - \alpha_i}{4K^2(n-1)}(2 + K - \sqrt{4 + K^2}).$$

We use the spectral transformation $y = c^{k-k_0}x$ ($k_0 \in \mathbb{Z}$) to transform (8) into

$$y' = c^{k+1-k_0} f(k, c^{k_0-k}y). \tag{9}$$

Then the following connection between the global pullback attractor A_1 of system (9) and the manifold $\mathcal{W}_{i+1,n}$ of system (8) holds:

$$A_1(k) = c^{k-k_0}\mathcal{W}_{i+1,n}(k) \quad \text{for all } k \in \mathbb{Z}.$$

If the system (8) is invertible, then we have the following connection between the global pullback attractor A_2 of the inverted version of (9),

$$y' = c^{-k-k_0-1} f^{-1}(-k-1, c^{k_0+k}y),$$

and the manifold $\mathcal{W}_{1,i}$ of system (8):

$$A_2(k) = c^{-k-k_0}\mathcal{W}_{1,i}(-k) \quad \text{for all } k \in \mathbb{Z}.$$

Proof. This result follows directly from Theorem 5.1 by putting the components $1, \ldots, i$ and $i+1, \ldots, n$ together as in the proof of Theorem 6.1.

Remark 6.3. *Provided the constant L of system (8) is small enough and system (8) is invertible, every manifold of the hierarchy can be approximated numerically. For the manifolds $\mathcal{W}_{1,i}$ and $\mathcal{W}_{i+1,n}$ ($i = 1, \ldots, n-1$) this follows from Theorem 6.2. The manifolds $\mathcal{W}_{i,j}$ for $1 < i \leq j < n$ can be obtained by intersecting $\mathcal{W}_{1,j}$ and $\mathcal{W}_{i,n}$ (see Theorem 6.1).*

References

[1] Aulbach, B., The fundamental existence theorem on invariant fiber bundles, *Journal of Difference Equations and Applications* **3** (1998), 501–537.

[2] Aulbach, B., Rasmussen, M., Siegmund, S., Approximation of attractors of nonautonomous dynamical systems, *Discrete and Continuous Dynamical Systems, Series B.* (in press).

[3] Aulbach, B., Wanner, T., Invariant foliations and decoupling of nonautonomous difference equations, *Journal of Difference Equations and Applications* **9** (2003), 459–472.

[4] Dellnitz, M., Hohmann, A., The Computation of unstable manifolds using subdivision and continuation, in: *Progress in Nonlinear Differential Equations and Their Applications* **19**, Birkhäuser (1996), 449–459.

[5] Dellnitz, M., Hohmann, A., A subdivision algorithm for the computation of unstable manifolds and global attractors, *Numerische Mathematik* **75** (1997), 293–317.

[6] Hadamard, J., Sur l'itération et les solutions asymptotiques des équations différentielles, *Bull. Soc. Math. France* **29** (1901), 224–228.

[7] Kloeden, P.E., Pullback attractors in nonautonomous difference equations, *Journal of Difference Equations and Applications* **6** (2000), 91–102.

[8] Poincaré, H., Mémoire sur les courbes définie par une équation différentielle, IV, *J. Math. Pures Appl.* **2** (1886), 151–217.

[9] Rasmussen, M., Approximation von Attraktoren und Mannigfaltigkeiten nichtautonomer Systeme, *Diploma Thesis*, University of Augsburg (2002).

[10] Siegmund, S., Computation of nonautonomous invariant manifolds, in: B. Aulbach, S. Elaydi, and G. Ladas (eds.), *Proceedings of the Sixth International Conference on Difference Equations*, 215–227, CRC, Boca Raton, FL, 2004.

Determination of Initial Data Generating Solutions of Bernoulli's Type Difference Equations with Prescribed Asymptotic Behavior

JAROMÍR BAŠTINEC

Department of Mathematics
Faculty of Electrical Engineering and Communication
Brno University of Technology
Brno, Czech Republic

and

JOSEF DIBLÍK[1]

Department of Mathematics and Descriptive Geometry
Faculty of Civil Engineering
Brno University of Technology
Brno, Czech Republic

Abstract The discrete Bernoulli's type equation

$$u(k+1) = q(k)(u(k))^r - v(k),$$

with variable $k \in N(a)$ and fixed $r \in \mathbb{R}^+$ is considered. A result concerning asymptotic behavior of its solutions for $k \to \infty$ is formulated and corresponding two-sided inequalities for them are given. In the paper we treat the problem of how to determine initial data generating solutions with such asymptotic behavior. It is shown that in some cases it is possible to get corresponding initial data with the aid of limits of two-number auxiliary sequences.

Keywords Bernoulli's type discrete equation, initial data, two-sided inequalities

AMS Subject Classification 39A10, 39A11

[1]Supported by Grant 201/01/0079 of Czech Grant Agency and by the Council of Czech Government MSM 2622000 13 of Czech Republic.

1 Introduction

In this paper we consider the discrete Bernoulli's type equation

$$u(k+1) = q(k)(u(k))^r - v(k), \tag{1}$$

with variable $k \in N(a)$, $N(a) = \{a, a+1, \dots\}$, $a \in \mathbb{N}$, $\mathbb{N} = \{1, 2, \dots\}$ and with fixed $r \in \mathbb{R}^+ := (0, \infty)$. Throughout this paper we suppose $q : N(a) \to \mathbb{R}^+$ and $v : N(a) \to \mathbb{R}_+ := [0, \infty)$. Define $\mathbb{N}^* := \{0\} \cup \mathbb{N}$. We will show that under certain conditions Equation (1) has at least one solution defined on $N(a)$ with prescribed asymptotic behavior. In the paper we treat the problem of how to determine initial data (a, u^*) generating solutions with such a behavior. It is shown that in some cases it is possible to get corresponding initial data (i.e., value u^*) with the aid of limits of two special auxiliary sequences.

1.1 Preliminaries

Put $\Delta u(k) := u(k+1) - u(k)$. Let us consider a scalar discrete equation containing one difference

$$\Delta u(k) = f(k, u(k)) \tag{2}$$

with a one-valued function $f : N(a) \times \mathbb{R} \to \mathbb{R}$ and $a \in \mathbb{N}$. Together with discrete Equation (2) we consider an initial problem. It is posed as follows: for a given $s \in \mathbb{N}^*$ we are seeking the solution $u = u(k)$ of (2) satisfying the initial condition

$$u(a+s) = u_s \in \mathbb{R} \tag{3}$$

with a prescribed constant u_s. Let us recall that the solution of initial problem (2), (3) is defined as an infinite sequence of numbers $\{u^k\}_0^\infty$ with $u^k = u(a+s+k)$, i.e.

$$u^0 = u_s = u(a+s), u^1 = u(a+s+1), \dots, u^n = u(a+s+n), \dots,$$

such that for any $k \in N(a+s)$ the equality (2) holds. Let us note that the existence and uniqueness of the solution of the initial problem (2), (3) is a consequence of properties of the function f. Suppose moreover that f is continuous with respect to the second argument. Then the initial problem (2), (3) depends continuously on its initial data.

Let b, c be real functions defined on $N(a)$ such that $b(k) < c(k)$ for every $k \in N(a)$. Let us define a set $\omega \subset N(a) \times \mathbb{R}$ as

$$\omega := \{(k, u) : k \in N(a),\ u \in \omega(k)\},$$

where

$$\omega(k) := \{(u) : b(k) < u < c(k)\}.$$

Let us define a closure of the set ω as

$$\overline{\omega} := \{(k, u) : k \in N(a),\ u \in \overline{\omega}(k)\},$$

with

$$\overline{\omega}(k) := \{(u) : b(k) \le u \le c(k)\}.$$

The following theorem, concerning an asymptotic behavior of solutions of Equation (2) is an improvement of Theorem 2 from [2] (see [3] as well). Tracing the proof of the mentioned theorem it is easy to see that the condition with respect to f concerning a Lipschitz-type condition with respect to the second argument can be weakened to the condition of continuity with respect to the second argument.

Theorem 1.1. *Let b, c be real functions defined on $N(a)$ such that $b(k) < c(k)$ for every $k \in N(a)$. Let $f : \overline{\omega} \to \mathbb{R}$ be continuous with respect to the second argument. If, moreover, for every $k \in N(a)$*

$$f(k, b(k)) - b(k+1) + b(k) < 0$$

and

$$f(k, c(k)) - c(k+1) + c(k) > 0$$

then there exists an initial condition

$$u^*(a) = u^* \in \omega(a) \tag{4}$$

such that the corresponding solution $u = u^*(k)$, $k \in N(a)$ of Equation (2) satisfies the relation

$$u^*(k) \in \omega(k) \tag{5}$$

for every $k \in N(a)$.

1.2 Application of Theorem 1.1 to Equation (1)

The Bernoulli's type Equation (1) is a partial case of Equation (2) with the right-hand side

$$f(k, u) := q(k)(u(k))^r - v(k) - u(k).$$

In the sequel we will suppose that b, c are real functions defined on $N(a)$ such that $0 < b(k) < c(k)$ for every $k \in N(a)$. Let us reformulate Theorem 1.1 with respect to Equation (1).

Theorem 1.2. *If for every $k \in N(a)$*

$$q(k)(b(k))^r - v(k) < b(k+1) \tag{6}$$

and

$$q(k)(c(k))^r - v(k) > c(k+1), \tag{7}$$

then there exists an initial condition (4) such that the corresponding solution $u = u^*(k)$, $k \in N(a)$ of Equation (1) satisfies the relation (5) for every $k \in N(a)$.

Let us underline that Theorem 1.2 (as well as Theorem 1.1) states only that there exists an initial condition (4) generating the solution $u = u^*(k)$ of the corresponding equation satisfying (5) for every $k \in N(a)$ without giving any concrete determination of the corresponding initial data u^* itself. In this contribution we fill this gap particularly in the case of Equation (1). A lot of investigations concern various asymptotic problems for discrete equations (e.g., in [1, 4]–[6]). The above formulated problem was not a topic of investigation yet.

2 Auxiliary operators and sequences

The operators introduced below will be defined on $\mathbb{R}^+$. Let us define with the aid of the right-hand side of equation (1) an operator T_k:

$$T_k(u) := q(k)u^r - v(k)$$

for every $k \in N(a)$, and corresponding inverse operator T_k^{-1} (the existence of which follows from conditions given above):

$$T_k^{-1}(u) := \left(\frac{u + v(k)}{q(k)}\right)^{1/r}.$$

Next we define for every $s \in \mathbb{N}^*$ operators L_{a+i}, $i = 0, 1, \ldots, s$ and inverse operators L_{a+i}^{-1}, $i = 0, 1, \ldots, s$ as:

$$\begin{aligned}
L_a(u) &= u, \\
L_{a+1}(u) &:= T_a(u), \\
L_{a+2}(u) &:= T_{a+1}\left(T_a(u)\right), \\
&\ldots \\
L_{a+s}(u) &:= T_{a+s-1}\left(T_{a+s-2}\left(\ldots\left(T_{a+1}\left(T_a(u)\right)\right)\ldots\right)\right)
\end{aligned}$$

and

$$\begin{aligned}
L_a^{-1}(u) &:= u, \\
L_{a+1}^{-1}(u) &:= T_a^{-1}(u), \\
L_{a+2}^{-1}(u) &:= T_a^{-1}\left(T_{a+1}^{-1}(u)\right), \\
&\ldots \\
L_{a+s}^{-1}(u) &:= T_a^{-1}\left(T_{a+1}^{-1}\left(\ldots\left(T_{a+s-2}^{-1}\left(T_{a+s-1}^{-1}(u)\right)\right)\ldots\right)\right).
\end{aligned}$$

Now we give some properties of introduced operators. The following property is obvious.

Lemma 2.1. *Operators T_k, T_k^{-1}, L_{a+i}, L_{a+i}^{-1}, $i = 0, 1, \ldots, s$, $s \in \mathbb{N}^*$ are increasing on $\mathbb{R}^+$ for every fixed $k \in N(a)$ and $i \in \{0, 1, \ldots, s\}$.*

Lemma 2.2. *For every $u \in \mathbb{R}^+$ and $s \in \mathbb{N}^*$:*

$$L_{a+s}^{-1}(T_{a+s}^{-1}(u)) = L_{a+s+1}^{-1}(u). \tag{8}$$

Proof. Directly from definitions we get

$$L_{a+s}^{-1}(u) = T_a^{-1}\left(T_{a+1}^{-1}\left(\dots\left(T_{a+s-2}^{-1}\left(T_{a+s-1}^{-1}(u)\right)\right)\dots\right)\right)$$

and

$$L_{a+s+1}^{-1}(u) = T_a^{-1}\left(T_{a+1}^{-1}\left(\dots\left(T_{a+s-1}^{-1}\left(T_{a+s}^{-1}(u)\right)\right)\dots\right)\right).$$

Consequently,

$$L_{a+s}^{-1}(T_{a+s}^{-1}(u)) =$$
$$T_a^{-1}\left(T_{a+1}^{-1}\left(\dots\left(T_{a+s-2}^{-1}\left(T_{a+s-1}^{-1}(T_{a+s}^{-1}(u))\right)\right)\dots\right)\right) = L_{a+s+1}^{-1}(u).$$

□

Lemma 2.3. *Let the inequalities (6), (7) be valid for every $k \in N(a)$. Then the sequence $\{u_{cs}\}_{s=0}^{\infty}$ with*

$$u_{cs} := L_{a+s}^{-1}(c(a+s)), \quad s \in \mathbb{N}^*, \tag{9}$$

is a decreasing convergent sequence, and the sequence $\{u_{bs}\}_{s=0}^{\infty}$ with

$$u_{bs} := L_{a+s}^{-1}(b(a+s)), \quad s \in \mathbb{N}^*, \tag{10}$$

is an increasing convergent sequence. Moreover $u_{cs} > u_{bs}$ holds for every $s \in \mathbb{N}^*$ and for corresponding limits

$$c^* = \lim_{s\to\infty} u_{cs}, \quad b^* = \lim_{s\to\infty} u_{bs}, \tag{11}$$

the inequality $c^* \geq b^*$ holds.

Proof. Let us divide the proof into several steps.

a) Property $u_{cs} > u_{bs}$, $s \in \mathbb{N}^*$.

Let us show that $u_{cs} > u_{bs}$ for every $s \in \mathbb{N}^*$. This follows from (9), (10), since $c(a+s) > b(a+s)$ for every $s \in \mathbb{N}^*$ and from Lemma 2.1.

b) Sequence $\{u_{cs}\}_{s=0}^{\infty}$ is a decreasing sequence.

Let us verify that $u_{cs} > u_{c,s+1}$ for every $s \in \mathbb{N}^*$. If $s = 0$ and $s = 1$ then (9) gives

$$u_{c0} = L_a^{-1}(c(a)) = c(a)$$

and

$$u_{c1} = L_{a+1}^{-1}(c(a+1)) = T_a^{-1}(c(a+1)) = \left(\frac{c(a+1)+v(a)}{q(a)}\right)^{1/r}.$$

The inequality $u_{c0} > u_{c1}$ is a consequence of (7) with $k = a$ since

$$q(a)(c(a))^r - v(a) > c(a+1),$$

i.e.,

$$c(a) > \left(\frac{c(a+1) + v(a)}{q(a)}\right)^{1/r}.$$

Let us consider the general case. For $k = a + s$, $s \in \mathbb{N}^*$, the inequality (7) gives

$$q(a+s)(c(a+s))^r - v(a+s) > c(a+s+1),$$

i.e.,

$$c(a+s) > \left(\frac{c(a+s+1) + v(a+s)}{q(a+s)}\right)^{1/r}. \tag{12}$$

Using (9) we are able to estimate the general term u_{cs}, $s \in \mathbb{N}^*$ of the sequence $\{u_{cs}\}_{s=0}^{\infty}$:

$$\begin{aligned}
u_{cs} &= L_{a+s}^{-1}(c(a+s)) > [\text{ due to (12) and Lemma 2.1 }] \\
&> L_{a+s}^{-1}\left(\left(\frac{c(a+s+1) + v(a+s)}{q(a+s)}\right)^{1/r}\right) \\
&= L_{a+s}^{-1}\left(T_{a+s}^{-1}(c(a+s+1))\right) = [\text{due to (8)}] \\
&= L_{a+s+1}^{-1}(c(a+s+1)) = u_{c,s+1}.
\end{aligned}$$

So, the inequality $u_{cs} > u_{c,s+1}$ holds for every $s \in \mathbb{N}^*$.

c) Sequence $\{u_{b,s}\}_{s=0}^{\infty}$ is an increasing sequence.

Let us show that $u_{bs} < u_{b,s+1}$ for $s \in \mathbb{N}^*$. If $s = 0$ and $s = 1$ then (10) gives

$$u_{b0} = L_a^{-1}(b(a)) = b(a)$$

and

$$u_{b1} = L_{a+1}^{-1}(b(a+1)) = T_a^{-1}(b(a+1)) = \left(\frac{b(a+1) + v(a)}{q(a)}\right)^{1/r}.$$

The inequality $u_{b0} < u_{b1}$ is a consequence of (6) with $k = a$ since

$$q(a)(b(a))^r - v(a) < b(a+1),$$

i.e.,

$$b(a) < \left(\frac{c(a+1) + v(a)}{q(a)}\right)^{1/r}.$$

Let us consider the general case. For $k = a + s$, $s \in \mathbb{N}^*$, the inequality (6) gives

$$q(a+s)(b(a+s))^r - v(a+s) < b(a+s+1),$$

i.e.,

$$b(a+s) < \left(\frac{b(a+s+1)+v(a+s)}{q(a+s)}\right)^{1/r}. \tag{13}$$

Then using (10) we are able to estimate the general term u_{bs}, $s \in \mathbb{N}^*$ of the sequence $\{u_{bs}\}_{s=0}^{\infty}$:

$$\begin{aligned}
u_{bs} &= L_{a+s}^{-1}(b(a+s)) < [\text{ due to (13) and Lemma 2.1 }] \\
&< L_{a+s}^{-1}\left(\left(\frac{b(a+s+1)+v(a+s)}{q(a+s)}\right)^{1/r}\right) \\
&= L_{a+s}^{-1}\left(T_{a+s}^{-1}(b(a+s+1))\right) = [\text{due to (8)}] \\
&= L_{a+s+1}^{-1}(b(a+s+1)) = u_{b,s+1}.
\end{aligned}$$

Consequently, the inequality $u_{bs} < u_{b,s+1}$ is verified for every $s \in \mathbb{N}^*$. The lemma is proved since all remaining affirmations are elementary consequences of the theory of number sequences. □

Let us formulate an elementary property of solutions. The corresponding proof is omitted.

Lemma 2.4. *Solutions $u(k)$, $U(k)$, $k \in N(a)$ of two initial conditions for equation* (1)*:*

$$u(a) = \alpha, \quad \text{and} \quad U(a) = \beta$$

with $0 < \alpha < \beta$ satisfy the inequalities

$$u(k) < U(k)$$

for every $k \in N(a)$.

Lemma 2.5. *Let the inequalities* (6), (7) *be valid for every $k \in N(a)$. Then*
a) *The solution $u = u_{cs}^*(k), k \in N(a)$ of the initial condition*

$$u_{cs}^*(a) = u_{cs}, \quad s \in \mathbb{N}^* \tag{14}$$

for the equation (1) *satisfies the relations*

$$u_{cs}^*(k) \in \omega(k) \tag{15}$$

for every $k = a, a+1, \ldots, a+s-1$ and

$$u_{cs}^*(a+s) = c(a+s). \tag{16}$$

Moreover,

$$u_{c,s+1}^*(k) < u_{cs}^*(k) \quad \text{if} \quad k = a, a+1, \ldots, a+s. \tag{17}$$

b) *The solution $u_{bs}^*(k), k \in N(a)$ of the initial condition*

$$u_{bs}^*(a) = u_{bs}, \quad s \in \mathbb{N}^* \tag{18}$$

for Equation (1) *satisfies the relations*

$$u^*_{bs}(k) \in \omega(k) \tag{19}$$

for every $k = a, a+1, \ldots, a+s-1$ *and*

$$u^*_{bs}(a+s) = b(a+s). \tag{20}$$

Moreover,

$$u^*_{b,s+1}(k) > u^*_{bs}(k) \quad \text{if} \quad k = a, a+1, \ldots, a+s. \tag{21}$$

Proof. **α)** Consider the initial condition (14). Then (as it follows directly from (1))

$$\begin{aligned} u^*_{cs}(a+1) &= q(a)(u^*_{cs}(a))^r - v(a) \\ &= q(a)(u_{cs})^r - v(a) = T_a(u_{cs}) = L_{a+1}(u_{cs}). \end{aligned} \tag{22}$$

Suppose that the formula

$$u^*_{cs}(a+k) = L_{a+k}(u^*_{cs}(a)) = L_{a+k}(u_{cs}) \tag{23}$$

holds for $k = 0, 1, \ldots, s-1$. For $k = 0$ and $k = 1$ it holds since we consequently get the relations (14) and (22). Moreover,

$$\begin{aligned} u^*_{cs}(a+s) &= q(a+s-1)(u^*_{cs}(a+s-1))^r - v(a+s-1) \\ &= q(a+s-1)\left[L_{a+s-1}(u_{cs})\right]^r - v(a+s-1) \\ &= T_{a+s-1}\left(L_{a+s-1}(u_{cs})\right) = L_{a+s}(u_{cs}) \end{aligned}$$

Comparing the last expression with (23), we conclude that (23) holds for $k = s$ and, moreover, for every $k \in \mathbb{N}^*$. Using the representation (9), we get

$$u^*_{cs}(a+s) = L_{a+s}(u_{cs}) = L_{a+s}\left(L^{-1}_{a+s}(c(a+s))\right) = c(a+s).$$

So, the formula (16) holds.

β) Consider the initial condition (18). Then (as it follows directly from (1))

$$\begin{aligned} u^*_{bs}(a+1) &= q(a)(u^*_{bs}(a))^r - v(a) \\ &= q(a)(u_{bs})^r - v(a) = T_a(u_{bs}) = L_{a+1}(u_{bs}). \end{aligned} \tag{24}$$

Suppose that the formula

$$u^*_{bs}(a+k) = L_{a+k}(u^*_{bs}(a)) = L_{a+k}(u_{bs})$$

holds for $k = 0, 1, \ldots, s-1$ (for $k = 0$ and $k = 1$ we consequently get relations (18) and (24)). Moreover,

$$\begin{aligned} u^*_{bs}(a+s) &= q(a+s-1)(u^*_{bs}(a+s-1))^r - v(a+s-1) \\ &= q(a+s-1)\left[L_{a+s-1}(u_{bs})\right]^r - v(a+s-1) \\ &= T_{a+s-1}\left(L_{a+s-1}(u_{bs})\right) = L_{a+s}(u_{bs}). \end{aligned}$$

Comparing the last expression with (24), we conclude that (24) holds for $k = s$ and, consequently, for every $k \in \mathbb{N}^*$. Using the representation (10) we get

$$u_{bs}^*(a+s) = L_{a+s}(u_{bs}) = L_{a+s}\left(L_{a+s}^{-1}(b(a+s))\right) = b(a+s).$$

So the formula (20) is proved.

γ) By Lemma 2.3 we have $u_{cs} > u_{bs}$, $s \in \mathbb{N}^*$. Then, by Lemma 2.4, for every $k \in N(a)$ inequalities $u_{cs}^*(k) > u_{bs}^*(k)$ holds. Since, by Lemma 2.3, $u_{cs} > u_{c,s+1}$, $s \in \mathbb{N}^*$ and $u_{bs} < u_{b,s+1}$, $s \in \mathbb{N}^*$, the properties (17) and (21) are a consequence of Lemma 2.3 and Lemma 2.4. Let us prove relations (15), (19). Since $u_{bs}^*(a) < u_{cs}^*(a)$ and $u_{b,s+1}^*(a) > u_{bs}^*(a)$, $u_{c,s+1}^*(a) < u_{cs}^*(a)$, Lemma 2.4 gives

$$u_{bs}^*(k) < u_{b,s+1}^*(k) < u_{c,s+1}^*(k) < u_{cs}^*(k).$$

For $k = a+s$ we get (due to (16) and (20))

$$b(a+s) = u_{bs}^*(a+s) < u_{b,s+1}^*(a+s) < u_{c,s+1}^*(a+s) < u_{cs}^*(a+s) = c(a+s).$$

The last inequalities can be rewritten as

$$b(a+\tilde{s}) = u_{b\tilde{s}}^*(a+\tilde{s}) < u_{b,\tilde{s}+1}^*(a+\tilde{s}) < u_{c,\tilde{s}+1}^*(a+\tilde{s}) < u_{c\tilde{s}}^*(a+\tilde{s}) = c(a+\tilde{s}).$$

Putting here consequently $\tilde{s} = 0, 1, 2 \ldots, s-1$ we see that (15) and (19) hold, too. □

3 Main results

In this final part we give results concerning the existence of at least one solution of the problem (1), (5).

Theorem 3.1 (Main Result). *Let the inequalities* (6), (7) *be valid for every* $k \in N(a)$*. Then every initial condition* (4) *with* $u^* \in [b^*, c^*]$*, where* b^* *and* c^* *are defined by* (11)*, determines a solution of Equation* (1) *satisfying relation* (5)*.*

Proof. The proof is a straightforward consequence of Lemmas 2.3, 2.4, 2.5. For solutions of the initial conditions $u(a) = b^*$, $U(a) = c^*$ we have $b(k) < u(k) \le U(k) < c(k)$ for every $k = a, a+1, \ldots$ and $u(k) \le \tilde{u}(k) \le U(k)$ for every $k = a, a+1, \ldots$ if $\tilde{u}(a) \in [b^*, c^*]$. □

Consequence 3.2. *If Theorem* 3.1 *holds then a solution of the problem* (1), (4) *satisfying relation* (5) *is defined by*

$$u(a+s) = L_{a+s}(u^*), \quad s \in \mathbb{N}^*$$

with $u^* \in [b^*, c^*]$*.*

Proof. This statement follows immediately from Theorem 3.1 and from the explicit form of the problem (1), (4) which can be derived from (1) directly. □

Theorem 3.3. *Let the inequalities* (6), (7) *be valid for every* $k \in N(a)$. *Then the initial condition* $u(a) = u^{\nabla}$ *with* $u^{\nabla} \in [b(a), c(a)] \setminus [b^*, c^*]$ *generate a solution* $u = u^{\nabla}(k)$, $k \in N(a)$ *of Equation* (1) *not satisfying relation* (5) *for all* $k \in N(a)$.

Proof. Let us suppose that $u^{\nabla} \in (c^*, c(a)]$. Then there exists a number $s = s^{\nabla} \in \mathbb{N}^*$ such that $u^{\nabla} > u_{cs^{\nabla}}$. By Lemma 2.3, the inequalities

$$u^{\nabla}(k) > u^*_{cs^{\nabla}}(k),$$

where $u^*_{cs^{\nabla}}(k)$ is solution with $u^*_{cs^{\nabla}}(a) = u_{cs^{\nabla}}$, hold for every $k \in N(a)$. In accordance with Lemma 2.4,

$$u^*_{cs^{\nabla}}(k) \in \omega(k), \quad k = a, a+1, \ldots, a + s^{\nabla} - 1$$

and

$$u^*_{cs^{\nabla}}(a + s^{\nabla}) = c(a + s^{\nabla}).$$

Moreover, due to (1) and (7),

$$\begin{aligned} u^*_{cs^{\nabla}}(a + s^{\nabla} + 1) &= q(a + s^{\nabla})(u^*_{cs^{\nabla}}(a + s^{\nabla}))^r - v(a + s^{\nabla}) \\ &= q(a + s^{\nabla})(c(a + s^{\nabla}))^r - v(a + s^{\nabla}) > c(a + s^{\nabla} + 1). \end{aligned}$$

Consequently,

$$u^{\nabla}(a + s^{\nabla} + 1) > c(a + s^{\nabla} + 1).$$

So the relation (5) does not hold for $k = a + s^{\nabla} + 1$. The case $u^{\nabla} \in [b(a), b^*)$ can be considered similarly. □

The following two corollaries follow obviously from Theorem 3.1 and Theorem 3.3.

Corollary 3.4. *Let the inequalities* (6), (7) *be valid for every* $k \in N(a)$. *Then a solution* $u = u(k)$, $k \in N(a)$ *of Equation* (1) *satisfies relation* (5) *for every* $k \in N(a)$ *if and only if* $u(a) \in [b^*, c^*]$.

Corollary 3.5. *Let the inequalities* (6), (7) *be valid for every* $k \in N(a)$. *Let, moreover,* $b^* = c^*$. *Then Equation* (1) *has a unique solution* $u = u^*(k)$, $k \in N(a)$ *satisfying for every* $k \in N(a)$ *relation* (5). *This solution is determined by initial data* $u^*(a) = u^* = b^*$.

Let us find sufficient conditions for the case $b^* = c^*$. Let the inequalities (6), (7) be valid for every $k \in N(a)$. Denote

$$\Delta(s) = u_{cs} - u_{bs}, \quad s = 0, 1, \ldots.$$

Then the length of the interval $[b^*, c^*]$ can be estimated (due to the monotonicity of sequences $\{u_{cs}\}_{s=0}^{\infty}$, $\{u_{bs}\}_{s=0}^{\infty}$) as

$$0 \leq c^* - b^* < \Delta(s), \quad s = 0, 1, \ldots.$$

From the definition of the expressions u_{cs}, u_{bs} we see that

$$\Delta(s) = u_{cs} - u_{bs} = \text{ [due to (9) and (10)] } = L_{a+s}^{-1}(c(a+s)) - L_{a+s}^{-1}(b(a+s)).$$

Then

$$\begin{aligned} 0 &< u_{cs} - c^* < \Delta(s), \quad s \in \mathbb{N} \\ 0 &< b^* - u_{bs} < \Delta(s), \quad s \in \mathbb{N}. \end{aligned}$$

Theorem 3.6. *Let the inequalities* (6), (7) *be valid for every* $k \in N(a)$. *Then* $b^* = c^*$ *if*

$$\lim_{s \to \infty} \Delta(s) = 0.$$

Proof. We have

$$c^* - b^* = \lim_{s \to \infty} \left(L_{a+s}^{-1}(c(a+s)) - L_{a+s}^{-1}(b(a+s)) \right) = \lim_{s \to \infty} \Delta(s) = 0.$$

□

References

[1] R.P. Agarwal, *Differential Equations and Inequalities,* Marcel Dekker, Inc., 2nd ed., 2000.

[2] J. Diblík, Retract principle for difference equations *Proceedings of the Fourth International Conference on Difference Equations*, Poznan, Poland, August 27–31, 1998. Eds. S. Elaydi, G. Ladas, J. Popenda, and J. Rakowski, Gordon and Breach Science Publ., 107–114, 2000.

[3] J. Diblík, Discrete retract principle for systems of discrete equations, *Comput. Math. Appl.* **42** (2001), 515–528.

[4] S.N. Elaydi, *An Introduction to Difference Equations,* 2nd ed., Springer, 1999.

[5] I. Györi, G. Ladas, *Oscillation Theory of Delay Differential Equations with Applications,* Clarendon, Oxford, 1991.

[6] I. Györi, M. Pituk, Asympotic formulae for the solutions of a linear delay difference equation, *J. Math. Anal. Appl.* **195** (1995), 376–392.

Solutions Bounded on the Whole Line for Perturbed Difference Systems

ALEXANDER BOICHUK
Institute of Mathematics, The National Academy of Science
Kyiv, Ukraine
and University of Žilina, Slovak Republic
and
MIROSLAVA RŮŽIČKOVÁ [1]
University of Žilina, Slovak Republic

Abstract Conditions for existence of solutions bounded on the whole line Z for a weakly perturbed linear difference systems $x(n+1) = A(n)x(n) + f(n) + \varepsilon A_1(n)x(n)$ are obtained under the assumption that the unperturbed ($\varepsilon = 0$) linear homogeneous system has a dichotomy on both half-lines Z_+ and Z_- and corresponding unperturbed linear nonhomogeneous system has no solutions bounded on the whole line Z for arbitrary nonhomogeneity.

Keywords Dichotomy, difference equation, bounded solution, Fredholm operator, pseudo-inverse matrix

AMS Subject Classification 39A05, 39A10

1 Preliminaries

Let us denote by $B(J)$ the Banach space of vector-valued functions $x : J \to R^N$ bounded on J with the norm $\|x\| = \sup_{n\in J} |x(n)|$, $|x(n)| := \|x\|_{R^N}$; J will usually denote the set of integers Z, or nonnegative integers Z_+, or nonpositive integers Z_-. Consider the system

$$x(n+1) = A(n)x(n) \tag{1}$$

with an invertible $N \times N$-matrix $A(n)$, $n \in Z$, whose elements are real-valued functions bounded on the whole line Z : $A(\cdot) \in B(Z)$. As in the differential case [1, 2], it is known [3, 4, 5] that system (1) has a dichotomy on J if there exists a projector P ($P^2 = P$) and constants $K \geq 1$ and $0 < \lambda < 1$, such that

$$\|X(n)PX^{-1}(m)\| \leq K\,\lambda^{n-m}, \quad n \geq m,$$

[1]Supported by grant VEGA 1/0026/03 of Slovak Grant Agency.

$$\|X(n)(I-P)X^{-1}(m)\| \leq K\,\lambda^{m-n}, \quad m \geq n$$

for all $m, n \in J$; $X(n)$ is the normal ($X(0) = I$) fundamental $N \times N$ matrix of system (1). Consider the problem of finding solutions $x : Z \to R^N$, $x(\cdot) \in B(Z)$ bounded on Z of the nonhomogeneous system

$$x(n+1) = A(n)x + f(n), \qquad A(\cdot), \quad f(\cdot) \in B(Z). \tag{2}$$

Lemma 1.1. *Let system (1) have a dichotomy on Z_+ and Z_- with projectors P and Q, respectively. Then:*
a) an operator

$$(L_0x)(n) \stackrel{def}{=} x(n+1) - A(n)x(n) : \; B(Z) \;\to\; B(Z) \tag{3}$$

is a Fredholm operator and

$$indL_0 = rank[PP_{N(D)}] - rank[P_{N(D^*)}(I-P)]$$

$$= rank[(I-Q)P_{N(D)}] - rank[P_{N(D^*)}Q] = r - d,$$

where $P_{N(D)}$ and $P_{N(D^*)}$ are orthoprojectors onto the kernel $N(D) = kerD$ and cokernel $N(D^*) = kerD^*$ of matrix $D = P - (I-Q)$;
b) the homogeneous system (1) has an r-parametric set of solutions $X_r(n)c_r$, $\forall c_r \in R^r$ bounded on Z, where

$$X_r(n) = X(n)[PP_{N(D)}]_r = X(n)[(I-Q)P_{N(D)}]_r,$$

$$(r = dimN(L_0) = rank[PP_{N(D)}] = rank[(I-Q)P_{N(D)}]\;)$$

and $[\,\bullet\,]_r$ is an $(N \times r)$-dimensional matrix whose columns make a complete set of r linearly independent columns of matrix in square bracket $[\,\bullet\,]$;
c) the system

$$x(n+1) = A^{*-1}(n)x(n), \tag{4}$$

adjoint to (1) has a d-parametric set of solutions $H_d(n)c_d$, $\forall c_d \in R^d$ bounded on Z, where

$$H_d(n) = X^{*-1}(n)[Q^*P_{N(D^*)}]_d = X^{*-1}(n)[(I-P^*)P_{N(D^*)}]_d,$$

$$(\,d = dimN(L_0^*) = rank[P_{N(D^*)}(I-P)] = rank[P_{N(D^*)}Q];)$$

d) $f \in Im(L_0)$ only if the condition

$$\sum_{k=-\infty}^{+\infty} H_d^*(k+1)f(k) = 0 \tag{5}$$

holds, where $H_d^*(n)$ is a $d \times N-$ matrix whose rows represent a complete set of d linearly independent solutions of system (4) bounded on Z;

e) the nonhomogeneous system (2) has an r-parametric set of solutions bounded on Z and the general solution bounded on Z of system (2) can be written as

$$x(n, c_r) = X_r(n)c_r + (G[f])(n) \quad \forall c_r \in R^r, \tag{6}$$

where $(G[f])(n)$ is the generalized Green operator [5] for the problem of solutions bounded on the whole line Z of the system (2.) Here $N(L_0), N(L_0^*)$ and $Im(L_0)$ are the kernel, cokernel, and image of the operator L_0, respectively; symbol $*$ means the operation of transposition.

2 Perturbed problems

Consider the perturbed nonhomogeneous linear problem in the form

$$x(n+1) = A(n)x(n) + f(n) + \varepsilon A_1(n)x(n), \quad A(n), A_1(n), f(n) \in B(Z) \tag{7}$$

We will assume that the generating problem (2) has no solutions bounded on the whole line Z for arbitrary nonhomogeneity $f(n) \in B(Z)$ and system (1) has a dichotomy on Z_+ and Z_-. **Is it possible to make the problem (2) to become solvable by means of linear perturbations and, if it is possible, then what kind should be the perturbation coefficient $A_1(n)$ in order to become the problem (7) everywhere solvable?** We can answer this question with the help of the $(d \times r)$-matrix

$$B_0 = \sum_{k=-\infty}^{+\infty} H_d^*(k+1)A_1(k)X_r(k), \tag{8}$$

the construction of which involves the perturbation term of problem (7). Using the method of [7] and method of generalized inverse operators [6] we can find conditions when solutions bounded on the whole line Z of problem (7) appear in the form of Laurent series in powers of a small parameter ε.

Lemma 2.1. *Let system (1) have a dichotomy on Z_+ and Z_- with projectors P and Q, respectively, and assume that problem (2) has no solutions bounded on the whole line Z for arbitrary nonhomogeneity $f(n) \in B(Z)$. If the following equivalent relations*

$$P_{N(B_0^*)} = 0 \qquad or \qquad rank\ B_0 = d \tag{9}$$

hold, then for arbitrary $f(n) \in B(Z)$ problem (7) has at least one solution bounded on Z in the form of the series

$$x(n, \varepsilon) = \sum_{i=-1}^{\infty} \varepsilon^i x_i(n), \tag{10}$$

converging for $\varepsilon \in (0, \varepsilon_0]$, where ε_0 is an appropriate constant characterizing the domain of the convergence of series (10).

Proof. Substitute (10) into (7) and equate the coefficients at equal powers of ε. For ε^{-1}, we obtain the homogeneous problem

$$x_{-1}(n+1) = A(n)x_{-1}(n), \tag{11}$$

which determines $x_{-1}(n)$. By the hypothesis of Lemma 1, the homogeneous problem (11) has an r-parametric ($r = n - n_1$) family of solutions $x_{-1}(n, c_{-1}) = X_r(n)c_{-1}$, bounded on Z, where the r-dimensional column vector $c_{-1} \in R^r$ can be determined from the solvability condition of the problem for $x_0(n)$. For ε^0, we get the problem

$$x_0(n+1) = A(n)x_0(n) + f(n) + A_1(n)x_{-1}(n), \tag{12}$$

which determines $x_0(n) \in B(Z)$. It is an implication of Lemma 1 that the solvability criterion for problem (12) has the form

$$\sum_{k=-\infty}^{+\infty} H_d^*(k+1)[f(k) + A_1(k)x_{-1}(k, c_{-1})] = 0,$$

from which we receive with respect to $c_{-1} \in R^r$ an algebraic system

$$B_0 c_{-1} = -\sum_{k=-\infty}^{+\infty} H_d^*(k+1)f(k). \tag{13}$$

System (13) is solvable for arbitrary $f(n) \in B(Z)$ if and only if the condition (9) is satisfied. The system (13) becomes resolvable with respect to $c_{-1} \in R^r$ up to an arbitrary constant vector $P_{N(B_0)}c$ $(\forall c \in R^r)$ from the null-space of matrix B_0 with

$$c_{-1} = -B_0^+ \sum_{k=-\infty}^{+\infty} H_d^*(k+1)f(k) + P_{N(B_0)}c,$$

where B_0^+ is a Moore-Penrose pseudo-inverse [6] to matrix B_0. This solution can be rewritten in the form

$$c_{-1} = \bar{c}_{-1} + P_{B_\rho}c_\rho, \ \forall\, c_\rho \in R^\rho, \ \ \bar{c}_{-1} = -B_0^+ \sum_{k=-\infty}^{+\infty} H_d^*(k+1)f(k) \tag{14}$$

and P_{B_ρ} is an $(r \times \rho)$-dimensional matrix, whose columns are a complete set of ρ linearly independent columns of $(r \times r)$-dimensional matrix $P_{N(B_0)}$, where

$$\rho = rank\ P_{N(B_0)} = r - rank\ B_0 = r - d.$$

So, for the solutions of problem (11) we have the following expression:

$$x_{-1}(n, c_\rho) = \bar{x}_{-1}(n, \bar{c}_{-1}) + X_r(n)P_{B_\rho}c_\rho, \ \forall\, c_\rho \in R^\rho; \ \bar{x}_{-1}(n, \bar{c}_{-1}) = X_r(n)\bar{c}_{-1}.$$

Assuming that (9) holds, problem (12) has the r-parametric family of solutions bounded on the whole line Z

$$x_0(n, c_0) = X_r(n)c_0 + (G[f(\cdot) + A_1(\cdot)x_{-1}(\cdot, c_{-1}])(n). \tag{15}$$

Here c_0 is an r-dimensional constant vector, which is determined at the next step from the solvability condition of the problem for $x_1(n)$. For ε^1, we get the problem

$$x_1(n+1) = A(n)x_1(n) + A_1(n)x_0(n), \tag{16}$$

which determines solution $x_1(n)$ bounded on Z. The solvability criterion for problem (16) has the form

$$B_0 c_0 = -\sum_{k=-\infty}^{+\infty} H_d^*(k+1)A_1(k)(G[f(\cdot) + A_1(\cdot)x_{-1}(\cdot, c_\rho)])(k) \tag{17}$$

$$= -\sum_{k=-\infty}^{+\infty} H_d^*(k+1)A_1(k)(G[f(\cdot) + A_1(\cdot)[\bar{x}_{-1}(\cdot, \bar{c}_{-1}) + X_r(\cdot)P_{B_\rho}c_\rho]])(k).$$

The algebraic system (17) has the following family of solutions:

$$c_0 = \bar{c}_0 + [\cdot, \cdot, \cdot]P_{B_\rho}c_\rho, \tag{18}$$

where

$$\bar{c}_0 = -B_0^+ \sum_{k=-\infty}^{+\infty} H_d^*(k+1)A_1(k)(G[f(\cdot) + A_1(\cdot)\bar{x}_{-1}(\cdot, \bar{c}_{-1})])(k), \tag{19}$$

$$[\cdot, \cdot, \cdot] = [I_r - B_0^+ \sum_{k=-\infty}^{+\infty} H_d^*(k+1)A_1(k)(G[A_1(\cdot)X_r(\cdot)])(k)].$$

So, for the ρ-parametric family of solutions bounded on Z of problem (12) we have the following expression

$$x_0(n, c_\rho) = \bar{x}_0(n, \bar{c}_0) + \bar{X}_0(n)P_{B_\rho}c_\rho, \quad \forall\ c_\rho \in R^\rho,$$

where

$$\bar{x}_0(n, \bar{c}_0) = X_r(n)\bar{c}_0 + (G[f(\cdot) + A_1(\cdot)\bar{x}_{-1}(\cdot, \bar{c}_{-1}])(n)$$

$$\bar{X}_0(n) = X_r(n)[I_r - B_0^+ \sum_{k=-\infty}^{+\infty} H_d^*(k+1)A_1(k)(G[A_1(\cdot)X_r(\cdot)])(k)] + (G[A_1(\cdot)X_r(\cdot)])(n).$$

Again, assuming that (9) holds, problem (16) has the r-parametric family of solutions bounded on Z

$$x_1(n, c_1) = X_r(n)c_1 + (G[A_1(\cdot)[\bar{x}_0(\cdot, \bar{c}_0 + \bar{X}_0(\cdot)P_{B_\rho}c_\rho)(n)]])(n). \tag{20}$$

Here c_1 is an r-dimensional constant vector, which is determined at the next step from the solvability condition of the problem for $x_2(n) \in B(Z)$

$$x_2(n+1) = A(n)x_2(n) + A_1(n)x_1(n). \tag{21}$$

The solvability criterion for problem (21) has the form

$$B_0 c_1 = -\sum_{k=-\infty}^{+\infty} H_d^*(k+1)A_1(k)G[A_1(\cdot)[\bar{x}_0(\cdot,\bar{c}_0) + \bar{X}_0(\cdot)P_{B_\rho}c_\rho]](n).$$

Under condition (9), the last equation has the ρ-parametric family of solutions

$$c_1 = \bar{c}_1 + [.,.,.]P_{B_\rho}c_\rho,$$

where

$$\bar{c}_1 = -B_0^+ \sum_{k=-\infty}^{+\infty} H_d^*(k+1)A_1(k)G[A_1(\cdot)\bar{x}_0(\cdot,\bar{c}_0)](k).$$

$$[.,.,.] = [I_r - B_0^+ \sum_{k=-\infty}^{+\infty} H_d^*(k+1)A_1(k)G[A_1(\cdot)\bar{X}_0(\cdot)](k)].$$

So, for the coefficient $x_1(n,c_1) = x_1(n,c_\rho)$ we have the following expression:

$$x_1(n,c_\rho) = \bar{x}_1(n,\bar{c}_1) + \bar{X}_1(n)P_{B_\rho}c_\rho, \quad \forall\, c_\rho \in R^\rho,$$

where

$$\bar{x}_1(n,\bar{c}_1) = X_r(n)\bar{c}_1 + (G[A_1(\cdot)\bar{x}_0(\cdot,\bar{c}_0])(n) \tag{22}$$

$$\bar{X}_1(n) = X_r(n)[I_r - B_0^+ \sum_{k=-\infty}^{+\infty} H_d^*(k+1)A_1(k)(G[A_1(\cdot)\bar{X}_0(\cdot)])(k)] + (G[A_1(\cdot)\bar{X}_0(\cdot)])(n).$$

Continuing this process, by assuming that (9) holds, it follows by induction that the coefficients $x_i(n,c_i) = x_i(n,c_\rho)$ of series (10) can be determined from the relevant problems as follows:

$$x_i(n,c_\rho) = \bar{x}_i(n,\bar{c}_i) + \bar{X}_i(n)P_{B_\rho}c_\rho, \quad \forall\, c_\rho \in R^\rho, \quad i = -1,0,1,2,..., \tag{23}$$

where

$$\bar{x}_i(n,\bar{c}_i) = X_r(n)\bar{c}_i + (G[A_1(\cdot)\bar{x}_{i-1}(\cdot,\bar{c}_{i-1}])(n), \quad i = 1,2,...;$$

$$\bar{c}_i = -B_0^+ \sum_{k=-\infty}^{+\infty} H_d^*(k+1)A_1(k)G[A_1(\cdot)\bar{x}_{i-1}(\cdot,\bar{c}_{i-1})](k), \quad i = 1,2,...;$$

$$\bar{X}_i(n) = X_r(n)[I_r - B_0^+ \sum_{k=-\infty}^{+\infty} H_d^*(k+1)A_1(k)(G[A_1(\cdot)\bar{X}_{i-1}(\cdot)])(k)]$$

$$+ (G[A_1(\cdot)\bar{X}_{i-1}(\cdot)])(n), \quad i = 0, 1, 2, ...; \quad \text{and} \quad \bar{X}_{-1}(t) = X_r(t).$$

Since the convergence of series (10) can be proved by traditional methods of majorization, the proof of the lemma is complete. □

¿From the above Lemma 2.1 we have the following conclusions. Problem (7) determines the bounded operator

$$(L_\varepsilon x)(n) \stackrel{def}{=} z(n+1) - A(n)x(n) - \varepsilon A_1(n)x(n), \tag{24}$$

which is acting from the space $B(Z)$ to the space $B(Z)$. Under the assumption (9), problem (7) is always solvable in the Banach spaces under consideration. This means that the image of the operator L_ε coincides with the whole space $B(Z)$, that is, Im $L_\varepsilon = B(Z)$. Therefore, L_ε is a normally solvable operator (see [6, 8]), while the problem adjoint to the homogeneous one

$$x(n+1) = A(n)x(n) + \varepsilon A_1(n)x(n), \tag{25}$$

has only trivial bounded solution, that is : $dimker\ L_\varepsilon^* = 0$, $\varepsilon \neq 0$, where the operator L_ε^* is the adjoint one to L_ε in the corresponding spaces. As is shown in the proof of Lemma 2, $dimker\ L_\varepsilon = \rho = r - d$. This, together with the above mentioned property $dimker L_\varepsilon^* = 0$, means that the normally-solvable operator L_ε is a Fredholm one. Now, it is not difficult to see that for the difference operator (24), the well-known fact from the theory of operators [8, 9], concerning the maintenance under small perturbations of the index of the Fredholm operator L_0 (3), is satisfied.

Theorem 2.2. *Consider the problem*

$$x(n+1) = A(n)x(n) + \varepsilon A_1(n)x(n) + f(n), \tag{26}$$

and assume that for arbitrary nonhomogeneity $f(n) \in B(Z)$ generating problem (2) has no solutions bounded on Z and system (1) has a dichotomy on Z_+ and Z_- with projectors P and Q, respectively. If the condition

$$rank\ [\ B_0 = \sum_{k=-\infty}^{+\infty} H_d^*(k+1)A_1(k)X_r(k)\ ] = d \tag{27}$$

holds, then, for sufficiently small $\varepsilon \in (0, \varepsilon_0]$:
1) the operator $L_\varepsilon : B(Z) \to B(Z)$ defined by the formula (24) is a Fredholm one with

$$ind\ L_\varepsilon = dimker\ L_\varepsilon - dimker\ L_\varepsilon^* = \rho = r - d,$$

where the operator L_ε^* is the adjoint one to L_ε,

$$(ind\ L_0 = \rho = r - d; \quad dimker\ L_0 = r, \quad dimker\ L_0^* = d);$$

2) the homogeneous problem (25) has a ρ*-parametric family of solutions*

$$x_0(n,\varepsilon,c_\rho) = \sum_{i=-1}^{\infty} \varepsilon^i \bar{X}_i(n) P_{B_\rho} c_\rho, \qquad \forall\ c_\rho \in R^\rho, \qquad (\rho = dimker L_\varepsilon) \quad (28)$$

with the properties $x_0(\cdot,\varepsilon,c_\rho) \in B(Z)$, $x_0(n,\cdot,c_\rho) \in C(0,\varepsilon_0]$;
3) the problem adjoint to (25) has only trivial bounded solution ($dimker\ L^*_\varepsilon = 0$, $\varepsilon \in C(0,\varepsilon_0]$)*;*
4) for arbitrary $f(n) \in B(Z)$ *problem (26) has a* ρ*-parametric set of solutions* $x(n,\varepsilon) = z(n,\varepsilon,c_\rho):\ x(\cdot,\varepsilon,c_\rho) \in B(Z),\ x(n,\cdot,c_\rho) \in C(0,\varepsilon_0]$, *in the form of the series*

$$x(n,\varepsilon,c_\rho) = \sum_{i=-1}^{\infty} \varepsilon^i [\bar{x}_i(n,\bar{c}_i) + \bar{X}_i(n) P_{B_\rho} c_\rho], \qquad \forall\ c_\rho \in R^\rho \qquad (29)$$

converging for $\varepsilon \in (0,\varepsilon_0]$, *where* ε_0 *is as in Lemma 2 and the coefficients* $\bar{x}_i(n,\bar{c}_i)$, $\bar{c}_i$ *and* $\bar{X}_i(n)$ *can be determined from (23),(19),(14).*

References

[1] Palmer K.J., Exponential dichotomies and transversal homoclinic points, *Journal of Differential Equations* **55** (1984), 225–256.

[2] Sacker R.J. Existence of dichotomies and invariant splittings for linear differential systems, *Journal of Differential Equations* **27** (1978), 106–137.

[3] Henry D., Geometric theory of semilinear parabolic equations, Lecture Notes in Math., **840**, Springer-Verlag, Berlin, 1981.

[4] Boichuk A.A., Solutions of linear and nonlinear difference equations bounded on the whole line, *Nonlinear Oscillations* **4** (2001), 16–27.

[5] Boichuk A.A. and Růžičková M., Solutions of nonlinear Difference E-quations Bounded on the Whole Line, in *Proceedings of the Colloquium on Differential and Difference Equations, Brno,* Folia FSN Universitatis Masarykianae, Brunensis, *Mathematica* **13** (2002), 45–60.

[6] Boichuk A.A., Zhuravlev V.F., and Samoilenko A.M., Generalized inverse operators and Noether boundary value problems, Inst. of Mathematics NAS of Ukraine, Kyiv (1995) (in Russian).

[7] Vishik M.I. and Lyusternik L.A., Solvability of some problems concerning perturbations in case of matrices and selfadjoint and non-selfadjoint differential equations, *Uspekhi Mat. Nauk* **15** (1960), 3–80 (in Russian).

[8] Kato T., *Perturbation Theory for Linear Operators*, Springer-Verlag, New York (1966).

[9] Samoilenko A.M., Boichuk A.A. and Boichuk An.A., Solutions of weakly perturbed linear systems bounded on the whole line, *Ukrainskij Math. Zh.* **54** (2002), 1517–1530.

Rational Third-Order Difference Equations

E. CAMOUZIS

Department of Mathematics
The American College of Greece, Deree College
Athens, Greece

Abstract The aim of this paper is to present some facts about rational third-order difference equations and to pose open problems and conjectures about the global behavior of their solutions.

Keywords Period-two convergence, period-five convergence, unbounded solutions

AMS Subject Classification 39A10, 37B55

In [4] several open problems and conjectures were posed about the behavior of solutions of the equation:

$$x_{n+1} = \frac{\alpha + \beta x_n + \gamma x_{n-1} + \delta x_{n-2}}{A + Bx_n + Dx_{n-2}}, \quad n = 0, 1, \ldots \tag{1}$$

with nonnegative parameters and nonnegative initial conditions.
One can see that Eq. (1) possesses period-two solutions if and only if

$$\gamma = \beta + \delta + A\,. \tag{2}$$

The following open problem was posed in [4].

Open Problem 1 Obtain necessary and sufficient conditions on $\alpha, \beta, \gamma, \delta, A$, B and D so that Eq. (1) possesses the following trichotomy:

(a) Every solution of Eq. (1) has a finite limit if and only if

$$\gamma < \beta + \delta + A.$$

(b) Every solution of Eq. (1) converges to a period-two solution if and only if

$$\gamma = \beta + \delta + A.$$

(c) Eq. (1) has positive unbounded solutions if and only if

$$\gamma > \beta + \delta + A.$$

The following trichotomy result was recently established about the rational third-order difference equation

$$x_{n+1} = \frac{\alpha + \gamma x_{n-1} + \delta x_{n-2}}{A + x_{n-2}}, \quad n = 0, 1, ... \tag{3}$$

with nonnegative parameters and nonnegative initial conditions such that the denominators are always positive.

Theorem A (see [1], [6], and [7]) *Assume $\gamma + \delta + A > 0$. Then the following statements are true:*

(a) Every solution of Eq. (3) has a finite limit if and only if

$$\gamma < \delta + A. \tag{4}$$

(b) Every solution of Eq. (3) converges to a period-two solution if and only if

$$\gamma = \delta + A. \tag{5}$$

(c) Eq. (3) has positive unbounded solutions if and only if

$$\gamma > \delta + A. \tag{6}$$

When we say, in part (b) of Theorem A, that every solution $\{x_n\}_{n=-1}^{\infty}$ of Eq. (3) converges to a period-two solution, we mean that the subsequences $\{x_{2n}\}_{n=0}^{\infty}$ and $\{x_{2n+1}\}_{n=-1}^{\infty}$ of even and odd terms have finite limits as $n \to \infty$, and these limits are not always equal, although sometimes they may be equal. In fact when (4) holds, Eq. (3) has infinitely many prime period-two solutions.

In the special case of Eq. (1) where

$$\beta = \gamma = A = D = 0 \text{ and } \alpha > 0$$

a necessary and sufficient condition for period-two trichotomy is $\delta = 0$. When $\delta > 0$ the resulting equation is

$$y_{n+1} = \frac{p + y_{n-2}}{y_n}, \quad n = 0, 1, ... \tag{7}$$

with positive initial conditions y_{-2}, y_{-1}, y_0 and $p > 0$.
The following result concerning the behavior of solutions of Eq. (7) has been recently established.

Theorem B (see [3] and [5])
(a) Every solution of Eq. (7) has a finite limit if

$$p \geq 2.$$

(b) Every solution of Eq. (7) converges to a period-five solution if and only if

$$p = 1.$$

(c) Eq. (7) has positive unbounded solutions if and only if

$$0 < p < 1.$$

Conjecture 1 Show that every solution of Eq. (7) has a finite limit when $1 < p < 2$.

In the special case of Eq. (1) where $\alpha = \beta = A = 0$, $B = D > 0$ and $\gamma, \delta > 0$ the resulting equation is

$$y_{n+1} = \frac{cy_{n-1} + y_{n-2}}{y_n + y_{n-2}}, \quad n = 0, 1, ... \tag{8}$$

with positive initial conditions y_{-2}, y_{-1}, y_0 and $c > 0$.
The following result concerning the behavior of solutions of (8) was recently established.

Theorem C (see [2]) *Let $c = 1$. Assume that $\{x_n\}_{n=-1}^{\infty}$ is a solution of Eq. (8) with initial conditions $x_{-2}, x_{-1}, x_0 \geq \frac{1}{2}$. Then every solution of Eq. (8) converges to a period-two solution.*

Open Problem 2 Prove or disprove the existence of unbounded solutions of Eq. (8) when $c = 1$.

Another special case of Eq. (1) is the following equation

$$x_{n+1} = \frac{1 + x_{n-1} + x_{n-2}}{x_n}, \quad n = 0, 1, ... \tag{9}$$

with positive initial conditions x_{-2}, x_{-1}, x_0.
Note that Eq. (9) follows from (1) with $\alpha = \beta = A = 0$ and $\gamma = \delta = B = D = 1$. On the other hand not all the solutions of Eq. (9) converge to a period-two solution of Eq. (9). In this case as we state in the theorem below there exist unbounded solutions.

Theorem 1 *Every positive solution $\{x_n\}_{n=-2}^{\infty}$ of Eq. (9) satisfies one of the following:*
Either

$$\lim_{n\to\infty} x_{2n} = x > 1 \text{ and } \lim_{n\to\infty} x_{2n+1} = \frac{1+x}{x-1} \tag{10}$$

or

$$\lim_{n\to\infty} x_{2n} = x \in [0, 1] \text{ and } \lim_{n\to\infty} x_{2n} = \infty. \tag{11}$$

The following identity follows from direct calculations using Eq. (9).

Lemma 1 *Let* $\{x_n\}_{n=-2}^{\infty}$ *be any positive solution of Eq. (9). Then the following is true:*

$$x_{n+1} - x_{n-1} = \frac{1}{x_n}(x_n - x_{n-2}), \quad n = 0, 1, \ldots \tag{12}$$

Proof of Theorem 1

Case 1 $0 < x_0 \leq x_{-2} \leq 1$ and $x_{-1} > 0$. From Eq. (9), we have

$$x_1 = \frac{1 + x_{-1} + x_{-2}}{x_0} > \max(1, x_{-1}).$$

In view of (12), the sequence x_{2n} is decreasing and less than one while x_{2n+1} is strictly increasing and greater than one. Suppose that x_{2n+1} converges to a finite limit l_1. Then x_{2n} also converges to a finite limit l_0, where

$$l_0 \in [0, 1] \text{ and } l_1 \in (1, \infty).$$

By taking limits in (9), as $n \to \infty$, we have

$$1 + l_1 + l_0 = l_1 l_0 \leq l_1,$$

which is a contradiction and so

$$l_1 = \infty.$$

Case 2 $0 < x_{-2} < x_0 < 1$ and $x_{-1} > 0$. From (9), we have

$$x_1 = \frac{1 + x_{-1} + x_{-2}}{x_0} > \max(x_{-1}, 1).$$

Subcase 1 With the appropriate choice of the initial conditions we can generate a solution such that $x_{2n} < 1$ for all n. Indeed we choose x_{-2}, x_{-1}, x_0 (for example, $x_{-2} = 0.8, x_0 = 0.9, x_{-1} > 0$), so that

$$x_1 - x_{-1} > 2. \tag{13}$$

In view of (12), x_{2n} is strictly increasing, and x_{2n+1} is also strictly increasing and greater than one for all $n \geq 1$. In view of (13), we have

$$x_1 - x_{-1} > 2 > 1 + x_0$$

and so $x_2 < 1$. Also in view of (12), we have

$$x_3 - x_1 = \frac{1}{x_2}(x_1 - x_{-1}) > x_1 - x_{-1} > 2 > 1 + x_2$$

and so $x_4 < 1$. With the use of induction it follows $x_{2n} < 1$. Following a similar argument as in Case 1 we show that

$$\lim_{n\to\infty} x_{2n} = l_0 \leq 1 \text{ and } \lim_{n\to\infty} x_{2n+1} = \infty.$$

Subcase 2 On the other hand assume that

$$x_2 > 1.$$

In view of (12), the sequences x_{2n}, x_{2n+1} are both strictly increasing and greater than one for all $n \geq 1$. With the use of (12) it can be easily shown that the solution x_n is bounded from above and below. Furthermore

$$\lim_{n\to\infty} x_{2n} = x \in (1,\infty) \text{ and } \lim_{n\to\infty} x_{2n+1} = \frac{1+x}{x-1} \in (1,\infty).$$

Case 3 $1 \leq x_{-2} \leq x_0$. In that case we will consider several subcases.
Subcase 1 If we choose the initial conditions (for example, $x_{-2} = 1.01, x_{-1} = 0.1, x_0 = 100.$), so that

$$0 < x_1 \leq x_{-1} \leq 1$$

the result follows from Case 1.
Subcase 2 Choose the initial conditions (for example, $x_{-2} = 1.1, x_{-1} = 0.1, x_0 = 3.2.$) so that

$$0 < x_{-1} < x_1 < 1.$$

If $x_{2n+1} < 1$ for all n, then x_{2n+1} is strictly increasing and smaller than one, while x_{2n} is strictly increasing and greater than one. Arguing as in Case 1 we show that

$$\lim_{n\to\infty} x_{2n} = \infty \text{ and } \lim_{n\to\infty} x_{2n+1} = l_1 \in (0,1].$$

On the other hand assume that

$$x_3 > 1.$$

In view of (12) it follows that x_{2n+1} is strictly increasing and greater than one while x_{2n} is strictly increasing and greater than one. The result follows from Subcase 2 of Case 2.
Subcase 3 Assume that

$$0 < x_1 < 1 < x_{-1}.$$

Then $x_3 < x_1 < 1$ and the result follows from Case 1.
Subcase 4 We choose the initial conditions ($x_{-2} = 2, x_{-1} = 1.01, x_0 = 3$), so that

$$1 < x_{-1} < x_1.$$

The result follows from Subcase 2 of Case 2.

Case 4 $1 < x_0 < x_{-2}$. If $0 < x_2 < 1$ the result follows from Case 1. Therefore we assume that $x_{2n} > 1$ for all n. Also in that case the subsequence x_{2n+1} cannot be smaller than one for any n, since if it does the two subsequences will have two finite limits l_0, l_1 and by taking limits in (9), as $n \to \infty$, we will get a contradiction. So we can also assume that $x_{2n+1} > 1$ for all n. Therefore if x_{2n} converges to 1, x_{2n+1} will have to be increasing to ∞, while if x_{2n} converges to a number $x > 1$, then if x_{2n+1} is increasing will converge to a finite limit, and if it is decreasing will also have to converge to a finite limit, which cannot be equal to one.
The proof of the Theorem is complete.

References

[1] A.M. Amleh, V. Kirk, and G. Ladas, On the Dynamics of $x_{n+1} = \frac{\alpha+\beta x_{n-1}}{A+Bx_{n-2}}$, *Math. Sci. Res. Hot-Line* **5** (2002), 1–15.

[2] E. Camouzis, R. Devault, and G. Papaschinopoulos, On the recursive sequence $x_{n+1} = \frac{\gamma x_{n-1}+x_{n-2}}{x_n+x_{n-2}}$ (in press).

[3] E. Camouzis, R. Devault, and W. Kosmala, On the period five trichotomy of all positive solutions $x_{n+1} = \frac{p+x_{n-2}}{x_n}$, J. Math. Anal. Appl. **291** (2004), 40–49.

[4] E. Camouzis, C.H. Gibbons, and G. Ladas, On period-two convergence in rational equations, *J. Difference Equ. Appl.* **9** (2003), 535–540.

[5] E. Camouzis and G. Ladas, Three trichotomy conjectures, *J. Difference Equ. Appl.* **8** (2002), 495–500.

[6] E. Camouzis, G. Ladas, and H.D. Voulov, On the dynamics of $x_{n+1} = \frac{a+\gamma x_{n-1}+\delta x_{n-2}}{A+x_{n-2}}$, *J. Difference Equ. Appl.* **9** (2003), 731–738.

[7] E.A. Grove, G. Ladas, M. Predescu, and M. Radin, On the global character of $y_{n+1} = \frac{py_{n-1}+y_{n-2}}{q+y_{n-2}}$, *Math. Sci. Res. Hot-Line* **5** (2001), 25–31.

Decaying Solutions for Difference Equations with Deviating Argument

MARIELLA CECCHI

Department of Electr. and Telecom., University of Florence
Florence, Italy

ZUZANA DOŠLÁ[1]

Department of Mathematics, Masaryk University
Brno, Czech Republic

and

MAURO MARINI

Department of Electr. and Telecom., University of Florence
Florence, Italy

Abstract We deal with the nonlinear difference equation with deviating argument

$$\Delta\big(a(n)\Phi_p(\Delta x(n))\big) = b(n)f\big(x(g(n))\big), \quad \Phi_p(u) = |u|^{p-2}u \quad p > 1,$$

where $\{a(n)\}$, $\{b(n)\}$ are positive real sequences for $n \geq n_0$, $f : [0, \infty) \to [0, \infty)$ is continuous with $f(0) = 0, f(u) > 0$ for $u > 0$ and $\{g(n)\}$ is a sequence of positive integers such that $\lim_n g(n) = \infty$. Necessary and sufficient conditions for the existence of positive solutions approaching zero are given. The role of the nonlinearity f and the effect of the deviating argument are enlightened and illustrated by examples.

Keywords Difference equations with deviating argument, decaying solutions

AMS Subject Classification 39A10

[1]Supported by the Czech Grant Agency, grant 201/01/0079.

1 Introduction

Consider the nonlinear difference equation with deviating argument

$$\Delta\big(a(n)\Phi_p(\Delta x(n))\big) = b(n)f\big(x(g(n))\big), \tag{1}$$

where $\{a(n)\}$,$\{b(n)\}$ are positive real sequences for $n \geq n_0$, $f : [0,\infty) \to [0,\infty)$ is continuous with $f(0) = 0, f(u) > 0$ for $u > 0$, $\{g(n)\}$ is a sequence of positive integers, defined for $n \geq n_0$, such that $\lim_n g(n) = \infty$ and $\Phi_p(u) = |u|^{p-2}u$ with $p > 1$.

Equation (1) includes the quasi-linear difference equation

$$\Delta\big(a(n)\Phi_p(\Delta x(n))\big) = b(n)\Phi_q\big(x(g(n))\big) \tag{2}$$

and, in particular, the equation without deviating argument

$$\Delta\big(a(n)\Phi_p(\Delta x(n))\big) = b(n)\Phi_q\big(x(n+1)\big). \tag{3}$$

Both equations have been deeply investigated: we refer to [3, 4, 5, 6, 7, 9, 10] and references therein. Other interesting contributions can be found in the recent monographies [1, 2].

Equation (1) comes out in the formulation of several physical problems, related to search of radial solutions for differential equations with p-Laplacian operator. In the discretization process of the quoted elliptic problems with free boundaries, a crucial role is played by the existence of the so-called *extremal solutions,* i.e. positive solutions approaching zero or infinity. By a *solution* of (1) we mean a real sequence $x = \{x(n)\}$ defined for $n \geq \min\{n_0, n_1\}$, where $n_1 = \min\{g(n), n \geq n_0\}$ and satisfying (1) for $n \geq n_0$. A positive solution x of (1) approaching zero is said to be *decaying solution* and a decaying solution x of (1) is called *regularly decaying solution* if its quasidifference

$$x^{[1]}(n) = a(n)\Phi_p(\Delta x(n))$$

satisfies $\lim_n x^{[1]}(n) = c_x < 0$.

Denote with Φ_{p^*} the inverse of the increasing map Φ_p. For the equation without deviating argument (3), the following holds.

Theorem A. [6, Corollary 3.3] *Equation (3) has regularly decaying solutions if and only if*

$$Y = \lim_m \sum_{k=1}^{m} b(k)\Phi_q\left(\sum_{i=k+1}^{m} \Phi_{p^*}\left(\frac{1}{a(i)}\right)\right) < \infty.$$

Our aim here is to continue the study in [6], by enlightening some discrepancies between (1) and (3), related to the existence of regularly decaying solutions. At first Equation (2) is considered and a necessary and sufficient condition, which extends Theorem A, is given. We note that the deviation

$n+1-g(n)$ is not assumed to be necessarily constant and may be unbounded. In the last section our results are restated for the more general case (1). Some examples, illustrating the role of the nonlinearity f and the effect of the deviating argument, complete the paper.

Other types of extremal solutions, i.e., decaying solutions x satisfying $\lim_n x^{[1]}(n) = 0$ and positive unbounded solutions, will be considered in a forthcoming paper.

2 Main result

When

$$Y_a = \sum_{i=1}^{\infty} \Phi_{p^*}\left(\frac{1}{a(i)}\right) = \infty,$$

then (1) does not have regularly increasing solutions. Indeed let $x = \{x(n)\}$ a regularly decaying solution of (1) and set $x_\infty^{[1]} = \lim_n x^{[1]}(n)$. Taking into account that $\{x^{[1]}(n)\}$ is eventually negative increasing, we have $x^{[1]}(n) < x_\infty^{[1]}(<0)$. Summing this inequality from n to N, $n<N$, we obtain

$$x(n) > x(N+1) - \Phi_{p^*}(x_\infty^{[1]}) \sum_{i=n}^{N} \Phi_{p^*}\left(\frac{1}{a(i)}\right), \tag{4}$$

which gives a contradiction as $N \to \infty$. Concerning Equation (2) the following stronger result holds.

Theorem 2.1. *Equation (2) has regularly decaying solutions if and only if*

$$W = \lim_m \sum_{k=1}^{m} b(k)\Phi_q \left[\sum_{i=g(k)}^{m} \Phi_{p^*}\left(\frac{1}{a(i)}\right)\right] < \infty. \tag{5}$$

Proof. Necessity. Let $x = \{x(n)\}$ be a regularly decaying solution of (2) and set $x_\infty^{[1]} = \lim_n x^{[1]}(n)$. As already claimed, necessarily $Y_a < \infty$. From (4) we have for every large n

$$x(g(n)) > -\Phi_{p^*}(x_\infty^{[1]}) \sum_{i=g(n)}^{\infty} \Phi_{p^*}\left(\frac{1}{a(i)}\right).$$

Then

$$\Phi_q(x(g(n))) > \Phi_q[-\Phi_{p^*}(x_\infty^{[1]})]\Phi_q\left(\sum_{i=g(n)}^{\infty} \Phi_{p^*}\left(\frac{1}{a(i)}\right)\right)$$

and from (2) we obtain

$$\Delta(a(n)\Phi_p(\Delta x(n))) > \Phi_q[-\Phi_{p^*}(x_\infty^{[1]})]b(n)\Phi_q\left(\sum_{i=g(n)}^{\infty} \Phi_{p^*}\left(\frac{1}{a(i)}\right)\right)$$

or $(m > n)$

$$x^{[1]}(m+1) - x^{[1]}(n) > \Phi_q[-\Phi_{p^*}(x^{[1]}_\infty)] \sum_{k=n}^{m} b(k)\Phi_q\left(\sum_{i=g(k)}^{\infty} \Phi_{p^*}\left(\frac{1}{a(i)}\right)\right),$$

which gives a contradiction as $m \to \infty$.

Sufficiency. Choose $k_0 \geq n_0$ so large that

$$\sum_{k=k_0}^{\infty} b(k)\Phi_q\left(\sum_{i=g(k)}^{\infty} \Phi_{p^*}\left(\frac{1}{a(i)}\right)\right) < \frac{1}{4}. \tag{6}$$

Denote by $\ell^\infty_{k_0}$ the Banach space of all bounded sequences defined for all integer $n \geq k_0$ and endowed with the topology of the supremum norm. Taking into account that (5) implies $Y_a < \infty$, let Ω be the nonempty subset of $\ell^\infty_{k_0}$ given by

$$\Omega = \left\{ \{u(n)\} \in \ell^\infty_{k_0} : \ 0 \leq u(n) \leq \sum_{i=n}^{\infty} \Phi_{p^*}\left(\frac{1}{a(i)}\right)\right\}.$$

Clearly Ω is bounded, closed, and convex. Because $\lim_n g(n) = \infty$, there exists $\overline{n}$ such that

$$g(i) \geq k_0 + 1 \text{ for } i \geq \overline{n}. \tag{7}$$

Now consider the operator $T : \Omega \longrightarrow \ell^\infty_{k_0}$, which assigns to any $u = \{u(n)\} \in \Omega$ the sequence $w = T(u) = \{w(n)\}$ given by

$$w(n) = \sum_{j=n}^{\infty} \Phi_{p^*}\Big(\frac{1}{a(j)}\Big)\Phi_{p^*}\left(\frac{1}{4} + \sum_{i=j}^{\infty} b(i)\Phi_q(u(g(i)))\right) \qquad \text{if } n \geq \overline{n},$$

$$w(n) = \frac{1}{4}\sum_{k=n}^{\infty} \Phi_{p^*}\Big(\frac{1}{a(k)}\Big) + w(\overline{n}) \qquad \text{if } k_0 \leq n < \overline{n}.$$

In view of (7), the operator T is well defined. Obviously, $w(n) \geq 0$ for $n \geq \overline{n}$. In view of (6) we have for $n \geq \overline{n}$

$$\begin{aligned} w(n) &\leq \sum_{j=n}^{\infty} \Phi_{p^*}\Big(\frac{1}{a(j)}\Big)\Phi_{p^*}\left(\frac{1}{4} + \sum_{i=j}^{\infty} b(i)\Phi_q\left(\sum_{k=g(i)}^{\infty} \Phi_{p^*}\Big(\frac{1}{a(k)}\Big)\right)\right) \\ &\leq \frac{1}{2}\sum_{j=n}^{\infty} \Phi_{p^*}\Big(\frac{1}{a(j)}\Big). \end{aligned}$$

For $k_0 \leq n < \overline{n}$ we have

$$\begin{aligned} w(n) &\leq \frac{1}{2}\sum_{k=n}^{\infty} \Phi_{p^*}\Big(\frac{1}{a(k)}\Big) + w(\overline{n}) \\ &\leq \frac{1}{2}\sum_{k=n}^{\infty} \Phi_{p^*}\Big(\frac{1}{a(k)}\Big) + \frac{1}{2}\sum_{j=\overline{n}}^{\infty} \Phi_{p^*}\Big(\frac{1}{a(j)}\Big) < \sum_{k=n}^{\infty} \Phi_{p^*}\Big(\frac{1}{a(k)}\Big). \end{aligned}$$

Then $T(\Omega) \subseteq \Omega$. In order to complete the proof it is sufficient to prove the relatively compactness of $T(\Omega)$, the continuity of T in Ω and to apply the Schauder fixed point theorem. Regarding the compactness of $T(\Omega)$, by a result in [8, Theorem 3.3], it is sufficient to show that for any $\varepsilon > 0$ there exists $N \geq k_0$ such that $|w(k) - w(\ell)| < \varepsilon$ whenever $k, \ell \geq N$ for any $w \in T(\Omega)$. Without loss of generality assume $k < \ell$. Then

$$|w(k) - w(\ell)| = \sum_{j=k}^{\ell-1} \Phi_{p^*}\left(\frac{1}{a(j)}\right) \Phi_{p^*}\left(\frac{1}{4} + \sum_{i=j}^{\infty} b(i)\Phi_q(u(g(i))\right) \leq \sum_{j=k}^{\ell-1} \Phi_{p^*}\left(\frac{1}{a(j)}\right).$$

Choosing N large, we obtain the assertion, because $Y_a < \infty$. The continuity of T in $T(\Omega)$ can be proved using the discrete analogous to the Lebesgue dominated convergence theorem and an argument similar to that given in [4, Theorem 2].

Now, by applying the Schauder theorem, we obtain the existence of at least one fixed point for the operator T, i.e., there exists a sequence $\{x(n)\} \in \Omega$ such that

$$x(n) = \sum_{j=n}^{\infty} \Phi_{p^*}\left(\frac{1}{a(j)}\left(\frac{1}{4} + \sum_{i=j}^{\infty} b(i)\Phi_q(x(g(i))\right)\right) \quad \text{if} \quad n \geq \overline{n}$$

$$x(n) = \frac{1}{4}\sum_{k=n}^{\infty} \Phi_{p^*}\left(\frac{1}{a(k)}\right) + x(\overline{n}) \quad \text{if} \quad k_0 \leq n < \overline{n}.$$

It is easy to verify that for $n \geq \overline{n}$ the nonnegative sequence $\{x(n)\}$ is a solution of (2) and

$$\Delta x(n) = -\Phi_{p^*}\left(\frac{1}{a(j)}\left(\frac{1}{4} + \sum_{i=j}^{\infty} b(i)\Phi_q(x(g(i))\right)\right) < 0. \tag{8}$$

Thus necessarily $\{x(n)\}$ is positive. Because $\{x(n)\} \in \Omega$, we have $0 \leq x(n) \leq \sum_{j=n}^{\infty} \Phi_{p^*}\left(\frac{1}{a(j)}\right)$ and therefore $\lim_n x(n) = 0$. From (8) we have $x^{[1]}(n) < -1/4$ and so $\{x(n)\}$ is a regularly decaying solution of (2). $\square$

Clearly if $g(n) \leq n + 1$ and (2) has regularly decaying solutions, then Equation (3) has regularly decaying solutions too, because $W < \infty \Longrightarrow Y < \infty$. The opposite implication can fail. In other words, if (3) has regularly decaying solutions, then the deviating argument may cause the lack of these types of solutions, as the following examples show.

Example 1. Consider the equations ($n \geq 2$)

$$\Delta\big(n(n+1)\Delta x(n)\big) = \frac{1}{\sqrt{n}}\, x(n)\,, \tag{9}$$

$$\Delta\big(n(n+1)\Delta x(n)\big) = \frac{1}{\sqrt{n}} x\big(g(n)\big), \tag{10}$$

where $g(n) = \lfloor n^{1/2} \rfloor + 1$ and $\lfloor y \rfloor$ denotes the integer part of the number indicated. Then $Y < \infty$ and $W = \infty$. Hence from Theorem A and Theorem 2.1, regularly decaying solutions exist for (9), but not for an equation with deviating argument (10). Observe that in this case the deviation $n+1-g(n)$ is unbounded.

Example 2. Consider the equations ($n \geq 2$)

$$\Delta\left(\frac{\exp(n^2)}{2n}\Delta x(n)\right) = \frac{\exp(n^2)}{n^2} x(n), \tag{11}$$

$$\Delta\left(\frac{\exp(n^2)}{2n}\Delta x(n)\right) = \frac{\exp(n^2)}{n^2} x(n+1). \tag{12}$$

We have

$$Y = \sum_{n=1}^{\infty} b(n) \sum_{i=n+1}^{\infty} \frac{1}{a(i)} < \sum_{n=1}^{\infty} \frac{1}{n^2} < \infty$$

$$W = \sum_{n=1}^{\infty} b(n) \sum_{i=n}^{\infty} \frac{1}{a(i)} > \sum_{n=1}^{\infty} \frac{b(n)}{a(n)} = \infty.$$

Then (11) does not have regularly decaying solutions, but, in view of Theorem A, the corresponding equation without deviating argument (12) has regularly decaying solutions. Observe that in this case the deviation is constant and it is 1.

When $g(n) > n+1$ the opposite situation can occur, as the following example shows.

Example 3. Consider the equation

$$\Delta(n(n+1)\Delta x(n)) = n^2 x(g(n)), \tag{13}$$

where $g(n) = \exp(n)$. It is easy to verify that, in view of Theorem 2.1 the "advanced" equation (13) has regularly decaying solutions. However, because $Y = \infty$, the corresponding equation without deviating argument does not have regularly decaying solutions.

3 The general case

Let $\Psi : [0,\infty) \to [0,\infty)$ a continuous nondecreasing function such that $\Psi(0) = 0$, $\Psi(u) > 0$ for $u > 0$. Theorem 2.1 can be generalized to (1) as follows.

Theorem 3.1. *Assume*

$$\liminf_{u\to 0} \frac{f(u)}{\Psi(u)} > 0.$$

If $Y_a = \infty$ *or for every positive constant* h *it holds that*

$$\lim_m \sum_{k=1}^{m} b(k)\Psi\Big(h \sum_{i=g(k)}^{m} \Phi_{p^*}\left(\frac{1}{a(i)}\right)\Big) = \infty,$$

then (1) does not have regularly decaying solutions.

Theorem 3.2. *Assume that*

$$\limsup_{u\to 0} \frac{f(u)}{\Psi(u)} < \infty.$$

If $Y_a < \infty$ *and there exists a positive constant* H *such that*

$$W_H = \lim_m \sum_{k=1}^{m} b(k)\Psi\Big(H \sum_{i=g(k)}^{m} \Phi_{p^*}\left(\frac{1}{a(i)}\right)\Big) < \infty,$$

then (1) has regularly decaying solutions.

The argument of the proof of Theorems 3.1, 3.2 is similar to that given in the proof of Theorem 2.1, with minor changes.

Observe that if Ψ is unbounded at infinity then $W_H < \infty \Longrightarrow Y_a < \infty$. However, when Ψ is bounded, it may happen that $W_H < \infty$ and $Y_a = \infty$. By other words, the convergence of the series W_H it is not sufficient for the existence of regularly decaying solutions of (1) as the following example illustrates.

Example 4. Consider the equation ($n \geq 1$)

$$\Delta^2 x(n) = \frac{1}{n^2} f(x(n)), \tag{14}$$

where $f(u) = arctg\ u$. We have $Y_a = \infty$ and $W_H \leq \frac{\pi}{2}\sum_{k=1}^{\infty} 1/k^2 < \infty$. Then, in view of Theorem 3.1, Equation (14) does not have regularly decaying solutions.

Summarizing Theorems 3.1, 3.2 we get the following result.

Corollary 3.3. *Let* f *be a nondecreasing function.*

(a) If f *is unbounded at infinity then (1) has regularly decaying solutions if and only if there exists a positive constant* H *such that*

$$\lim_m \sum_{k=1}^{m} b(k) f\left(H \sum_{i=g(k)}^{m} \Phi_{p^*}\left(\frac{1}{a(i)}\right)\right) < \infty.$$

(b) If f *is bounded at infinity then (1) has regularly decaying solutions if and only if* $Y_a < \infty$ *and* $\sum_{k=1}^{\infty} b(k) < \infty$.

References

[1] R.P. Agarwal, *Difference Equations and Inequalities*, 2nd Edition, Pure Appl. Math. 228, Marcel Dekker, New York, 2000.

[2] R.P. Agarwal, S.R. Grace, D. O'Regan, *Oscillation Theory for Difference and Functional Differential Equations*, Kluwer, Dordrecht, 2000.

[3] R.P. Agarwal, W.T. Li, P.Y.H. Pang, *Asymptotic behavior of nonlinear difference systems,* Appl. Mathematics Comput.**140** (2003), 307-316.

[4] M. Cecchi, Z. Došlá, M. Marini, *Positive decreasing solutions of quasilinear difference equations*, Computers Math. Appl. **42** (2001), 1401–1410.

[5] M. Cecchi, Z. Došlá, M. Marini, *Unbounded solutions of quasilinear difference equations*, Computer Math. Appl., **45** (2003), 1113–1123.

[6] M. Cecchi, Z. Došlá, M. Marini, *Limit behavior for quasilinear difference equations*, in *Proceeding of Int. Conf. of Difference Equations and Appl.*, Augsburg, August 1-5, 2001, B. Aulbach, S. Elyadi, G. Ladas Eds., Taylor&Francis Publ., London, 2003, 381–388.

[7] S. Cheng, H.J. Li, W.T. Patula, *Bounded and zero convergent solutions of second order difference equations*, J. Math. Anal. Appl. **141** (1989), 463–483.

[8] S.S. Cheng, W.T. Patula, *An existence theorem for a nonlinear difference equation, Nonlinear Anal.* **20** (1993), 193–203.

[9] E. Thandapani, K. Ravi, *Bounded and monotone properties of solutions of second-order quasilinear difference equations*, Computers Math. Appl. **38** (1999), 113–121.

[10] P. J. Y. Wong, R.P. Agarwal, *Oscillations and nonoscillations of half-linear difference equations generated by deviating arguments*, Comp. Math. Appl., 10-12, (1998) 11-26.

A Differential and Difference Equation with a Constant Delay

JAN ČERMÁK[1]

Institute of Mathematics, Faculty of Mechanical Engineering
Brno University of Technology
Brno, Czech Republic

Abstract This paper presents some resemblances between the asymptotic behavior of solutions of the delay differential equation

$$\dot{x}(t) = a(t)x(t) + b(t)x(t-1)$$

and the corresponding delay difference equation

$$a(t)\varphi(t) = |b(t)|\varphi(t-1).$$

Keywords Differential and difference equation, delay, asymptotic behavior

AMS Subject Classification 34K25, 39A11

1 Introduction and preliminaries

We consider the delay differential equation

$$\dot{x}(t) = a(t)x(t) + b(t)x(t-1), \quad t \in I := [t_0, \infty), \tag{1}$$

where a, b are continuous functions on I such that $a(t) > 0$ and $b(t) \neq 0$ for all $t \in I$. Our goal is to establish conditions under which solutions of (1) can be estimated via a solution of the delay difference equation

$$a(t)\varphi(t) = |b(t)|\varphi(t-1), \quad t \in I. \tag{2}$$

There are numerous interesting results that unify the qualitative theory of differential and difference equations. The oscillation theory for delay difference equations that parallels the oscillation theory for delay differential equations has been developed in the book by Györi and Ladas [6] (for some later relevant results see also Shen [15] or Baštinec and Diblík [1]). The related asymptotic properties of both types of these equations have been described, e.g., in the papers of Györi and Pituk [7, 8], Liu [11] or Péics [13].

[1]Supported by grant # 201/01/0079 of the Czech Grant Agency.

The difference equations occurring in these analogies are usually obtained via replacing the differential term by the difference one. However, it is not the only way to convert Equation (1) into the difference case having similar qualitative properties. In [4] the asymptotics of all solutions of (1) has been related to the behavior of a solution φ of the difference equation

$$0 = a(t)\varphi(t) + b(t)\varphi(t-1).$$

Similarly, the asymptotic behavior of some differential equations with a proportional delay

$$\dot{x}(t) = a(t)x(t) + b(t)x(\lambda t), \quad 0 < \lambda < 1, \quad t \geq 0$$

can be approached by the behavior of a solution of the λ-difference equation

$$0 = a(t)\varphi(t) + b(t)\varphi(\lambda(t)), \quad 0 < \lambda < 1, \quad t \geq 0$$

(for some particular cases see, e.g., Makay and Terjéki [12] or Kato and McLeod [10]). In the general case, the related asymptotics of

$$\dot{x}(t) = a(t)x(t) + b(t)x(\tau(t)), \quad t \in I$$

and

$$0 = a(t)\varphi(t) + b(t)\varphi(\tau(t)), \quad t \in I$$

has been discussed, e.g., in Heard [9] or in [2, 3].

All these results have been derived under the assumption of the negativity of the coefficient a. If a is positive, then we cannot expect that all solutions can asymptotically behave in this way. However, we show that under certain assumptions there exists a class of solutions of Equation (1) such that asymptotic bounds of these solutions can be given in terms of a continuous positive solution of the difference equation (2). We note that due to our assumptions such a solution does exist.

2 Results and proofs

Throughout this section we assume that a, b are continuous functions on I, $a(t) > 0$, $b(t) \neq 0$ for all $t \in I$, $|b(t)|/a(t)$ is nonincreasing as $t \to \infty$ and $\limsup_{t\to\infty} a(t-1)/a(t) < 1$.

First we show that under these assumptions every solution x of (1) is either an asymptotic equivalent as $t \to \infty$ to a solution y of

$$\dot{y}(t) = a(t)y(t) \tag{1}$$

or it is of order less than $y(t)$ as $t \to \infty$.

Lemma 2.1. *Consider Equation (1), where $a, b \in C(I)$, $a(t) > 0$, $b(t) \neq 0$ for all $t \in I$, $\limsup_{t\to\infty} a(t-1)/a(t) < 1$ and $|b(t)|/a(t)$ is nonincreasing as $t \to \infty$. If x is a solution of (1), then there exists a constant L (depending on x) such that*

$$\lim_{t\to\infty} x(t) \exp\left\{-\int_{t_0}^{t} a(s)\mathrm{d}s\right\} = L.$$

Proof. Let σ_0 be such that $a(t-1) \leq \alpha a(t)$ for all $t \geq \sigma_0$ and a suitable $\alpha < 1$. Put $z(t) = x(t)\exp\left\{-\int_{t_0}^{t} a(s)\mathrm{d}s\right\}$, $t \geq \sigma_0$ to obtain

$$\dot{z}(t) = p(t)z(t-1), \quad t \geq \sigma_0, \tag{2}$$

where $p(t) = b(t)\exp\left\{-\int_{t-1}^{t} a(s)\mathrm{d}s\right\}$. It is well known (see, e.g., Pituk [14]) that every solution z of (2) tends to a finite limit provided $\int_{\sigma_0}^{\infty} |p(t)|\mathrm{d}t$ converges. To estimate this integral we write

$$\begin{aligned}
\int_{\sigma_0}^{\infty} |p(t)|\mathrm{d}t &\leq \int_{\sigma_0}^{\infty} a(t)\frac{|b(t)|}{a(t)} \exp\left\{-\int_{t-1}^{t} a(s)\mathrm{d}s\right\}\mathrm{d}t \\
&\leq \frac{|b(\sigma_0)|}{a(\sigma_0)}\int_{\sigma_0}^{\infty} a(t)\exp\left\{-\int_{t-1}^{t} a(s)\mathrm{d}s\right\}\mathrm{d}t \\
&= \frac{|b(\sigma_0)|}{a(\sigma_0)}\int_{\sigma_0}^{\infty} \frac{-a(t)}{a(t)-a(t)\frac{a(t-1)}{a(t)}}\frac{\mathrm{d}}{\mathrm{d}t}\exp\left\{-\int_{t-1}^{t} a(s)\mathrm{d}s\right\}\mathrm{d}t \\
&\leq \frac{|b(\sigma_0)|}{a(\sigma_0)}\frac{1}{1-\alpha}\exp\left\{-\int_{\sigma_0-1}^{\sigma_0} a(s)\mathrm{d}s\right\}\mathrm{d}t < \infty.
\end{aligned}$$

Substituting back we get

$$\lim_{t\to\infty} z(t) = \lim_{t\to\infty} x(t)\exp\left\{-\int_{t_0}^{t} a(s)\mathrm{d}s\right\} = L$$

for a suitable $L \in \mathbb{R}$. □

Remark 2.2. Given $L \in \mathbb{R}$, we denote by F_L the set of all solutions x of (1) such that either $x(t) \sim L\exp\left\{\int_{t_0}^{t} a(s)\mathrm{d}s\right\}$ as $t \to \infty$ (if $L \neq 0$) or $x(t) = o\left(\exp\left\{\int_{t_0}^{t} a(s)\mathrm{d}s\right\}\right)$ as $t \to \infty$ (if $L = 0$). Under the assumptions of Lemma 2.1, for any solution x of (1) there exists $L \in \mathbb{R}$ such that $x \in F_L$. If $x \in F_L$ with $L \neq 0$, then x is asymptotically comparable with the solution of (1); hence the behavior of x at infinity cannot be described by means of a solution φ of (2). We show that if $x \in F_0$, then the situation becomes different.

Theorem 2.3. *Consider Equation (1), where $a, b \in C(I)$, $a(t) > 0$, $b(t) \neq 0$ for all $t \in I$, $\limsup_{t\to\infty} a(t-1)/a(t) < 1$, $|b(t)|/a(t)$ is nonincreasing as $t \to \infty$ and let φ be a positive continuous solution of (2). If x is a solution of (1) such that $x \in F_0$, then $x(t) = O(\varphi(t))$ as $t \to \infty$.*

Proof. Rewrite Equation (1) as

$$\begin{aligned}\dot{x}(t)\exp\left\{-\int_{t_0}^{t}a(s)\mathrm{d}s\right\} &= a(t)x(t)\exp\left\{-\int_{t_0}^{t}a(s)\mathrm{d}s\right\}\\ &\quad + b(t)\exp\left\{-\int_{t_0}^{t}a(s)\mathrm{d}s\right\}x(t-1).\end{aligned}$$

The integration over $[t,\infty)$ yields

$$\left[x(t)\exp\left\{-\int_{t_0}^{t}a(s)\mathrm{d}s\right\}\right]_t^{\infty} = \int_t^{\infty}b(u)\exp\left\{-\int_{t_0}^{u}a(s)\mathrm{d}s\right\}x(u-1)\mathrm{d}u,$$

i.e.,

$$x(t) = -\exp\left\{\int_{t_0}^{t}a(s)\mathrm{d}s\right\}\int_t^{\infty}b(u)\exp\left\{-\int_{t_0}^{u}a(s)\mathrm{d}s\right\}x(u-1)\mathrm{d}u$$

by using the fact that $x \in F_0$.

Let $\sigma_0 \geq t_0$ be such that $a(t-1) \leq \alpha a(t)$ for all $t \geq \sigma_0$ and a suitable $\alpha < 1$, and set $\sigma_j := \sigma_0 + j$, $I_j = [\sigma_j, \sigma_{j+1}]$, $j = 0, 1, 2, \ldots$. Now let $t \in I_1$. Since $|x(t)| \leq M\exp\left\{\int_{t_0}^{t}a(s)\mathrm{d}s\right\}$ for all $t \in I$ and a suitable $M > 0$, we have

$$\begin{aligned}|x(t)| &\leq M\exp\left\{\int_{t_0}^{t}a(s)\mathrm{d}s\right\}\int_t^{\infty}|b(u)|\exp\left\{-\int_{t_0}^{u}a(s)\mathrm{d}s\right\}\\ &\quad\times\exp\left\{\int_{t_0}^{u-1}a(s)\mathrm{d}s\right\}\mathrm{d}u\\ &= M\exp\left\{\int_{t_0}^{t}a(s)\mathrm{d}s\right\}\int_t^{\infty}|b(u)|\exp\left\{-\int_{u-1}^{u}a(s)\mathrm{d}s\right\}\mathrm{d}u\\ &= M\exp\left\{\int_{t_0}^{t}a(s)\mathrm{d}s\right\}\\ &\quad\times\int_t^{\infty}\frac{a(u)\frac{|b(u)|}{a(u)}}{-a(u)+\frac{a(u-1)}{a(u)}a(u)}\frac{\mathrm{d}}{\mathrm{d}u}\left[\exp\left\{-\int_{u-1}^{u}a(s)\mathrm{d}s\right\}\right]\mathrm{d}u\\ &= M\exp\left\{\int_{t_0}^{t}a(s)\mathrm{d}s\right\}\int_t^{\infty}\frac{\frac{|b(u)|}{a(u)}}{1-\frac{a(u-1)}{a(u)}}\frac{\mathrm{d}}{\mathrm{d}u}\left[-\exp\left\{-\int_{u-1}^{u}a(s)\mathrm{d}s\right\}\right]\mathrm{d}u\\ &\leq M\exp\left\{\int_{t_0}^{t}a(s)\mathrm{d}s\right\}\frac{|b(t)|}{a(t)}\frac{1}{1-\alpha}\exp\left\{-\int_{t-1}^{t}a(s)\mathrm{d}s\right\}\\ &\leq M\frac{|b(\sigma_1)|}{a(\sigma_1)}\frac{1}{1-\alpha}\exp\left\{\int_{t_0}^{\sigma_1}a(s)\mathrm{d}s\right\}.\end{aligned}$$

The repeated application of the above calculations yields

$$|x(t)| \leq M\prod_{k=1}^{j}\frac{|b(\sigma_k)|}{a(\sigma_k)}\frac{1}{1-\alpha^k}\exp\left\{\int_{t_0}^{\sigma_1}a(s)\mathrm{d}s\right\} \qquad \text{for all} \quad t \in I_j,$$

i.e.,

$$|x(t)| \le M^* \prod_{k=1}^{j} \frac{|b(t_k)|}{a(t_k)} \qquad \text{for all} \quad t \in I_j \tag{3}$$

and a suitable $M^* > 0$. Now let φ be a solution of the difference equation (2). If $\varphi(t) \ge N$ for all $t \in I_0$, then

$$\varphi(t) = \frac{|b(t)|}{a(t)}\varphi(t-1) \ge \frac{|b(t)|}{a(t)}N \ge N\frac{|b(\sigma_1)|}{a(\sigma_1)}$$

for all $t \in I_1$. The repetition leads to

$$\varphi(t) \ge N \prod_{k=1}^{j} \frac{|b(t_k)|}{a(t_k)} \qquad \text{for} \quad t \in I_j. \tag{4}$$

Comparing (3) and (4) we obtain

$$x(t) = O(\varphi(t)) \qquad \text{as} \quad t \to \infty.$$

□

3 Some consequences

In this section, we apply the previous results to some particular cases of (1).

Corollary 3.1. *Consider the equation*

$$\dot{x}(t) = a(t)[x(t) - Qx(t-1)], \qquad t \in I, \tag{1}$$

where $Q \in \mathbb{R}$, $Q \neq 0$, $a \in C^1(I)$, $a(t) > 0$ *for all* $t \in I$, $a(t-1)/a(t)$ *is nonincreasing and* $\limsup_{t\to\infty} a(t-1)/a(t) < 1$. *If* x *is a solution of (3.1) such that* $x \in F_0$, *then*

$$x(t) = g(t)\varphi(t) + O\left(\frac{|\varphi(t)|}{a(t)}\right) \qquad \text{as} \quad t \to \infty,$$

where g *is a suitable continuous function satisfying* $g(t) = g(t-1)$ *for all* $t \in I$ *and* $\varphi(t) = Q^t$ *is a solution of the difference equation*

$$\varphi(t) = Q\varphi(t-1). \tag{2}$$

Proof. By Theorem 2.3, $x(t) = O(|Q|^t)$ as $t \to \infty$ for any solution $x \in F_0$ of (1). Further, we introduce the function $y(t) = \dot{x}(t)/a(t)$, which is a solution of

$$\dot{y}(t) = a(t)y(t) - Qa(t-1)y(t-1).$$

The repeated application of Theorem 2.3 yields $y(t) = O(\omega(t))$ as $t \to \infty$, where ω is a solution of

$$a(t)\omega(t) = |Q|a(t-1)\omega(t-1).$$

Then we can easily verify that the function $a(t)\omega(t) = |Q|^t$ is fulfilling this equation; hence $\dot{x}(t) = O(|Q|^t)$ as $t \to \infty$.

Now set $z(t) = \frac{x(t)}{\varphi(t)}$, where $\varphi(t) = Q^t$ is a solution of (2). Then (1) becomes

$$\dot{z}(t) = \left[a(t) - \frac{\dot{\varphi}(t)}{\varphi(t)}\right] z(t) - Qa(t)\frac{\varphi(t-1)}{\varphi(t)} z(t-1),$$

i.e.,

$$\dot{z}(t) = [a(t) - \log Q]\, z(t) - a(t)z(t-1).$$

Rewrite this equation into the difference form

$$z(t) - z(t-1) = \frac{1}{a(t)}\left[\dot{z}(t) + z(t)\log Q\right] = O\left(\frac{1}{a(t)}\right) \qquad \text{as} \quad t \to \infty$$

using the asymptotic estimate on x and $\dot{x}$. Consequently,

$$z(t) = g(t) + O\left(\frac{1}{a(t)}\right) \qquad \text{as} \quad t \to \infty$$

for a continuous periodic function g with a period 1. □

Remark 3.2. Using the substitution $z(t) = x(t)\exp\left\{-\int_{t_0}^t a(s)\mathrm{d}s\right\}$ we can immediately deduce from Pituk [14] that there exists a solution x^* of (1) defined for all sufficiently large t, which is asymptotic equivalent to $\exp\left\{\int_{t_0}^t a(s)\right.$ $\left.\mathrm{d}s\right\}$ as $t \to \infty$ (for a similar situation see the proof of Lemma 2.1). Then using Lemma 2.1 and Corollary 3.1 we get the following asymptotic representation valid for any solution x of (1):

Let all the assumptions of Corollary 3.1 be fulfilled. Then for any solution x of (1) there exists a constant $L \in \mathbb{R}$ and a continuous periodic function g of a period 1 such that

$$x(t) = Lx^*(t) + g(t)Q^t + O\left(|Q|^t/a(t)\right) \qquad \text{as} \quad t \to \infty$$

(for related results see Diblík [5] and the references cited therein).

Corollary 3.3. *Consider the equation:*

$$\dot{x}(t) = a(t)[x(t) - \frac{1}{t}x(t-1)], \qquad t \geq t_0 > 0, \tag{3}$$

where $a \in C(I)$, $a(t) > 0$ for all $t \in I$ and $\limsup_{t\to\infty} a(t-1)/a(t) < 1$. If x is a solution of (3) belonging to F_0, then

$$x(t) = O(\varphi(t)) \qquad \text{as} \quad t \to \infty, \tag{4}$$

where $\varphi(t) = 1/\Gamma(t)$ (Γ stands for the Euler gamma function) is the function fulfilling the auxiliary difference equation:

$$\varphi(t) = \frac{1}{t}\varphi(t-1).$$

Proof. The proof follows immediately from Theorem 2.3. □

Example 3.4. *Consider the equation*

$$\dot{x}(t) = a^t x(t) - b^t x(t-1), \qquad t \in I, \tag{5}$$

where $a \geq b > 1$ are reals. Then, by Theorem 2.3, the estimate (4), where

$$\varphi(t) = \left(\frac{b}{a}\right)^{\binom{t+1}{2}}$$

is a solution of the difference equation

$$a^t \varphi(t) = b^t \varphi(t-1),$$

holds for any solution x of (5) such that $x \in F_0$.

References

[1] Baštinec, J. and Diblík, J., Subdominant positive solutions of discrete equation $\triangle u(k+n) = -p(k)u(k)$, *Abstr. Appl. Anal.* (in press).

[2] Čermák, J., A change of variables in the asymptotic theory of differential equations with unbouded delays, *J. Comp. Appl. Math.* **143** (2002), 81–93.

[3] Čermák, J., The asymptotic of solutions for a class of delay differential equations, *Rocky Mountain J. Math.* (in press).

[4] Čermák, J., On the related asymptotics of delay differential and difference equations, *Dynam. Systems Appl.* (in press).

[5] Diblík, J., Asymptotic representation of solutions of equation $\dot{y}(t) = \beta(t)[y(t) - y(t - \tau(t))]$, *J. Math. Anal. Appl.* **217** (1998), 200–215.

[6] Györi, I. and Ladas, G., *Oscillation Theory of Delay Differential Equations with Applications*, Clarendon Press, Oxford, 1991.

[7] Györi, I. and Pituk, M., Asymptotic formulae for the solutions of a linear delay difference equations, *J. Math. Anal. Appl.* **195** (1995), 376–392.

[8] Györi, I. and Pituk, M., Comparison theorems and asymptotic equilibrium for delay differential and difference equations, *Dynam. Systems Appl.* **5** (1996), 277–302.

[9] Heard, M. L., A change of variables for functional differential equations, *J. Differential Equations* **18** (1975), 1–10.

[10] Kato, T. and McLeod, J.B., The functional-differential equation $y'(x) = ay(\lambda x) + by(x)$, *Bull. Am. Math. Soc.* **77** (1971), 891–937.

[11] Liu, Y., Numerical investigation of the pantograph equation, *Appl. Numer. Math.* **24** (1997), 309–317.

[12] Makay, G. and Terjéki, J., On the asymptotic behavior of thc pantograph equations, *Electron. J. Qual. Theory Differ. Equ.* **2** (1998), 1–12.

[13] Péics, H., On the asymptotic behavior of solutions of a system of functional equations, *Mathematica Perodica Hungarica*, to appear.

[14] Pituk, M., On the limits of solutions of functional differential equations, *Math. Bohemica* **118** No. 1 (1993), 53–66.

[15] Shen, J. and Stavroulakis, I.P., Oscillation criteria for delay difference equations, *Electron. J. Diff. Eqns* **2001** No. 10 (2001), 1–15.

Some Discrete Nonlinear Inequalities and Applications to Boundary Value Problems[1]

WING-SUM CHEUNG
Department of Mathematics
University of Hong Kong

Abstract In this note, we present some new discrete Gronwall-Bellman-Ou-Iang-type inequalities over 2-dimensional lattices. These on the one hand generalize some existing results in the literature and on the other hand provide a handy tool for the study of qualitative properties of solutions of difference equations. We illustrate this by applying these new results to certain boundary value problems for difference equations.

Keywords Discrete Gronwall-Bellman-Ou-Iang-type inequalities, boundary value problems

2000 AMS Subject Classification 26D15, 39A10, 39A70, 47J20

1 Introduction

It is well recognized that integral inequalities in general provide a very useful and handy device for the study of qualitative as well as quantitative properties of solutions of differential equations. Among various types of integral inequalities, the Gronwall-Bellman type (see, e.g. [3–6, 8, 10–12, 14–16]) is particularly useful in that they provide explicit bounds for the unknown functions. A specific branch of this type of integral inequalities is originated by Ou-Iang. In [13], in order to study the boundedness behavior of the solutions of certain 2nd-order differential equations, Ou-Iang established the following integral inequality which is now known as Ou-Iang's inequality in the literature.

Theorem (Ou-Iang [13]). *If u and f are nonnegative functions on $[0, \infty)$ satisfying*

$$u^2(x) \leq k^2 + 2 \int_0^x f(s)u(s)ds \ , \quad x \in [0, \infty),$$

[1]This work has been supported in part by the Research Grants Council of the Hong Kong SAR, China (Project No. HKU7040/03P).

for some constant $k \geq 0$, *then*

$$u(x) \leq k + \int_0^x f(s)ds \ , \quad x \in [0, \infty).$$

While Ou-Iang's inequality has a neat form and is interesting in its own right as an integral inequality, its importance lies equally heavily on its many beautiful applications in differential equations (see, e.g., [2,3,9,11,12]). Over the years, many generalizations of Ou-Iang's inequality to various situations have been established. Among these, the discretization is an interesting direction because one naturally expects that discrete versions of these inequalities should play an important role in the study of difference equations, just as the continuous versions play a fundamental role in the study of differential equations. Results in this direction can be found in, e.g., [1,16,17] and the references cited there.

The purpose of this note is to outline some new discrete Gronwall-Bellman-Ou-Iang-type inequalities with explicit bounds. Besides generalizing some existing results in the literature, these inequalities also provide a handy tool for the study of qualitative properties of solutions of difference equations. We illustrate this by applying these new inequalities to study the boundedness, uniqueness, and continuous dependence of the solutions of a boundary value problem for difference equations. Details of the work involved will appear in [7].

2 Discrete Gronwall-Bellman-Ou-Iang-type inequalities

Throughout this paper, $I := [m_0, M) \cap \mathbb{Z}$ and $J := [n_0, N) \cap \mathbb{Z}$ are two fixed lattices of integral points in $\mathbb{R}$, where $m_0, n_0 \in \mathbb{Z}$, $M, N \in \mathbb{Z} \cup \{\infty\}$. Let $\Omega := I \times J \subset \mathbb{Z}^2$, $\mathbb{R}_+ := [0, \infty)$, and for any $(s,t) \in \Omega$, the sublattice $[m_0, s] \times [n_0, t] \cap \Omega$ of Ω will be denoted as $\Omega_{(s,t)}$.

If U is a lattice in $\mathbb{Z}$ (or in $\mathbb{Z}^2$), the collection of all $\mathbb{R}$-valued functions on U is denoted by $\mathcal{F}(U)$, and that of all $\mathbb{R}_+$-valued functions by $\mathcal{F}_+(U)$. As usual, the collection of all continuous functions of a topological space X into a topological space Y will be denoted by $C(X, Y)$, the forward difference operator with stepsize 1 on 1-dimensional lattices will be denoted by Δ, and the partial forward difference operators with stepsize 1 on 2-dimensional lattices with respect to the first and second arguments by Δ_1 and Δ_2, respectively.

We start with a 2-dimensional discrete Gronwall-Bellman-type inequality (for a proof, see [7]).

Theorem 2.1. *Suppose $u \in \mathcal{F}_+(\Omega)$. If $c \geq 0$ is a constant and $b \in \mathcal{F}_+(\Omega)$, $w \in C(\mathbb{R}_+, \mathbb{R}_+)$ are functions satisfying*
(i) *w is nondecreasing with $w(r) > 0$ for $r > 0$; and*
(ii) *for any $(m,n) \in \Omega$,*

$$u(m,n) \leq c + \sum_{s=m_0}^{m-1} \sum_{t=n_0}^{n-1} b(s,t) w\big(u(s,t)\big),$$

then

$$u(m,n) \leq \Phi^{-1}\big[\Phi(c) + B(m,n)\big] \tag{1}$$

for all $(m,n) \in \Omega_{(m_1,n_1)}$, where

$$B(m,n) := \sum_{s=m_0}^{m-1} \sum_{t=n_0}^{n-1} b(s,t),$$
$$\Phi(r) := \int_1^r \frac{ds}{w(s)}, \quad r > 0,$$
$$\Phi(0) := \lim_{r \to 0^+} \Phi(r),$$

Φ^{-1} is the inverse of Φ, and $(m_1,n_1) \in \Omega$ is chosen such that $\Phi(c) + B(m,n) \in$ Dom(Φ^{-1}) for all $(m,n) \in \Omega_{(m_1,n_1)}$.

With Theorem 2.1 at hand, it is not too difficult to arrive at the following 2-dimensional discrete Ou-Iang-type inequality (for a proof, see [7]).

Theorem 2.2. *Suppose $u \in \mathcal{F}_+(\Omega)$. If $k \geq 0$ is a constant and $a, b \in \mathcal{F}_+(\Omega)$, $w \in C(\mathbb{R}_+, \mathbb{R}_+)$ are functions satisfying*
(i) *w is nondecreasing with $w(r) > 0$ for $r > 0$; and*
(ii) *for any $(m,n) \in \Omega$,*

$$u^2(m,n) \leq k^2 + \sum_{s=m_0}^{m-1} \sum_{t=n_0}^{n-1} a(s,t) u(s,t) + \sum_{s=m_0}^{m-1} \sum_{t=n_0}^{n-1} b(s,t) u(s,t) w\big(u(s,t)\big),$$

then

$$u(m,n) \leq \Phi^{-1}\Big[\Phi\big(k + A(m,n)\big) + B(m,n)\Big] \tag{2}$$

for all $(m,n) \in \Omega_{(m_1,n_1)}$, where

$$A(m,n) := \sum_{s=m_0}^{m-1} \sum_{t=n_0}^{n-1} a(s,t),$$

Φ and $B(m,n)$ are defined as in Theorem 2.1, and $(m_1,n_1) \in \Omega$ is chosen such that $\Phi\big(k + A(m,n)\big) + B(m,n) \in$ Dom Φ^{-1} for all $(m,n) \in \Omega_{(m_1,n_1)}$.

Remark. In many cases the nondecreasing function w satisfies $\int_1^\infty \frac{ds}{w(s)} = \infty$. For example, $w =$ constant > 0, $w(s) = s$, $w(s) = \sqrt{s}$, etc., are such functions. In such cases $\Phi(\infty) = \infty$ and so we may take $m_1 = M$, $n_1 = N$. In particular, inequalities (1) and (2) hold for all $(m,n) \in \Omega$.

In case Ω degenerates into a 1-dimensional lattice, Theorem 2.2 takes the following simpler form, which is equivalent to a result of Pachpatte in [16].

Corollary 2.3. *Suppose $u \in \mathcal{F}_+(I)$. If $k \geq 0$ is a constant and $a, b \in \mathcal{F}_+(I)$, $w \in C(\mathbb{R}_+, \mathbb{R}_+)$ are functions satisfying*

(i) *w is nondecreasing with $w(r) > 0$ for $r > 0$; and*

(ii) *for any $m \in I$,*

$$u^2(m) \leq k^2 + \sum_{s=m_0}^{m-1} a(s)u(s) + \sum_{s=m_0}^{m-1} b(s)u(s)w\big(u(s)\big),$$

then

$$u(m) \leq \Phi^{-1}\left[\Phi\Big(k + \sum_{s=m_0}^{m-1} a(s)\Big) + \sum_{s=m_0}^{m-1} b(s)\right]$$

for all $m \in [m_0, m_1] \cap I$, where Φ is defined in Theorem 2.1, and $m_1 \in I$ is chosen such that $\Phi\Big(k + \sum\limits_{s=m_0}^{m-1} a(s)\Big) \in Dom\ \Phi^{-1}$ for all $m \in [m_0, m_1] \cap I$.

By choosing $w =$ the identity mapping, Theorem 2.2 gives the following useful consequence.

Corollary 2.4. *Suppose $u \in \mathcal{F}_+(\Omega)$. If $k \geq 0$ is a constant and $a, b \in \mathcal{F}_+(\Omega)$ are functions such that for any $(m,n) \in \Omega$,*

$$u^2(m,n) \leq k^2 + \sum_{s=m_0}^{m-1}\sum_{t=n_0}^{n-1} a(s,t)u(s,t) + \sum_{s=m_0}^{m-1}\sum_{t=n_0}^{n-1} b(s,t)u^2(s,t),$$

then

$$u(m,n) \leq \big[k + A(m,n)\big] \exp B(m,n)$$

for all $(m,n) \in \Omega$, where $A(m,n)$ and $B(m,n)$ are as defined in Theorem 2.2.

In case Ω degenerates into a 1-dimensional lattice, Corollary 2.4 takes the following simpler form, which is equivalent to another result of Pachpatte in [16].

Corollary 2.5. *Suppose $u \in \mathcal{F}_+(I)$. If $k \geq 0$ is a constant and $a, b \in \mathcal{F}_+(I)$ are functions such that for any $m \in I$,*

$$u^2(m) \leq k^2 + \sum_{s=m_0}^{m-1} a(s)u(s) + \sum_{s=m_0}^{m-1} b(s)u^2(s),$$

then

$$u(m) \leq \Big[k + \sum_{s=m_0}^{m-1} a(s)\Big] \prod_{s=m_0}^{m-1} \exp b(s)$$

for all $m \in I$.

Another special situation of Corollary 2.4 is the following 2-dimensional discrete version of Ou-Iang's inequality, which is readily seen by taking $a \equiv 0$ there.

Corollary 2.6. *Suppose $u \in \mathcal{F}_+(\Omega)$. If $k \geq 0$ is a constant and $b \in \mathcal{F}_+(\Omega)$ is a function such that for any $(m, n) \in \Omega$,*

$$u^2(m,n) \leq k^2 + \sum_{s=m_0}^{m-1} \sum_{t=n_0}^{n-1} b(s,t)u^2(s,t),$$

then

$$u(m,n) \leq k \exp B(m,n)$$

for all $(m,n) \in \Omega$, where $B(m,n)$ is as defined in Theorem 2.1.

By taking $a \equiv 0$ in Corollary 2.5, or by considering in Corollary 2.6 the case where Ω degenerates into a 1-dimensional lattice, we have the following 1-dimensional discrete analog of Ou-Iang's inequality.

Corollary 2.7. *Suppose $u \in \mathcal{F}_+(I)$. If $k \geq 0$ is a constant and $b \in \mathcal{F}_+(I)$ is a function such that for any $m \in I$,*

$$u^2(m) \leq k^2 + \sum_{s=m_0}^{m-1} b(s)u^2(s),$$

then

$$u(m) \leq k \prod_{s=m_0}^{m-1} \exp b(s)$$

for all $m \in I$.

Remark. It is evident that the results above can be generalized to obtain explicit bounds for functions satisfying certain discrete sum inequalities involving more retarded arguments. It is also clear that these results can be extended to functions on higher dimensional lattices in the obvious way.

3 Applications to boundary value problems

In this section, we shall illustrate how the results obtained in §2 can be applied to study the boundedness, uniqueness, and continuous dependence of the solutions of certain boundary value problems for difference equations involving 2 independent variables. We consider the following:
Boundary Value Problem (BVP):

$$\begin{cases} \Delta_{12}z^2(m,n) = f(m,n,z(m,n)) \\ z(m,n_0) = p(m), \; z(m_0,n) = q(n), \; p(m_0) = q(n_0) = 0, \end{cases} \tag{BVP}$$

where $\Delta_{12} = \Delta_2\Delta_1$, $f \in \mathcal{F}(\Omega \times \mathbb{R})$, $p \in \mathcal{F}(I)$, and $q \in \mathcal{F}(J)$ are given.

Our first result deals with the boundedness of solutions.

Theorem 3.1. *Consider (BVP). For any $(m,n) \in \Omega$ and any $v \in \mathbb{R}$, if*

$$\left|f(m,n,v)\right| \le b(m,n)|v|^2$$

and

$$p^2(m) + q^2(n) \le k^2$$

for some $k \ge 0$, where $b \in \mathcal{F}_+(\Omega)$, then all solutions of (BVP) satisfy

$$|z(m,n)| \le k \exp B(m,n), \quad (m,n) \in \Omega,$$

where $B(m,n)$ is defined as in Theorem 2.1. In particular, if $B(m,n)$ is bounded on Ω, then every solution of (BVP) is bounded on Ω.

The next result is about uniqueness.

Theorem 3.2. *Consider (BVP). For any $(m,n) \in \Omega$ and any $v_1, v_2 \in \mathbb{R}$, if*

$$\left|f(m,n,v_1) - f(m,n,v_2)\right| \le b(m,n)\left|v_1^2 - v_2^2\right|$$

for some $b \in \mathcal{F}_+(\Omega)$, then (BVP) has at most one solution on Ω.

Finally, we investigate the continuous dependence of the solutions of (BVP) on the function f and the boundary data p and q. For this we consider the following variation of (BVP):
($\overline{\text{BVP}}$):

$$\Delta_{12}z^2(m,n) = \overline{f}(m,n,z(m,n))$$

with

$$z(m,n_0) = \overline{p}(m), \; z(m_0,n) = \overline{q}(n), \; \overline{p}(m_0) = \overline{q}(n_0) = 0,$$

where $\overline{f} \in \mathcal{F}(\Omega \times \mathbb{R})$, $\overline{p} \in \mathcal{F}(I)$, and $\overline{q} \in \mathcal{F}(J)$ are given.

Theorem 3.3. *Consider (BVP) and ($\overline{BVP}$). For any $(m,n) \in \Omega$ and any $v_1, v_2 \in \mathbb{R}$, if*

(i) $|f(m,n,v_1) - f(m,n,v_2)| \leq b(m,n)|v_1^2 - v_2^2|$ *for some* $b \in \mathcal{F}_+(\Omega)$;

(ii) $\left|\left(p^2(m) - \overline{p}^2(m)\right) + \left(q^2(n) - \overline{q}^2(n)\right)\right| \leq \frac{\varepsilon}{2}$ *for some* $\varepsilon > 0$; *and*

(iii) *for all solutions* $\overline{z}(m,n)$ *of* ($\overline{BVP}$),

$$\sum_{s=m_0}^{m-1} \sum_{t=n_0}^{n-1} \left| f\left(s,t,\overline{z}(s,t)\right) - \overline{f}\left(s,t,\overline{z}(s,t)\right) \right| \leq \frac{\varepsilon}{2},$$

then for any solution z of (BVP),

$$\left| z^2(m,n) - \overline{z}^2(m,n) \right| \leq \varepsilon \exp\left(2B(m,n)\right),$$

where $B(m,n)$ is as defined in Theorem 2.1. Hence z^2 depends continuously on f, p, and q. In particular, if z does not change sign, it depends continuously on f, p and q.

Remark. The boundary value problem (BVP) is clearly not the only problem for which the boundedness, uniqueness, and continuous dependence of its solutions can be studied by using the results in §2. For example, one can arrive at similar results for the following variation of the (BVP):

$$\Delta_{12} z^2(m,n) = f\Big(m, n, z(m,n),\ z(m,n) \cdot w\big(|z(m,n)|\big)\Big)$$

with

$$z(m,n_0) = p(m),\ z(m_0,n) = q(n),\ p(m_0) = q(n_0) = 0,$$

where $f \in \mathcal{F}(\Omega \times \mathbb{R}^2)$, $p \in \mathcal{F}(I)$, $q \in \mathcal{F}(J)$, and $w \in C(\mathbb{R}_+, \mathbb{R}_+)$ are given.

References

[1] Agarwal, R.P., *Difference Equations and Inequalities*, Marcel Dekker, New York, 2000.

[2] Bainov, D. and Simeonov, P., *Integral Inequalities and Applications*, Kluwer Academic Publishers, Dordrecht, 1992.

[3] Beckenbach, E.F. and Bellman, R., *Inequalities*, Springer-Verlag, New York, 1961.

[4] Bellman, R., The stability of solutions of linear differential equations, *Duke Math. J.* **10** (1943), 643–647.

[5] Bihari, I., A generalization of a lemma of Bellman and its application to uniqueness problems of differential equations, *Acta Math. Acad. Sci. Hungar.* **7** (1956), 71–94.

[6] Cheung, W.S., On some new integrodifferential inequalities of the Gronwall and Wendroff type, *J. Math. Anal. Appl.* **178** (1993), 438–449.

[7] Cheung, W.S., Some discrete nonlinear inequalities and applications to boundary value problems for difference equations, *J. Difference Equ. Appl.* **10**, No.2 (2004), 213–223.

[8] Cheung, W.S. and Ma, Q.H., Nonlinear retarded integral inequalities for functions in two variables, *J. Concrete Appl. Math.* (in press).

[9] Haraux, H., *Nonlinear Evolution Equation: Global Behavior of Solutions*, Lecture Notes in Mathematics, vol. 841, Springer-Verlag, Berlin, 1981.

[10] Ma, Q.M. and Yang, E.H., On some new nonlinear delay integral inequalities, *J. Math. Anal. Appl.* **252** (2000), 864–878.

[11] Mitrinović, D.S., *Analytic Inequalities*, Springer-Verlag, New York, 1970.

[12] Mitrinović, D.S., Pečarić, J.E., and Fink, A.M., *Inequalities Involving Functions and Their Integrals and Derivatives*, Kluwer Academic Publishers, Dordrecht, 1991.

[13] Ou-Iang, L., The boundedness of solutions of linear differential equations $y'' + A(t)y = 0$, *Shuxue Jinzhan* **3** (1957), 409–415.

[14] Pachpatte, B.G., Explicit bounds on certain integral inequalities, *J. Math. Anal. Appl.* **267** (2002), 48–61.

[15] Pachpatte, B.G., *Inequalities for Differential and Integral Equations*, Academic Press, New York, 1998.

[16] Pachpatte, B.G., On some new inequalities related to certain inequalities in the theory of differential equations, *J. Math. Anal. Appl.* **189** (1995), 128–144.

[17] Pang, P.Y.H. and Agarwal, R.P., On an integral inequality and discrete analogue, *J. Math. Anal. Appl.* **194** (1995), 569–577.

The Asymptotic Stability of $\mathbf{x(n+k)+ax(n)+bx(n-\mathit{l})=0}$

F.M. DANNAN

Mathematics Department

Faculty of Science

Quatar University, Doha-Qatar

Abstract We established necessary and sufficient conditions for the asymptotic stability of the difference equation

$$x(n+k)+ax(n)+bx(n-l)=0, n=0,1,2,...,$$

where the coefficients a and b are real numbers and k and l are nonnegative integers.

Keywords Fibonacci numbers, Lucas numbers, Pell numbers

AMS Subject Classification 11B39

1 Introduction

Nonlinear difference equations arise in mathematical models in biology, economics, dynamical systems, statistics, computing, electrical circuits analysis, and other fields.

The question of local stability of nonlinear difference equation can be reduced to the corresponding linearized equations. The asymptotic stability of the linear difference equations is reduced to a discussion of the location of the roots of the corresponding characteristic (polynomial) equations. Consider the difference equation

$$x(n+k)+ax(n)+bx(n-l)=0, \qquad n=0,1,2,\dots, \tag{1}$$

where k, l are nonnegative integers and a, b are real.

It is well known that (1) is asymptotically stable if and only if all the roots of its characteristic equation

$$\lambda^{\ell+k}+a\lambda^{\ell}+b=0 \tag{2}$$

are inside the unit disk.

There are several theorems (Marden [6], Theorems 42.1 and 43.2) that provide necessary and sufficient conditions for the roots of a polynomial to be inside the unit disk. However, the conditions established by such theorems are extremely complicated. For this reason, a series of recent papers established necessary and sufficient conditions for the asymptotic stability of Eq. (1) in particular cases. For $k = 1$ and $l > 1$, Clark [1] showed that $|a| + |b| < 1$ is a sufficient condition for asymptotic stability. For the case $k = 1$, $a = -1$, $l > 0$, Levin and May [5] showed that

$$0 < b < 2\cos\frac{l\pi}{2l+1} \tag{3}$$

is a necessary and sufficient condition for asymptotic stability. Kuruklis [4] later discussed the location of the roots of Eq. (2) for $k = 1$ and established necessary and sufficient conditions for the asymptotic stability of

$$x(n+1) + ax(n) + bx(n-l) = 0. \tag{4}$$

Gyori and Ladas ([3] p. 163) have described the equation

$$x(n) - ax(n+1) + bx(n+k) = 0, \qquad n = 0, 1, 2, \dots, \tag{5}$$

where a and b are arbitrary reals and $k > 1$ is an integer, as "advanced". Dannan [2] obtained necessary and sufficient conditions for all the roots of its characteristic equation

$$b\lambda^k - a\lambda + 1 = 0 \tag{6}$$

to lie inside the unit disk. This implies necessary and sufficient conditions for asymptotic stability of Eq. (5). Notice that Eq. (6) is a particular case of Eq. (2),with $l = 1$.

In this article we establish necessary and sufficient conditions for the asymptotic stability of Eq. (1).

2 Auxiliary lemmas

The trivial cases when $a = b = 0$ or $a = 0, b \neq 0$ or $a \neq 0, b = 0$ are easily treated. Thus we assume that $ab \neq 0$. Without loss of generality we may assume from now on that ℓ and k are relatively prime. This follows from

Lemma 2.1. *If $l = sl'$ and $k = sk'$ for some positive integers s, l' and k', then all roots of Eq. (2) lie inside the unit disk if and only if all roots of*

$$u^{\ell'+k'} + au^{\ell'} + b = 0 \tag{7}$$

lie inside the unit disk.

Proof. Equation (2) can be written as

$$v^{\ell'+k'} + av^{\ell'} + b = 0 \tag{8}$$

where $v = \lambda^s$. Assume that any root λ of Eq. (2) satisfies $|\lambda| < 1$. Then $|v| = |\lambda^s| < 1$. Therefore all roots of Eq. (7) lie inside the unit disk. If any root u of Eq. (7) satisfies $|u| < 1$, then $|v| = |\lambda^s| < 1$, so $|\lambda| < 1$ and all roots of Eq. (2) lie inside the unit disk.
Note that for $a > 0$ Eq. (2) can be written as

$$z^{\ell+k} + z^{\ell} + p = 0, \tag{9}$$

where $z = \lambda/(a)^{\frac{1}{k}}$ and $p = b/(a)^{\frac{\ell+k}{k}}$. For $a < 0$ Eq. (2) can be written as

$$z^{\ell+k} - z^{\ell} + p = 0, \tag{10}$$

where $z = \lambda/\,|a|^{\frac{1}{k}}$ and $p = b/\,|a|^{\frac{\ell+k}{k}}$. Whenever $a > 0$ or $a < 0$, determining asymptotic stability of Eq. (1), and thus the location of the roots of Eq.(2) with respect to the unit circle, is reduced to determining the location of the roots of Eqs.(9) and (10) with respect to the circle $|z| = h$, where $h = 1/\,|a|^{\frac{1}{k}}$. The possible cases are summarized as follows: For Eq.(10), where l and k are positive integers, we have three main cases:

Case A: l and k are odd $(a < 0)$.

Case B: l is odd and k is even $(a < 0)$.

Case C: l is even and k is odd $(a < 0)$.

For Eq. (9), we have

Case D: l is odd and k is even $(a > 0)$.

If we replace z by $-z$ in Eq. (9), then it reduces to Eq. (10): Case A) Eq. (10) with $-p$ instead of p: Case B) when l and k are odd (l is even and k is odd). If $h = 1/\,|a|^{\frac{1}{k}}$ we will consider the subcases: $h = 1$, $h > 1$, and $h < 1$.

3 Main results

The following Theorems give the necessary and sufficient conditions for the roots of Eq. (2) to lie inside the unit disk. The proofs of these theorems will be published later.

Theorem 3.1. *Let $l \geqslant 1$ and $k > 1$ be relatively prime odd integers. Then the roots of Eq. (2) lie inside the unit disk if and only if $|a| < 1$ and*

$$|a| - 1 < b < \min_{\theta \in S}(1 + a^2 - 2\,|a| \cos k\theta)^{\frac{1}{2}}, \tag{11}$$

where S is the solution set of

$$\frac{1}{|a|} = \frac{\sin l\theta}{\sin(l+k)\theta} \tag{12}$$

on the interval $(0, \pi)$.

Theorem 3.2. *Let $l \geqslant 1$ be an odd integer, k be an even integer with l and k relatively prime. Then the roots of Eq. (2) lie inside the unit disk if and only if*

$$|b| < 1 - |a| \qquad for \qquad -1 < a < 0 \tag{13}$$

and

$$|b| < \min_{\theta \in S^*}(1 + a^2 + 2a\cos k\theta)^{\frac{1}{2}} \qquad for \qquad 0 < a < 1, \tag{14}$$

where S^* is the solution set of

$$-\frac{1}{a} = \frac{\sin l\theta}{\sin(l+k)\theta} \tag{15}$$

on the interval $(0, \pi)$.

Theorem 3.3. *Let l be an even integer and $k > 1$ be an odd integer, with l and k relatively prime. Then the roots of Eq. (2) lie inside the unit disk if and only if*

$$|a| - 1 < b < \min_{\theta \in S}(1 + a^2 - 2|a|\cos k\theta)^{\frac{1}{2}} \qquad for \qquad -1 < a < 0 \tag{16}$$

and

$$a - 1 < -b < \min_{\theta \in S}(1 + a^2 - 2a\cos k\theta)^{\frac{1}{2}} \quad for \quad 0 < a < 1. \tag{17}$$

where S is the solution set of

$$\frac{1}{|a|} = \frac{\sin l\theta}{\sin(l+k)\theta} \tag{18}$$

in $(0, \pi)$.

Now we present our main results on the asymptotic stability of the difference equation (1), $x(n+k) + ax(n) + bx(n-l) = 0, \quad n = 0, 1, ...,$ which which follows from Theorems 3.1–3.3.

Theorem 3.4. *Assume that $l \geqslant 1$ and $k > 1$ are relatively prime odd integers. Then Eq. (1) is asymptotically stable if and only if $|a| < 1$ and*

$$|a| - 1 < b < \min_{\theta \in S}(1 + a^2 - 2|a|\cos k\theta)^{\frac{1}{2}},$$

where S is the solution set of $\frac{1}{|a|} = \frac{\sin l\theta}{\sin(l+k)\theta}$ on the interval $(0, \pi)$.

Theorem 3.5. *Assume that $l \geq 1$ is an odd integer, k is an even integer, and l and k are relatively prime. Then Eq. (1) is asymptotically stable if and only if*

$$|b| < 1 - |a| \quad for \qquad -1 < a < 0$$

and

$$|b| < \min_{\theta \in S^*} \min(1 + a^2 + 2a\cos k\theta)^{\frac{1}{2}} \quad for \quad 0 < a < 1,$$

where S^* is the solution set of $-\frac{1}{a} = \frac{\sin \ell\theta}{\sin(\ell+k)\theta}$ on the interval $(0, \pi)$.

Theorem 3.6. *Assume that l is an even integer, $k > 1$ is an odd integer and l and k are relatively prime. Then Eq. (1) is asymptotically stable if and only if*

$$|a| - 1 < b < \min_{\theta \in S}(1 + a^2 - 2|a| \cos k\theta)^{\frac{1}{2}} \quad \textit{for} \quad -1 < a < 0$$

and

$$a - 1 < -b < \min_{\theta \in S}(1 + a^2 - 2a \cos k\theta)^{\frac{1}{2}} \quad \textit{for} \quad 0 < a < 1,$$

where S^* is the solution set of $\frac{1}{|a|} = \frac{\sin l\theta}{\sin(l+k)\theta}$ on the interval $(0, \pi)$.

Consider the difference equation

$$x(n + 25) + a\, x(n) + b\, x(n - 15) = 0, \quad n = 0, 1, 2, \ldots$$

The corresponding characteristic equation is

$$\lambda^{40} + a\lambda^{25} + b = 0,$$

and in the reduced form is

$$\lambda^8 + a\lambda^5 + b = 0.$$

Here we have $\ell = 5$, and $k = 3$. Therefore Theorem 3.5 is applicable and the given equation is asymptotically stable if and only if $|a| < 1$, and

$$|a| - 1 < b < \min_{\theta \in S}(1 + a^2 - 2|a| \cos 3\theta)^{\frac{1}{2}},$$

where S is the solution set of $\frac{1}{|a|} = \frac{\sin 5\theta}{\sin 8\theta}$ on the interval $(0, \pi)$. If we let $a = 0.6$, then the given equation is asymptotically stable if and only if $-0.4 < b < 0.4477703541$.

References

[1] C.W. Clark, A delay-recruitment model of population dynamics with application to baleen whale populations , *J. Math. Biol.* **3** (1976), 381–391.

[2] F.M. Dannan, S. Elaydi, Asymptotic stability of linear difference equation of advanced type (in press).

[3] I. Gyori, G.Ladas, *Oscillation Theory of Delay Differential Equations with Applications* , Clarendon Press, Oxford, 1991.

[4] S.A. Kuruklis, The asymptotic stability of $x(n+1) - ax(n) + bx(n-k) = 0$, J. Math. Anal. Appl. **188** (1994), 719–731.

[5] S.A. Levin and R.M. May, A note on difference delay equations, Theoret Popul. Biol., **9** (1976), 178–187 .

[6] M. Marden, *Geometry of Polynomials*, American Mathematical Society, Providence, RI, 1966.

Estimates of the Spectral Radius and Oscillation Behavior of Solutions to Functional Equations

A. DOMOSHNITSKY
College of Judea and Samaria, Ariel, Israel

and

M. DRAKHLIN[1]
College of Judea and Samaria, Ariel, Israel

Abstract In this paper it will be demonstrated that sign behavior of solutions of functional equations can be described through the spectral radius of a corresponding operator acting in the space of essentially bounded functions.

Keywords Integro-functional equation, spectral radius, upper and lower solution
AMS Subject Classification 39A10

1 Introduction

The main object in this paper is the following functional equation

$$y(t) = (Ty)(t) + f(t), \quad t \in [0, +\infty), \tag{1}$$

where $T : L^{\infty}_{[0,+\infty)} \to L^{\infty}_{[0,+\infty)}$ is a linear positive operator ($L^{\infty}_{[0,+\infty)}$ is the space of measurable essentially bounded functions $y : [0, +\infty) \to (-\infty, +\infty)$), $f \in L^{\infty}_{[0,+\infty)}$ is a given function. The integro-functional equation

$$y(t) = \sum_{i=1}^{m} p_i(t) y(g_i(t)) + \int_{h_1(t)}^{h_2(t)} k(t, s) y(s) ds, \quad t \in [0, +\infty), \tag{2}$$

$$y(\xi) = \varphi(\xi), \quad \xi < 0. \tag{3}$$

[1]The research of both authors was supported by the KAMEA and Giladi Programs of the Ministry of Absorption of the State of Israel.

is a particular case of equation (1). Let us assume that $p_i : [0, +\infty) \to [0, +\infty)$ $(i = 1, ..., m)$ and $k(\cdot, \cdot) : [0, +\infty) \times (-\infty, +\infty) \to [0, +\infty)$ are measurable bounded in essential nonnegative functions, g_i $(i = 1, ..., m)$ are measurable functions, h_1 and h_2 are continuous functions such that the differences $t - h_j(t)$ $(j = 1, 2)$ and $t - g_i(t)$ $(i = 1, ..., m)$ are bounded on $[0, +\infty)$, and among these functions there are both delays $(g_i(t) \leq t,\ i = 1, ..., n,\ h_1(t) \leq t)$ and advances $(g_i(t) \geq t,\ i = n + 1, ..., m,\ h_2(t) \geq t)$.

The following equation

$$y(t) = \sum_{i=1}^{m} p_i(t) y(g_i(t)), \quad t \in [0, +\infty), \tag{4}$$

$$y(\xi) = \varphi(\xi), \quad \xi < 0. \tag{5}$$

is an important particular case of equation (2), (3).

If we set in equation (4) $g_i(t) = t - i$, where i is an integer number, coefficients $p_i(t)$ and the initial function $\varphi(t)$ are constants on each interval $[k, k + 1)$ we obtain that corresponding solutions $y(t)$ are constants on each interval $[k, k + 1) : y(t) = y_k$. In this case we actually have a difference equation

$$y_n = \sum_{k=-m}^{m} p_k y_{n-k}, \quad n \in [1, +\infty), \tag{6}$$

$$y(n) = \varphi_n, \quad n < 0. \tag{7}$$

Our approach is a way to a corresponding analog for functional equations of the method of upper and lower solutions known for difference equations. This method can be combined in a future with the monotone iterative technique. For difference equations we can note in this connection the monograps by R.Agarwal [1], V.Lakshmikantham and D.Trigante [7], Y.Domshlak A.Matakajev, A.Askiev [3] and the papers by D.Jiang, Lili Zhang and R.Agarwal [5], Y.M.Wang [10] and W.Zhuang, Y.Chen and S.S.Cheng [11].

Note that oscillation of difference equations was considered, for example, in the following papers [3,4,6,9] and of functional equations- in [8].

Definition 1.1. A function $y \in L^{\infty}_{[0,+\infty)}$ satisfying equation (1) for almost every $t \in [0, +\infty)$ is called a solution of (1)

Considering solutions in the space $L^{\infty}_{[0,+\infty)}$ we should define the notions of zeros, oscillation and nonoscillation for non-continuous functions.

Definition 1.2. A point ν *is a zero of a function* $y \in L^{\infty}_{[0,+\infty)}$ *if either product*

$$\lim_{t \to \nu-} vrai \sup_{s \in [t,\nu]} y(s) \cdot \lim_{t \to \nu+} vrai \inf_{s \in [\nu,t]} y(s)$$

or

$$\lim_{t \to \nu-} vrai \inf_{s \in [t,\nu]} y(s) \cdot \lim_{t \to \nu+} vrai \sup_{s \in [\nu,t]} y(s)$$

is nonpositive.

Obviously, if y is a continuous function, then a point ν is its zero if and only if $y(\nu) = 0$. Zeros determined by this definition include "traditional" zeros as well as points of sign change. Note, for example, that the function $y(t) = \sin\frac{1}{t}$, $t \in [-1, 1]$ has a zero at the point $t = 0$ according to Definition 1.2.

Definition 1.3. We say that a function $y \in L^{\infty}_{[0,+\infty)}$ *nonoscillates* if there exists ν such that $y(t)$ does not have zeros for $t \in (\nu, +\infty)$. If there exists a sequence of zeros of a function y tending to infinity, we say that a function $y \in L^{\infty}_{[0,+\infty)}$ *oscillates.*

Let $L^{\infty}_{[\nu,\mu]}$, where $0 \leq \nu < \mu \leq \infty$, be the space of measurable essentially bounded functions $x : (\nu, \mu) \to (-\infty, +\infty)$. In order to study oscillation properties of Equation (1) let us introduce the following operator $T_{\nu,\mu}$: $L^{\infty}_{[\nu,\mu]} \to L^{\infty}_{[\nu,\mu]}$ as follows

$$(T_{\nu,\mu}x)(t) = (Ty)(t), \tag{8}$$

where $y \in L^{\infty}_{[0,+\infty)}$ is such that

$$y(t) = \begin{cases} x(t), & \nu \leq t \leq \mu, \\ 0, & \text{elsewhere}, \end{cases}$$

Note that for Equation (2) this operator $T_{\nu\mu} : L^{\infty}_{[\nu,\mu]} \to L^{\infty}_{[\nu,\mu]}$ is determined by the equality

$$(T_{\nu\mu}x)(t) = \sum_{i=1}^{m} p_i(t)x(g_i(t)) + \int_{h_1(t)}^{h_2(t)} k(t,s)x(s)ds, \quad t \in [\nu, \mu], \tag{9}$$

where $x(\xi) = 0$ if either $\xi < \nu$ or $\xi > \mu$.

The problems of oscillation are reduced to estimates of the spectral radius of the operator $T_{\nu\mu}$. Note that the result of an act of the operator $T_{\nu\mu}$: $L^{\infty}_{[\nu,\mu]} \to L^{\infty}_{[\nu,\mu]}$, determined by equality (9), on a continuous functions can be generally speaking a noncontinuous function. That is a reason to consider Equation (1) in the space $L^{\infty}_{[\nu,\mu]}$.

Relations between lower estimates of the spectral radius of the operator $T_{\nu\mu} : L^{\infty}_{[\nu,\mu]} \to L^{\infty}_{[\nu,\mu]}$ and oscillation will be obtained in Part 2, and between upper estimates of the spectral radius of the operator $T_{\nu\mu}$ and nonoscillation - in Part 3.

2 Lower estimates of the spectral radius and oscillation of solutions

Let us fix an interval (ν, μ) and define

$$\sigma_{\nu,\mu}(t) = \begin{cases} 1, & \nu \leq t \leq \mu, \\ 0, & \text{elsewhere}, \end{cases}$$

and $\theta^-(\nu)$, $\theta^+(\mu)$ such that

$$(Ty)(t) = (T\sigma_{\theta^-(\nu),\theta^+(\mu)}y)(t), \; t \in [\nu, \mu], \tag{10}$$

for each function $y \in L^\infty_{[0,+\infty)}$. Obviously, $\theta^-(\nu) = 0, \;\; \theta^+(\mu) = +\infty$ satisfy this condition. It will be clear below that we are interested in choosing the maximal possible (for equality (10)) number $\theta^-(\nu)$ and the minimal possible $\theta^+(\mu)$. The values $\theta^-(\nu)$ and $\theta^+(\mu)$ describe bounds of the "memory" of the operator T on the interval (ν, μ).

Theorem 2.1 [2]. *Let there exist h such that $f(t) = 0$ for $t > h$. If the spectral radius $r_{\nu\mu}$ of the operator $T_{\nu\mu}$ satisfies the inequality $r_{\nu\mu} > 1$, where $\nu > h$, then there is no solution of Equation* (1) *without zero in at least one of the intervals $[\theta^-(\nu), \mu]$ or $[\nu, \theta^+(\mu)]$.*

In order to formulate a corresponding result for Equation (2), let us denote

$$\begin{aligned} h &= \min\{t : h_1(t) \geq 0, \, g_i(t) \geq 0, \;\; i = 1, \ldots, n\}, \\ \theta^-(\nu) &= \min_{t\in[\nu,\mu]} \{h_1(t), \; g_i(t), \; i = 1, ..., m\}, \\ \theta^+(\mu) &= \max_{t\in[\nu,\mu]} \{h_2(t), \; g_i(t), \;\; i = 1, ..., m\}. \end{aligned}$$

Corollary 2.1. *Let $\nu > h$ and the spectral radius $r_{\nu\mu}$ of the operator $T_{\nu\mu}$ determined by* (9) *satisfy the inequality $r_{\nu\mu} > 1$, then there are no solutions of Equation* (2) *without zero in at least one of the intervals $[\theta^-(\nu), \mu]$ or $[\nu, \theta^+(\mu)]$.*

Theorem 2.2. *There are no nonoscillating solutions of homogeneous Equation* (1) *(i.e., $f = 0$ in* (1)*) if for each ν there exists μ $(\nu < \mu < \infty)$ such that the spectral radius $r_{\nu\mu}$ of the operator $T_{\nu\mu}$ satisfies the inequality $r_{\nu\mu} > 1$*

3 Upper estimates of the spectral radius and positivity of solutions

Theorem 3.1 [2]. *The following assertions are equivalent:*

(1) *The spectral radius r of the operator $T : L^\infty_{[0,+\infty)} \to L^\infty_{[0,+\infty)}$ is less than one;*

(2) *For each positive* $f \in L^{\infty}_{[0,+\infty)}$ *there exists a unique positive solution* $y \in L^{\infty}_{[0,+\infty)}$ *of Equation* (1);

(3) *There exists a positive function* $v \in L^{\infty}_{[0,+\infty)}$ *without zeros such that the function* $\psi(t) \equiv v(t) - (Tv)(t)$ *is positive without zeros for* $t \in [0, +\infty)$.

Corollary 3.1. *If the spectral radius* r *of the operator* $T : L^{\infty}_{[0,+\infty)} \to L^{\infty}_{[0,+\infty)}$ *is less than one and functions* $\varphi_1, \varphi_2 \in L^{\infty}_{[0,+\infty)}$ *satisfy the inequality* $\varphi_2(t) \geq f(t) \geq \varphi_1(t)$ *for* $t \in [0, +\infty)$, *then a solution* $y \in L^{\infty}_{[0,+\infty)}$ *of Equation* (1) *and solutions* $z_i \in L^{\infty}_{[0,+\infty)}$ *of the equations*

$$y(t) = (Ty)(t) + \varphi_i(t), \quad t \in [0, +\infty), \ i = 1, 2, \tag{11}$$

satisfy the inequality $z_1(t) \leq y(t) \leq z_2(t)$ *for* $t \in [0, +\infty)$.

For each fixed ν we can write

$$(Ty)(t) = (T_{0,\nu}y)(t) + (T_{\nu,+\infty}y)(t), \ t \in [\nu \ , +\infty), \tag{12}$$

Definition 3.1. We say that operator $T_{0,\nu}$ is *strongly positive* if $(T_{0,\nu}y)(t) > 0, \ t \in [\nu \ , +\infty)$ for each $y \in L^{\infty}_{[0,+\infty)}$ such that $y(t) > 0$, $t \in [0, \nu)$.

Corollary 3.2 [2]. *If* $\sup_{t \in [0,+\infty)} (Ty)(t) < 1$ *and* $f \in L^{\infty}_{[0,+\infty)}$ *is nonnegative (nonpositive), then a solution* $y \in L^{\infty}_{[0,+\infty)}$ *of Equation* (1) *is nonpositive (nonnegative). If, in addition,* f *is positive (negative) for* $t \in [0, \varepsilon]$ *and the operators* $T_{0,\nu}$ *are strongly positive for each* $\nu > \varepsilon$, *then each solution is positive (negative) for* $t \in [0, +\infty)$.

Theorem 3.2 [2]. *Let the following inequality be fulfilled*

$$\sup_{t \in [0,+\infty)} \sum_{i=1}^{m} p_i(t) + \int_{h_1(t)}^{h_2(t)} k(t,s)\sigma_{0,\infty}(s)ds < 1, \tag{13}$$

then

(1) *The spectral radius of the operator* $T : L^{\infty}_{[0,+\infty)} \to L^{\infty}_{[0,+\infty)}$ *determined by the formula* (9) *(with* $\nu = 0$ *and* $\mu = \infty$*) is less than one;*

(2) *If, in addition,* φ *is a positive (negative) initial function and the inequality*

$$\sum_{i=1}^{n} p_i(t) + \int_{h_1(t)}^{t} k(t,s)\sigma_{0,\infty}(s)ds > 0, \quad t \in [0, +\infty), \tag{14}$$

holds, then the corresponding solution $y \in L^{\infty}_{[0,+\infty)}$ *of* (2), (3) *is positive (negative) for* $t \in [0, +\infty)$.

Let us now consider the following equations

$$b(t)y(t) = a(t)y(t+1) + c(t)y(t-1), \quad t \in [0, +\infty), \tag{15}$$

and

$$b_0 y(t) = a_0 y(t+1) + c_0 y(t-1), \quad t \in [0, +\infty), \tag{16}$$

in which a, b and c are positive functions from the space $L^{\infty}_{[0,+\infty)}$, and a_0, b_0, and c_0 are positive constants.

If we substitute $v = \beta^t$ in the condition 3) of Theorem 3.1, then the following inequality

$$a_0\beta^2 - b_0\beta + c_0 < 0 \tag{17}$$

is obtained. In this case the spectral radius of the operator $T_0 : L^{\infty}_{[0,+\infty)} \to L^{\infty}_{[0,+\infty)}$ determined by

$$(T_0 y)(t) = \frac{a_0}{b_0} y(t+1) + \frac{c_0}{b_0} y(t-1), \tag{18}$$

$$y(\xi) = 0, \ \xi < 0, \tag{19}$$

is less than one. We have proved the following assertion for Equation (16).

Theorem 3.3. *Let the following inequality be fulfilled*

$$b_0^2 > 4a_0c_0, \tag{20}$$

then the spectral radius of the operator T_0 is less than one and a solution of (16), (5) *is positive for each positive initial function φ.*

Remark 3.1. It is known (see, for example, [9]) that the inequality

$$b_0^2 < 4a_0c_0,$$

implies oscillation of all solutions of Equation (16). Thus, inequality (20) cannot be improved.

Theorem 3.4 [2]. *Let inequalities* (20) *and $a(t) \le a_0$, $b(t) \ge b_0$, $c(t) \le c_0$ be fulfilled, then a solution of* (15), (5) *is positive for each positive initial function φ.*

Remark 3.2. Sufficient conditions for nonnegativity of solutions of difference equations on a *finite* interval were obtained in Theorems 2.1 – 2.4 of the recent paper [5], in which a corresponding hypothesis on a size of this interval (see the condition (iv)) was assumed.

References

[1] R.P. Agarwal. *Difference Equations and Inequalities*, Marcel Dekker, New York, 1992.

[2] A. Domoshnitsky, M. Drakhlin, and I.P. Stavroulakis, Distribution of zeros of solutions to functional equations, submitted to *Advances in Difference Equations.*

[3] Y. Domshlak, A. Matakajev, and A.Askiev, *Oscillatory Properties of Difference Equations*, "Ilim", Frunze, USSR, 1990 (in Russian).

[4] L.H. Erbe and B.G. Zhang, Oscillation of discrete analogues of delay equations, *Differential Integral Equations* **2** (1989), 300–309.

[5] D. Jiang, L. Zhang, and R.P. Agarwal, Monotone method for first order periodic boundary value problems and periodic solutions of delay difference equations, *Memoirs on Differential Equations and Mathematical Physics* **28** (2003), 75–88.

[6] G. Ladas, Ch. Philos, and Y.G. Sfikas, Sharp conditions for the oscillation for delay difference equations, *J. Appl. Math. Simulation* **2** (1989), 101–112.

[7] V. Lakshmikantham and D. Trigante, *Theory of Difference Equations: Numerical Methods and Applications,* Mathematics in Science and Engineering, 181, Academic Press, Boston, 1988.

[8] W. Nowakowska and J. Werbowski, Oscillatory behavior of solutions of functional equations, *Nonlinear Analysis TMA* **44** (2001), 767–775.

[9] J. Shen and I.P. Stavroulakis, Oscillation criteria for delay difference equations, *Electron. J. Differential Equations*, 2001, No. 10, 1–15.

[10] Y.M. Wang, Monotone methods for a boundary value problem of second order discrete equations, *Comput. Math. Appl.* **36** (1998), 77–92.

[11] W. Zhang, Y. Chen, and S.S. Cheng, Monotone methods for a discrete boundary value problems, *Comput. Math. Appl.* **32** (1996), 41–49.

Nonoscillatory Solutions of Third-Order Difference Equations

ZUZANA DOŠLÁ

Department of Mathematics, Masaryk University Brno
Brno, Czech Republic

and

ALEŠ KOBZA[1]

Department of Mathematics, Masaryk University Brno
Brno, Czech Republic

Abstract We deal with the third-order nonlinear difference equations

$$\Delta(p_n\Delta(r_n\Delta x_n)) - q_n f(x_{n+\sigma}) = 0,$$

where (p_n), (r_n) and (q_n) are sequences of positive real numbers for $n \in \mathbb{N}$, $\sigma \in \{0,1,2\}$, and $f : \mathbb{R} \to \mathbb{R}$ is a continuous function such that $f(u)u > 0$ for $u \neq 0$. All nonoscillatory solutions are classified according to the sign of their quasidifferences and their asymptotic behavior is investigated.

Keywords Third-order nonlinear difference equation, nonoscillatory solution, strongly monotone solutions, asymptotic properties

AMS Subject Classification 39A10

1 Introduction

We study the nonoscillatory solutions of third-order nonlinear difference equations

$$\Delta\left(p_n\Delta\left(r_n\Delta x_n\right)\right) - q_n f\left(x_{n+\sigma}\right) = 0, \tag{1}$$

where Δ is the forward difference operator $\Delta x_n = x_{n+1} - x_n$, (p_n), (r_n) and (q_n) are sequences of positive real numbers for $n \in \mathbb{N}$, $\sigma \in \{0,1,2\}$ and $f : \mathbb{R} \to \mathbb{R}$ is a continuous function such that $f(u)u > 0$ for $u \neq 0$.

[1]Research supported by the Czech Grant Agency, grant 201/01/0079.

Special cases of (1) have been widely considered in the literature. For instance, oscillatory and asymptotic properties for the third-order linear difference equation

$$\Delta(\Delta^2 x_n - p_{n+1}x_{n+1}) - q_n x_{n+2} = 0, \tag{2}$$

and for the nonlinear difference equation (1) with $r_n = 1$ have been studied in [8, 9] and [6], respectively. The nonoscillatory solutions of (1) with $q_n < 0$ has been studied in [4, 5]. Similar problems for the differential equations of the third-order with quasiderivatives have been studied in [2, 3].

A solution x of (1) is a real sequence (x_n) defined for all $n \in \mathbb{N}$ and satisfying (1) for all $n \in \mathbb{N}$. A solution of (1) is called a *nontrivial* if for any $n_0 \geq 1$ there exists $n > n_0$ such that $x_n \neq 0$. Otherwise, the solution is called a *trivial.* A nontrivial solution x of (1) is said to be *oscillatory* if for any $n_0 \geq 1$ there exists $n > n_0$ such that $x_{n+1}x_n \leq 0$. Otherwise, the nontrivial solution is said to be *nonoscillatory.*

Denote quasidifferences $x^{[i]}$, $i = 0, 1, 2$, of a solution x of (1) as follows

$$x_n^{[0]} = x_n, \quad x_n^{[1]} = r_n \Delta x_n, \quad x_n^{[2]} = p_n \Delta x_n^{[1]}, \quad x_n^{[3]} = \Delta x_n^{[2]}.$$

Our aim is to study all possible types of nonoscillatory solutions of (1) and to give conditions ensuring that all nonoscillatory solutions (1) satisfy $\lim |x_n^{[i]}| = \infty$ for $i = 0, 1, 2$.

2 Classification of nonoscillatory solutions

Proposition 2.1. *Let x be a solution of* (1). *If there exists $a \in \mathbb{N}$ such that $x_a \neq 0$, then x is a nontrivial solution.*

Proof. Rewritting (1) to the recurrent formula, we have that any solution such that $x_{n_0} = x_{n_0+1} = x_{n_0=2} = 0$ satisfies $x_n = 0$ for all $n \in N$.

Suppose that a solution x of (1) satisfying $x_a \neq 0$ is trivial. Then there exists $n_0 > a$ such that $x_n = 0$ for all $n > n_0$. This together with the above property gives a contradiction.

Lemma 2.2. *If x is a nonoscillatory solution of* (1), *then there exists $N \in \mathbb{N}$ such that $x_n^{[i]} \neq 0$ for $i = 0, 1, 2$ and $n \geq N$.*

Proof. Without loss of generality, assume that $x_{n+\sigma} > 0$ for $n \geq n_0$. Then $x_n^{[3]} > 0$ for $n \geq n_0$ and therefore $x^{[2]}$ is increasing for $n \geq n_0$. Thus there exists $n_1 \geq n_0$ such that $x^{[1]}$ is either increasing or decreasing for $n \geq n_1$. Hence there exists $n_2 \geq n_1$ such that $x^{[1]}$ is either positive or negative for $n \geq n_2$ and letting $N = n_2$ we get the conclusion.

In view of Lemma 1, all nonoscillatory solutions x of (1) belong to the following classes:

$$\begin{aligned} N_0 &= \{x : \exists n_x \text{ such that } x_n x_n^{[1]} < 0,\ x_n x_n^{[2]} > 0 \text{ for all } n \ge n_x\}, \\ N_1 &= \{x : \exists n_x \text{ such that } x_n x_n^{[1]} > 0,\ x_n x_n^{[2]} < 0 \text{ for all } n \ge n_x\}, \\ N_2 &= \{x : \exists n_x \text{ such that } x_n x_n^{[1]} > 0,\ x_n x_n^{[2]} > 0 \text{ for all } n \ge n_x\}, \\ N_3 &= \{x : \exists n_x \text{ such that } x_n x_n^{[1]} < 0,\ x_n x_n^{[2]} < 0 \text{ for all } n \ge n_x\}. \end{aligned}$$

In the sequel we give conditions to be the class N_i $i \in \{0,1,3\}$ empty.

Theorem 2.3. *a) The class N_0 is empty if*

$$\sum_{i=1}^{\infty} \frac{1}{p_i} = \infty. \tag{3}$$

b) The class N_3 is empty if

$$\sum_{j=1}^{\infty} \frac{1}{r_j} = \infty. \tag{4}$$

Proof. Claim a). Let $x \in N_0$. Without loss of generality we can suppose that there exists $n_0 \in \mathbb{N}$ such that $x_n^{[1]} < 0$, $x_n^{[i]} > 0$ for $i = 0,2,3$ and for $n \ge n_0$. Since $x_n^{[3]} > 0$, $x^{[2]}$ is increasing for $n \ge n_0$. Denote $c_2 = x_{n_0}^{[2]}$ $(c_2 > 0)$. Then we have $p_n \Delta x_n^{[1]} > c_2$ for $n > n_0$. Summing the last inequality from n_0 to n, we obtain

$$x_{n+1}^{[1]} > x_{n_0}^{[1]} + c_2 \sum_{i=n_0}^{n} \frac{1}{p_i}.$$

We get a contradiction with the negativeness of $x^{[1]}$ for n large, because the right-hand side tends to ∞ for $n \to \infty$.

Claim b). Let $x \in N_3$. Without loss of generality assume that there exists $n_0 \in \mathbb{N}$ such that $x_n > 0$, $x_n^{[i]} < 0$ for $i = 1,2$ and for $n \ge n_0$. Since $x_n^{[2]} < 0$, $x^{[1]}$ is decreasing for $n \ge n_0$. Denote $c_1 = x_{n_0}^{[1]}$ $(c_1 < 0)$. Then we have $r_n \Delta x_n < c_1$ for $n > n_0$ and by summation from n_0 to n, we obtain

$$x_{n+1} < x_{n_0} + c_1 \sum_{j=n_0}^{n} \frac{1}{r_j}.$$

Passing $n \to \infty$, we get a contradiction with the positiveness of x for n large.

Theorem 2.4. *Let*

$$\liminf_{|u| \to \infty} |f(u)| > 0. \tag{5}$$

The class N_1 is empty if

$$\sum_{i=1}^{\infty} q_i = \infty. \tag{6}$$

Proof. Let $x \in N_1$. Without loss of generality assume that there exists $n_0 \in \mathbb{N}$ such that $x_n^{[i]} > 0$, $x_n^{[2]} < 0$ for $i = 0, 1, 3$ and for $n \geq n_0$. Since x is increasing for $n \geq n_0$, there exists $\lim_{n\to\infty} x_n = c$, where $c \in (0, \infty]$. In view of this, (5) and the sign condition posed on f, there exists positive constant K such that $f(x_n) \geq K$ for $n \geq n_0$. By summation of (1) from n_0 to n we get

$$\sum_{i=n_0}^{n} \Delta x_i^{[2]} = \sum_{i=n_0}^{n} q_i f(x_{i+\sigma}) \geq K \sum_{i=n_0}^{n} q_i$$

and so

$$x_{n+1}^{[2]} \geq x_{n_0}^{[2]} + K \sum_{i=n_0}^{n} q_i.$$

Letting $n \to \infty$ we get $x^{[2]} \to \infty$, which is a contradiction with the negativeness of $x^{[2]}$ for n large.

3 Strongly monotone solutions

If (3) and (4) hold then by Theorem 2.3 all nonoscillatory solutions belong to the set $N_1 \cup N_2$. In this section we study solutions of the class N_2. Such solutions are usually called *strongly monotone solutions.*

Lemma 3.1. *Let x be a solution of* (1) *satisfying*

$$x_m \geq 0, \quad x_m^{[1]} \geq 0, \quad x_m^{[2]} > 0$$

for some integer $m \geq 1$, then

$$x_k \geq 0, \quad x_k^{[1]} > 0, \quad x_k^{[2]} > 0 \tag{7}$$

for each $k \in \mathbb{N}$ such that $k > m$. And moreover $x_k > 0$ for $k > m + 1$.

Proof. We will prove (7) for $k = m + 1$. Using the assumption $x_m^{[1]} \geq 0$ we get $\Delta x_m \geq 0$, from where $x_{m+1} \geq x_m \geq 0$. Similarly, using $x_m^{[2]} > 0$, we obtain $\Delta x_m^{[1]} > 0$, and so $x_{m+1}^{[1]} > x_m^{[1]} \geq 0$. Since $x_{m+1}^{[1]} > 0$, we get $x_{m+2} > 0$. Because $x_{m+\sigma} \geq 0$ for $\sigma \in \{0, 1, 2\}$, we have

$$\Delta x_m^{[2]} = q_m f(x_{m+\sigma}) \geq 0.$$

Therefore $x_{m+1}^{[2]} \geq x_m^{[2]} > 0$. We have proved $x_{m+1} \geq 0$, $x_{m+1}^{[1]} > 0$, $x_{m+1}^{[2]} > 0$ and $x_{m+2} > 0$. Repeating this process we get a conclusion.

Theorem 3.2. *The class N_2 is always nonempty.*

Proof. According to Lemma 3.1, any solution of (1) with initial values $x_1^{[i]} > 0$ for $i = 0, 1, 2$ satisfies, $x_n^{[i]} > 0$ for $i = 0, 1, 2$ for each $n \in \mathbb{N}$. This solution belongs to the class N_2.

Theorem 3.3. *Let $x \in N_2$ and let (5) hold.*

a) *If (6) holds, then $\lim_{n\to\infty} |x_n^{[2]}| = \infty$.*

b) *If $\sum_{j=1}^{\infty} \frac{1}{p_{j+1}} \sum_{i=1}^{j} q_i = \infty$, then $\lim_{n\to\infty} |x_n^{[1]}| = \infty$.*

c) *If*

$$\sum_{k=1}^{\infty} \frac{1}{r_{k+2}} \sum_{j=1}^{k} \frac{1}{p_{j+1}} \sum_{i=1}^{j} q_i = \infty, \tag{8}$$

then $\lim_{n\to\infty} |x_n| = \infty$.

Proof. Let $x \in N_2$. Without loss of generality suppose that there exists $n_0 \in \mathbb{N}$ such that $x_n > 0$, $x_n^{[i]} > 0$ for $i = 1, 2, 3$ and for $n \geq n_0$. By the same argument as in the proof of the Theorem 2.4 there exists constant K such that $f(x_n) \geq K > 0$ for $n \geq n_0$.

Claim a). Since

$$\Delta x_n^{[2]} = q_n f(x_{n+\sigma}) \geq K q_n$$

for $n \geq n_0$, by summation from n_0 to n we obtain

$$x_{n+1}^{[2]} - x_{n_0}^{[2]} \geq K \sum_{i=n_0}^{n} q_i.$$

Denote $x_{n_0}^{[2]} = c_2$ ($c_2 > 0$). We have

$$x_{n+1}^{[2]} \geq c_2 + K \sum_{i=n_0}^{n} q_i, \tag{9}$$

thus passing $n \to \infty$ and applying (6), we get $\lim_{n\to\infty} x_n^{[2]} = \infty$.

Claim b). According to (9)

$$p_{n+1} \Delta x_{n+1}^{[1]} \geq c_2 + K \sum_{i=n_0}^{n} q_i.$$

By summation from n_0 to n we get

$$x_{n+2}^{[1]} \geq x_{n_0+1}^{[1]} + c_2 \sum_{j=n_0}^{n} \frac{1}{p_{j+1}} + K \sum_{j=n_0}^{n} \frac{1}{p_{j+1}} \sum_{i=n_0}^{j} q_i. \tag{10}$$

Because the right-hand side tends to ∞ for $n \to \infty$, we get $\lim_{n\to\infty} x_n^{[1]} = \infty$.

Claim c). According to (10)

$$r_{n+2} \Delta x_{n+2} \geq c_1 + c_2 \sum_{j=n_0}^{n} \frac{1}{p_{j+1}} + K \sum_{j=n_0}^{n} \frac{1}{p_{j+1}} \sum_{i=n_0}^{j} q_i,$$

where $c_1 = x^{[1]}_{n_0+1}$ $(c_1 > 0)$ and $c_2 = x^{[2]}_{n_0}$ $(c_2 > 0)$. By summation from n_0 to n we obtain

$$x_{n+3} \geq x_{n_0+2} + c_1 \sum_{k=n_0}^{n} \frac{1}{r_{k+2}} + c_2 \sum_{k=n_0}^{n} \frac{1}{r_{k+2}} \sum_{j=n_0}^{k} \frac{1}{p_{j+1}} + K \sum_{k=n_0}^{n} \frac{1}{r_{k+2}} \sum_{j=n_0}^{k} \frac{1}{p_{j+1}} \sum_{i=n_0}^{j} q_i. \tag{11}$$

Letting $n \to \infty$, the right-hand side tends to ∞, we get $\lim_{n\to\infty} x_n = \infty$. The proof is complete.

Under the stronger assumption on the nonlinearity f, the condition (6) in Theorem 3.3 -a) can be relaxed.

Theorem 3.4. *Assume that*

$$\liminf_{|u|\to\infty} \frac{f(u)}{u} > 0 \tag{12}$$

and

$$\sum_{k=1}^{\infty} q_k \sum_{j=2}^{k+\sigma-1} \frac{1}{r_j} \sum_{i=1}^{j-1} \frac{1}{p_i} = \infty. \tag{13}$$

Then any solution $x \in N_2$ satisfies $\lim_{n\to\infty} |x^{[2]}_n| = \infty$.

Proof. Let $x \in N_2$. Without loss of generality suppose that there exists $n_0 \in \mathbb{N}$ such that $x^{[i]}_n > 0$ for $i = 0, 1, 2$ and for $n \geq n_0$. Denote $L_0 = \lim_{n\to\infty} x_n$. Clearly, $L_0 > 0$. Assume $\lim_{n\to\infty} |x^{[2]}_n| < \infty$. By summation of (1) from n_0 to n we have

$$x^{[2]}_{n+1} - x^{[2]}_{n_0} = \sum_{i=n_0}^{n} q_i f(x_{i+\sigma}),$$

from where

$$\sum_{k=n_0}^{\infty} q_k f(x_{k+\sigma}) < \infty. \tag{14}$$

Using (11) we have

$$x_{n+3} > c_2 \sum_{k=n_0}^{n} \frac{1}{r_{k+2}} \sum_{j=n_0}^{k} \frac{1}{p_{j+1}} = c_2 \sum_{k=n_0+2}^{n+2} \frac{1}{r_k} \sum_{j=n_0+1}^{k-1} \frac{1}{p_j},$$

where $c_2 = x^{[2]}_{n_0}$. This implies that

$$x_{n+\sigma} > c_2 \sum_{k=n_0+2}^{n+\sigma-1} \frac{1}{r_k} \sum_{j=n_0+1}^{k-1} \frac{1}{p_j}. \tag{15}$$

Let $L_0 < \infty$. Then (15) yields

$$\sum_{k=n_0+2}^{\infty} \frac{1}{r_k} \sum_{j=n_0+1}^{k-1} \frac{1}{p_j} < \infty. \tag{16}$$

Since x is positive increasing for $n \geq n_0$, there exists $K > 0$ such that $f(x_n) \geq K$ for $n \geq n_0$. ¿From here, (14) and (16) we have

$$\lim_{n\to\infty} \frac{q_n \sum_{j=2}^{n+\sigma-1} 1/r_j \sum_{i=1}^{j-1} 1/p_i}{q_n f(x_{n+\sigma})} < \infty,$$

i.e.,

$$\sum_{k=1}^{\infty} q_k \sum_{j=2}^{k+\sigma-1} \frac{1}{r_j} \sum_{i=1}^{j-1} \frac{1}{p_i} < \infty,$$

which is a contradiction with (13).

Let $L_0 = \infty$. By (12) there exists $K > 0$ such that $f(x_n) \geq Kx_n$ for large n. Applying (15) we get

$$f(x_{n+\sigma}) \geq Kx_{n+\sigma} > Kc_2 \sum_{k=n_0+2}^{n+\sigma-1} \frac{1}{r_k} \sum_{j=n_0+1}^{k-1} \frac{1}{p_j}$$

and using (14)

$$\infty > \sum_{k=n_0}^{\infty} q_k f(x_{k+\sigma}) \geq Kc_2 \sum_{k=n_0}^{\infty} q_k \sum_{j=n_0+2}^{k+\sigma-1} \frac{1}{r_j} \sum_{i=n_0+1}^{j-1} \frac{1}{p_i}.$$

This is a contradiction to (13).

The following examples illustrate our results.

Example 3.5. *Consider the equation*

$$\Delta\left[n\Delta\left(n\Delta x_n\right)\right] - \frac{1}{n}\left|x_{n+\sigma}\right|^{\lambda} sgn x_{n+\sigma} = 0,$$

where $\lambda > 0$ *and* $\sigma \in \{0, 1, 2\}$. *By Theorems 2.3, 2.4, and 3.3 every nonoscillatory solutions is from the class* N_2 *and satisfies*

$$\lim_{n\to\infty} |x_n^{[i]}| = \infty \quad for \quad i = 0,\,1,\,2.$$

Example 3.6. *Consider the equation*

$$\Delta\left[(n+1)\Delta\left(\frac{1}{n}\Delta x_n\right)\right] - \frac{1}{n^3+3n^2+2n} x_{n+2} = 0.$$

By Theorem 2.3, $N_0 = N_3 = \emptyset$ *and by Theorem 3.2,* $N_2 \neq \emptyset$. *However, not all solutions belong to the class* N_2. *Indeed, one can check that this equation has solution* (x_n) *where* $x_n = n$ *and this solution is from the class* N_1.

Open problem. For linear Equation (1) with $\sigma = 2$ and $p_n = r_n = 1$ it holds that $N_1 = \emptyset$ if and only if there exist oscillatory solutions, see [8]. It would be interesting to extend this result to Equation (1).

References

[1] Agarwal, R.P., *Difference Equations and Inequalities*, 2nd Edition, Pure Appl. Math. 228, Marcel Dekker, New York, 2000.

[2] Bartušek M. and Osička J., Asymptotic behaviour of solutions of a third order nonlinear differential equation, *Nonlinear Analysis*, **34** (1998), 653–664.

[3] Cecchi M., Došlá Z. and Marini M., On nonlinear oscillations for equations associated to disconjugate operators, *Nonlinear Analysis*, **30** (1997), 1583–1594.

[4] Došlá Z. and Kobza A., Global asymptotic properties of third order difference equations, *J. Comput. Appl. Math.*, **48** (2004), 191–200.

[5] Kobza A., Property A for third order difference equations, *Stud. Univ. Žilina Math. Ser.* **17** (2003), 109-114.

[6] Migda M., Schmeidel E., and Drozdowicz A., Nonoscillation results for some third order nonlinear difference equations, *Folia FSN Univ. Masaryk. Brunensis Math.* **13** (2003), Proceedings of Colloquium on Differential and Difference Eqs., Brno, 2002, 185–192.

[7] Smith B., Oscillatory and asymptotic behavior in certain third order difference equations, *Rocky Mountain J. Math.* **17** (1987), 597–606.

[8] Smith B., Oscillation and nonoscillation theorems for third order quasi-adjoint difference equations, *Portugal. Math.*, **45** (1988), 229–243.

[9] Smith B., Linear third order difference equations: oscillatory and asymptotic behavior, *Rocky Mountain J. Math.*, **22** (1992), 1559–1564.

[10] Smith B. and Taylor W. E., Asymptotic behavior of solutions of a third order difference equation, *Portugaliae Mathematica*, **44** (1987), 113–117.

Global Stability of Periodic Orbits of Nonautonomous Difference Equations in Population Biology and the Cushing-Henson Conjectures[1]

SABER ELAYDI

Department of Mathematics, Trinity University
San Antonio

and

ROBERT J. SACKER[2]

Department of Mathematics
University of Southern California
Los Angeles, California

Abstract Elaydi and Yakubu showed that a globally asymptotically stable (GAS) periodic orbit in an autonomous difference equation must in fact be a fixed point whenever the phase space is connected. In this paper we extend this result to periodic nonautonomous difference equations via the concept of skew-product dynamical systems. We show that for a k-periodic difference equation, if a periodic orbit of period r is GAS, then r must be a divisor of k. In particular subharmonic, or long periodic, oscillations cannot occur. Moreover, if r divides k we construct a nonautonomous dynamical system having a minimum period k and which has a GAS periodic orbit with minimum period r. Our methods are then applied to prove two conjectures of J. Cushing and S. Henson concerning a nonautonomous Beverton-Holt equation that arises in the study of the response of a population to a periodically fluctuating environmental force such as seasonal fluctuations in carrying capacity

[1]The authors thank Jim Cushing for many helpful comments on the manuscript.

[2]Supported by the University of Southern California, Letters Arts and Sciences Faculty Development Grant.

or demographic parameters like birth or death rates. We show that the periodic fluctuations in the carrying capacity always have a deleterious effect on the average population, thus answering in the affirmative the second of the conjectures.

Keywords Difference equation, global asymptotic stability, population biology

AMS Subject Classification 39A11, 92D25

1 Nonautonomous difference equations

A periodic difference equation with period k,

$$x_{n+1} = F(n, x_n), \quad F(n+k, x) = F(n, x) \quad x \in \mathbb{R}^n,$$

may be treated in the setting of skew-product dynamical systems [10], [9] by considering mappings

$$f_n(x) = F(n, x) \qquad f_i : \mathcal{F}_i \to \mathcal{F}_{i+1 \bmod k},$$

where $\mathcal{F}_i$, the "fiber" over f_i, is just a copy of $\mathbb{R}^n$ residing over f_i and consisting of those x on which f_i acts (Fig.1). Then the unit time mapping

$$(x, f_i) \longrightarrow (f_i(x), f_{i+1 \bmod k})$$

generates a semidynamical system on the product space

$$X \times Y \quad \text{where} \quad Y = \{f_0, \ldots, f_{k-1}\} \subset \mathbf{C}, \qquad X = \mathbb{R}^n, \tag{1}$$

where $\mathbf{C}$ is the space of continuous functions, fig.1. We thus study the k-periodic mapping system

$$x_{n+1} = f_n(x_n), \qquad f_{n+k} = f_n. \tag{2}$$

It is then not difficult to see that an *autonomous* equation f is one that leaves the fiber over f invariant, or put another way, f is a fixed point of the mapping $f_i \longrightarrow f_{i+1 \bmod k}$.

In Elaydi and Sacker [6] the concept of a "geometric r-cycle" was introduced and defined. The definition says essentially that a geometric r-cycle is the projection onto the factor X in the product space (1) of an r-cycle in the skew-product flow.

A geometric cycle is called globally asymptotically stable if the corresponding periodic orbit in the skew-product flow is globally asymptotically stable in the usual sense. The example in Fig. 1 is clearly *not* globally asymptotically stable. Globally asymptotically stable geometric r-cycles may be constructed using the following simple device. On $\mathbb{R}$ define $g(x) = 0.5x$. Then for $r = 3$ any $k \geq 5$ define

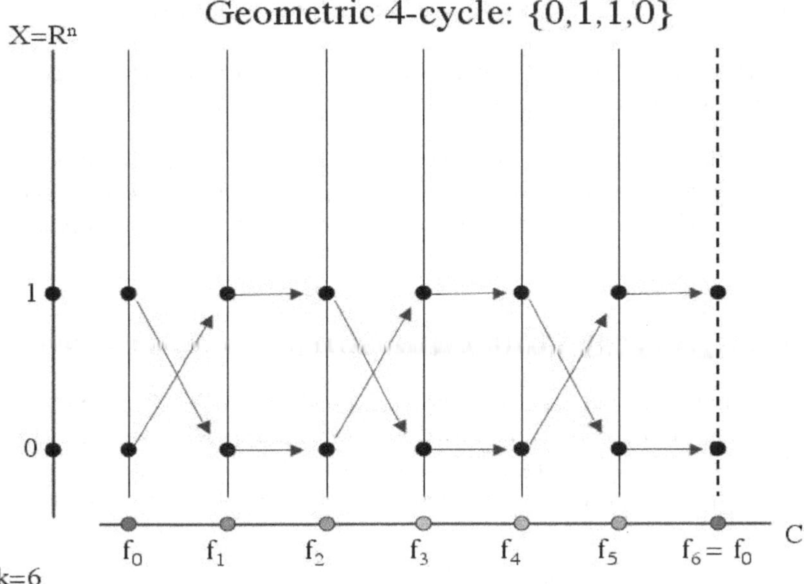

Figure 1: Skew-product flow. Example shown with n=1.

$$\begin{aligned} f_0 = f_1 &= \cdots = f_{k-4} \\ f_{k-3} &= g(x) + 1 \\ f_{k-2} &= g(x-1) + 2 \\ f_{k-1} &= g(x-2). \end{aligned}$$

The geometric 3-cycle consists of $\{0, 1, 2\}$. However if one watches the progress of "0" in $\mathbb{R}$ alone one will observe the (minimum period) k-cycle

$$x_0 = 0 \to 0 \to \cdots \to 0 \to 0 \to 0 \to 1 \to 2 \to 0 = x_0,$$

even though "0" seems at first to be fixed (imagine k very large). This can easily be generalized to the following theorems.

Theorem 1.1 *Given $r \geq 1$ and $k > r + 1$ there exists a k periodic mapping system having a globally asymptotically stable geometric k-cycle one of whose points "appears" fixed, i.e., it is fixed for $k - r$ iterations.*

In general, when we have a geometric r-cycle with $r \leq k$, one has the following.

Theorem 1.2 [6] *Assume that X is a connected metric space and each $f_i \in Y$ is a continuous map on X. Let $c_r = \{\bar{x}_0, \bar{x}_1, \ldots, \bar{x}_{r-1}\}$ be a geometric r-cycle of Equation (2). If c_r is globally asymptotically stable then $r|k$, i.e., r divides k.*

Thus the geometric 4-cycle in Fig. 1 cannot be globally asymptotically stable.

The next theorem shows how to construct such a dynamical system given any two positive integers r and k with $r|k$.

Theorem 1.3 **[6]** *Given any two positive integers r and k with $r|k$ then there exists a nonautonomous dynamical system having a minimum period k and which has a globally asymptotically stable geometric r-cycle with minimum period r.*

2 The Beverton-Holt equation

The Beverton-Holt equation has been studied extensively by Jim Cushing and Shandelle Henson [3, 4]. Also known as the *Beverton-Holt stock-recruitment equation* [1], it is a model for density dependent growth that exhibits *compensation* (Neave [8]) as opposed to *overcompensation* (Clark [2]), see also (Kot [7]). The equation takes the form

$$x_{n+1} = \frac{\mu K x_n}{K + (\mu - 1)x_n}, \qquad x_0 \geq 0 \quad K > 0,$$

where μ is the per-capita growth rate and K is the carrying capacity. It is easily shown that for $0 < \mu < 1$ the equilibrium (fixed point) $x = 0$ s globally asymptotically stable whereas for $\mu > 1$ the fixed point K is globally asymptotically stable.

In [3] the authors considered a periodic carrying capacity $K_{n+k} = K_n$ caused by a periodically (seasonally) fluctuating environment

$$x_{n+1} = \frac{\mu K_n x_n}{K_n + (\mu - 1)x_n}.$$

Defining

$$f_i(x) \doteq \frac{\mu K_i x}{K_i + (\mu - 1)x}$$

we have an equation of the form (2) with period k.

Although this is not always desirable from a qualitative point of view, we will compute a "solution" in closed form of the periodic Beverton-Holt equation.

After two iterations

$$x_2 = f_1 \circ f_0(x_0) = \frac{\mu^2 K_1 K_0 x_0}{K_1 K_0 + (\mu - 1)M_1 x_0},$$

and inductively after k iterations,

$$x_k = f_{k-1} \circ f_{k-2} \circ \cdots \circ f_0(x_0) = \frac{\mu^k K_{k-1} K_{k-2} \cdots K_0 x_0}{K_{k-1} K_{k-2} \cdots K_0 + (\mu - 1)M_{k-1} x}, \tag{3}$$

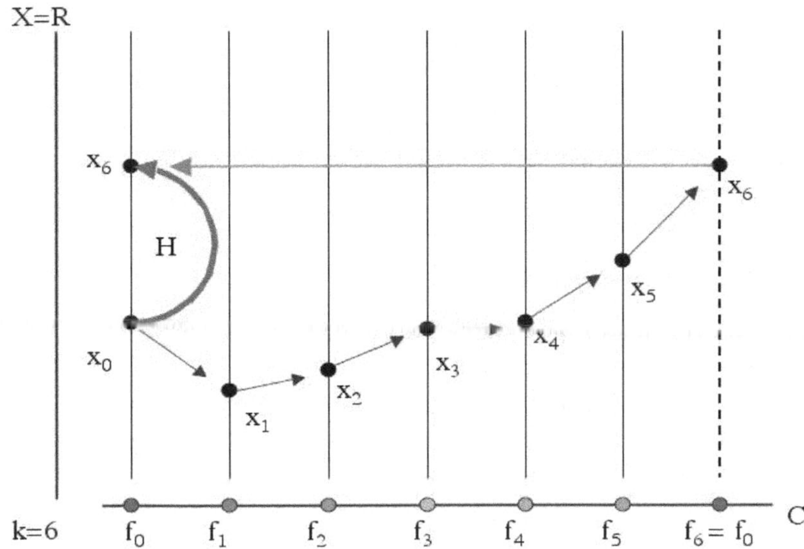

Figure 2: $k = 6$

where M_n satisfies the $2nd$-order *linear* difference equation:

$$M_{n+1} = K_{n+1}M_n + \mu^{n+1}K_nK_{n-1}\dots K_0, \quad M_0 = 1.$$

Thus

$$M_{k-1} = \prod_{j=0}^{k-2} K_{j+1} + \sum_{m=0}^{k-2} \left(\prod_{i=m+1}^{k-2} K_{i+1} \right) \mu^{m+1} K_m K_{m-1} \cdots K_0.$$

Letting $L_{k-1} = K_{k-1}K_{k-2}\dots K_0$, we finally obtain (defining H)

$$H(x) \doteq f_{k-1} \circ f_{k-2} \circ \cdots \circ f_0(x_0) = \frac{\mu^k L_{k-1} x_0}{L_{k-1} + (\mu - 1)M_{k-1}x_0}.$$

But then the difference equation, $x_{n+1} = H(x_n)$ leaves the fiber (copy of $\mathbb{R}$) invariant and is thus independent of n, i.e., *autonomous*! (See Fig. 2 for $k = 6$).

While this may or may not be easy to glean from (3) we will nevertheless use (3) later, but only in the case $k = 2$:

$$\bar{x} = (\mu + 1)\frac{K_1 K_0}{K_1 + \mu K_0}. \tag{4}$$

The mapping $x_{n+1} = H(x_n)$ thus has the unique positive fixed point

$$\bar{x} = \frac{\mu^k - 1}{\mu - 1} \frac{L_{k-1}}{M_{k-1}},$$

which is globally asymptotically stable with respect to positive initial conditions. By Theorem 1.2 we have further that either $\bar{x}$ is of minimal period k or of minimal period r where $r|k$.

3 The Ricatti equation

We next consider the more general autonomous Ricatti equation:

$$x_{n+1} = f(x_n), \qquad f(x) \doteq \frac{ax+b}{cx+d}$$

where we assume the following conditions:

$$\begin{array}{ll} 1. & a, c, d > 0,\ b \geq 0 \\ 2. & ad - bc \neq 0 \\ 3. & bc > 0 \quad \text{or} \quad a > d \end{array} \tag{5}$$

$$\begin{array}{lll} 1 & \Longrightarrow & f : \mathbb{R}^+ \to \mathbb{R}^+ \\ 2 & \Longrightarrow & f \quad \text{not a constant function} \\ 3 & \Longrightarrow & f \quad \text{has a positive fixed point (Beverton Holt if } b = 0) \end{array}$$

Under composition, letting

$$g(x) = \frac{\alpha x + \beta}{\gamma x + \delta},$$

one easily obtains

$$g \circ f(x) = \frac{(a\alpha + c\beta)x + (b\alpha + d\beta)}{(a\gamma + c\delta)x + (b\gamma + d\delta)}$$

from which 1, 2, and 3 in (5) easily follow.

The following change of variables, Elaydi [5, p. 87]

$$cx_n + d = \frac{y_{n+1}}{y_n}$$

reduces the Ricatti equation to a $2nd$-order *linear* equation,

$$y_{n+2} - py_{n+1} - qy_n = 0,$$

where $p = a + d$ and $q = bc - ad$. The general solution is then

$$y_n = c_1\lambda_{max}^n + c_2\lambda_{min}^n,$$

where λ_{max} and λ_{min} are the roots of the characterstic equation $\lambda^2 - p\lambda - q = 0$,

$$\lambda_{max} = \frac{a + d + \sqrt{(a-d)^2 + 4bc}}{2}, \tag{6}$$

$$\lambda_{min} = \frac{a + d - \sqrt{(a-d)^2 + 4bc}}{2}. \tag{7}$$

One can show that $c_1 \neq 0$ for solutions through $x_0 > 0$ from which it follows that

$$\frac{y_{n+1}}{y_n} \to \lambda_{max}.$$

From this it follows that

$$x_n \to x^* \doteq \frac{\lambda_{max} - d}{c},$$

and therefore x^* is a globally asymptotically stable fixed point in $\mathbb{R}^+$.

We now consider the **periodic Ricatti equation:**

$$x_{n+1} = f_n(x_n) \doteq \frac{a_n x_n + b_n}{c_n x_n + d_n},$$

where the coefficients satisfy 1, 2, and 3 in (5) and have period $k > 0$. Again, the function H defined by

$$H(x) \doteq f_{k-1} \circ f_{k-2} \circ \cdots \circ f_1 \circ f_0$$

has the same Ricatti form and satisfies 1, 2, and 3 in (5). Thus we conclude that the periodic Ricatti equation has a globally asymptotically stable geometric r-cycle and by Theorem 1.2, $r|k$.

4 The general case

In the previous sections we based our analysis on the special form the difference equations had. In this section we extract the salient properties that makes it all work. Recall that $h : \mathbb{R}^+ \to \mathbb{R}^+$ is **concave** if

$$h(\alpha x + \beta y) \geq \alpha h(x) + \beta h(y) \qquad \text{for all } x, y \in \mathbb{R}^+,$$

where $\alpha, \beta \geq 0, \alpha + \beta = 1$. The following property is easily verified: **If f, g are concave and f is increasing then $f \circ g$ is concave.** Note however that by requiring our maps to take values in $\mathbb{R}^+$ *and* to be defined on *all* of $\mathbb{R}^+$, a concave function is automatically increasing.

Define the class $\mathcal{K}$ to be all functions that satisfy

(1) $f : \mathbb{R}^+ \to \mathbb{R}^+$ is continuous

(2) f is concave (and therefore increasing)

(3) There exist x_1 and x_2 such that $f(x_1) > x_1$ and $f(x_2) < x_2$, i.e. the graph of f crosses the "diagonal."

$\mathcal{K}$ generalizes the class "A1" of Cushing and Henson [3].
Properties of $\mathcal{K}$.

(a) $\mathcal{K}$ is closed under the operation of composition, i.e., $f, g \in \mathcal{K}$ implies $f \circ g \in \mathcal{K}$. Thus $\mathcal{K}$ is a semigroup under composition.

(b) Each f has a unique globally asymptotically stable fixed point $x_f > 0$

(c) If $f, g \in \mathcal{K}$ with $x_f < x_g$ then $x_f < x_{f \circ g} < x_g$ and $x_f < x_{g \circ f} < x_g$

Thus, for the k-periodic difference equation

$$x_{n+1} = F(n, x_n), \qquad x \in \mathbb{R} \tag{8}$$

if for all n, $f_n \in \mathcal{K}$, where $f_n(x) = F(n, x)$, then g defined by

$$g(x) \doteq f_{k-1} \circ f_{k-2} \circ \cdots \circ f_1 \circ f_0 \in \mathcal{K},$$

represents an autonomous equation

$$x_{n+1} = g(x_n),$$

having a unique globally asymptotically stable fixed point. Therefore the difference equation (8) has a globally asymptotically stable geometric r-cycle and by Theorem 1.2, $r|k$.

5 The Cushing and Henson conjectures

In [4], Cushing and Henson conjectured that for the periodic k-Beverton-Holt equation, $k \geq 2$

$$x_{n+1} = \frac{\mu K_n x_n}{K_n + (\mu - 1)x_n}, \quad \mu > 1, \ K_n > 0,$$

1. **[Conjecture 1]** There is a positive k-periodic solution $\{\bar{x}_0. \ldots, \bar{x}_{k-1}\}$ and it globally attracts all positive solutions

2. **[Conjecture 2]** The average over n values $\{y_0, y_1, \ldots, y_{n-1}\}, av(y_n) \doteq \frac{1}{k}\sum_{i=0}^{k-1} y_i$ satisfies

$$av(\bar{x}_n) < av(K_n).$$

In Conjecture 2 it is implicit that the minimal period of the cycle $\{\bar{x}_0.\ldots,\bar{x}_{k-1}\}$ exceeds one, i.e., it is not a fixed point. The truth of Conjecture 2 implies that a fluctuating habitat has a deleterious effect on the population in the sense that the average population is less in a periodically oscillating habitat than it is in a constant habitat with the same average. Earlier [3] they proved both statements for $k = 2$. By our remarks in the previous section, Conjecture 1 is now completely solved: There exists a positive r-periodic globally asymptotically stable solution, with respect to $(-\infty,\infty)$, and moreover $r|k$.

We now answer Conjecture 2 in the affirmative for all $k \geq 2$ and in the process give a much simpler proof in the $k = 2$ case.

Comment: Without loss of generality we will now assume that k is the *minimal* period. Then for the periodic sequence $\{K_0, K_1, \ldots, K_{k-1}, K_k = K_0\}$, it follows that $K_i \neq K_{i+1}$ for at least one $i \in \{0, 1, \ldots, k-2\}$.
Everything then follows from an elementary algebraic lemma.

Lemma 5.1 *Define, for* $\alpha, \beta, x, y \in (0,\infty), \alpha+\beta = 1$,

$$g(x,y) = \frac{xy}{\alpha x + \beta y} - \beta x - \alpha y.$$

Then $g(x,y) \leq 0$ *with* "$=$" $\iff x = y$.
Equivalently,

$$\frac{xy}{\alpha x + \beta y} < \beta x + \alpha y, \qquad \iff x \neq y.$$

Proof

$$\begin{aligned} & (\alpha x + \beta y) g(x,y) \\ = \; & \{(1-\alpha^2-\beta^2)xy - \alpha\beta(x^2+y^2)\} \\ = \; & -\alpha\beta(x-y)^2. \end{aligned}$$

QED

We first derive a formula for a fixed point, x_{fog} of the composition of 2 Beverton-Holt functions using the formula (4):

$$\begin{aligned} f(x) &= \frac{\mu K x}{K + (\mu-1)x}, \quad g(x) = \frac{\mu L x}{L + (\mu-1)x} \\ x_{fog} &= (1+\mu)\frac{KL}{K+\mu L} = \frac{KL}{\alpha K + \beta L} = \frac{x_f x_g}{\alpha x_f + \beta x_g} \end{aligned} \tag{9}$$

where $\alpha = 1/(\mu+1)$ and $\beta = \mu/(\mu+1)$. From the previous comment and the lemma it follows that

$$x_{fog} = \frac{x_f x_g}{\alpha x_f + \beta x_g} \leq \beta x_g + \alpha x_f, \tag{10}$$

with strict inequality for at least one pair $f = f_i,\ g = f_{i+1},\ i \in \{0, 1, \ldots, k-2\}$.

Proof of Conjecture 2 for k=2:

$$f_0(x) = \frac{\mu K_0 x}{K_0 + (\mu - 1)x}, \quad f_1(x) = \frac{\mu K_1 x}{K_1 + (\mu - 1)x}$$

$$\begin{aligned} f_1 \circ f_0(x_0) &= x_0 = x_{f_1 \circ f_0} = \frac{K_1 K_0}{\alpha K_1 + \beta K_0} \\ f_0 \circ f_1(x_1) &= x_1 = x_{f_0 \circ f_1} = \frac{K_0 K_1}{\alpha K_0 + \beta K_1}. \end{aligned}$$

Now add and use (10) and $\alpha + \beta = 1$,

$$\begin{aligned} x_0 + x_1 &= \frac{K_1 K_0}{\alpha K_1 + \beta K_0} + \frac{K_0 K_1}{\alpha K_0 + \beta K_1} \\ &< \alpha K_0 + \beta K_1 + \alpha K_1 + \beta K_0 \\ &= K_0 + K_1. \end{aligned}$$

QED

Proof for k > 2:

I. **For** k **odd**, assume, inductively that Conjecture 2 is true for $(k+1)/2$. Take **k=3**. Then

$$f_2 \circ (f_1 \circ f_0)(x_0) = x_0, \qquad f_0 \circ (f_2 \circ f_1)(x_1) = x_1$$
$$f_1 \circ (f_0 \circ f_2)(x_2) = x_2.$$

Using the period 2 result,

$$\begin{aligned} 1.\quad f_1 \circ f_0(x_0) &= x_2, \quad \text{and} \quad f_2(x_2) = x_0 \implies \\ x_0 + x_2 &\leq x_{f_1 \circ f_0} + x_{f_2} = \frac{K_1 K_0}{\alpha K_1 + \beta K_0} + K_2 \\ &\leq \alpha K_0 + \beta K_1 + K_2 \\ 2.\quad f_2 \circ f_1(x_1) &= x_0, \quad \text{and} \quad f_0(x_0) = x_1 \implies \\ x_0 + x_1 &\leq x_{f_2 \circ f_1} + x_{f_0} \leq \alpha K_1 + \beta K_2 + K_0 \\ 3.\quad f_0 \circ f_2(x_2) &= x_1, \quad \text{and} \quad f_1(x_1) = x_2 \implies \\ x_1 + x_2 &\leq x_{f_0 \circ f_2} + x_{f_1} \leq \alpha K_2 + \beta K_0 + K_1, \end{aligned}$$

where at least one of the inequalities is strict. Adding, using (10) and $\alpha + \beta = 1$ we get

$$2(x_0 + x_1 + x_2) < 2(K_0 + K_1 + K_2).$$

Sketch for k=5:(Fig. 3)

$$f_1 \circ f_0(x_0) = x_2, \quad f_3 \circ f_2(x_2) = x_4,$$
$$f_4(x_4) = x_0 \implies$$
$$x_0 + x_2 + x_4 \leq x_{f_1 \circ f_0} + x_{f_3 \circ f_2} + x_{f_4}.$$

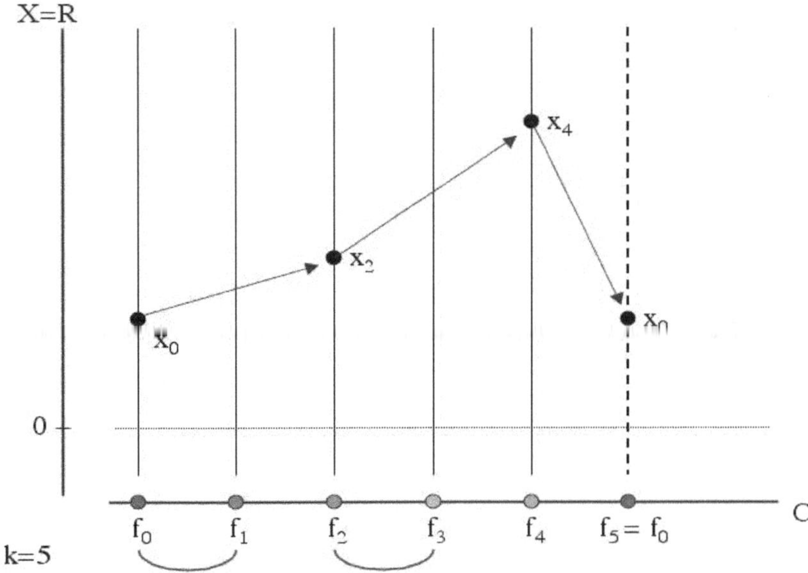

Figure 3: $k = 5$

After one cyclic permutation:

$$x_1 + x_3 + x_0 \leq x_{f_2 \circ f_1} + x_{f_4 \circ f_3} + x_{f_0}.$$

After 3 more cyclic permutation:

$$\begin{aligned} x_2 + x_4 + x_1 &\leq x_{f_3 \circ f_2} + x_{f_0 \circ f_4} + x_{f_1} \\ x_3 + x_0 + x_2 &\leq x_{f_4 \circ f_3} + x_{f_1 \circ f_0} + x_{f_2} \\ x_4 + x_1 + x_3 &\leq x_{f_0 \circ f_4} + x_{f_2 \circ f_1} + x_{f_3} \end{aligned}$$

where at least one of the inequalities is strict. Adding gives the result

$$3(x_0 + x_1 + \cdots + x_4) < 3(K_0 + K_1 + \cdots + K_4).$$

Sketch for k even, k=6: (Fig. 4) Inductively assume the conjecture true for $k/2$.

$$\begin{aligned} f_1 \circ f_0(x_0) &= x_2, \quad f_3 \circ f_2(x_2) = x_4, \quad f_5 \circ f_4(x_4) = x_0 \\ &\Longrightarrow \quad x_0 + x_2 + x_4 \leq x_{f_1 \circ f_0} + x_{f_3 \circ f_2} + x_{f_5 \circ f_4} \qquad (11) \\ &\leq \quad \alpha K_0 + \beta K_1 + \alpha K_2 + \beta K_3 + \alpha K_4 + \beta K_5. \end{aligned}$$

After one (and only one) cyclic permutation: (Fig. 5)

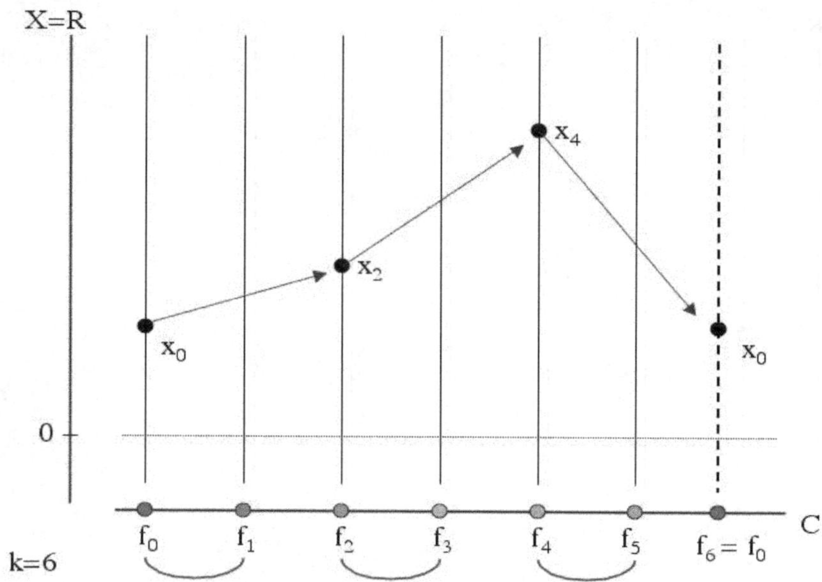

Figure 4: $k = 6$

$$\begin{aligned} f_2 \circ f_1(x_1) &= x_3, \quad f_4 \circ f_3(x_3) = x_5, \quad f_0 \circ f_5(x_5) = x_1 \\ &\Longrightarrow \quad x_1 + x_3 + x_5 \leq x_{f_2 \circ f_1} + x_{f_4 \circ f_3} + x_{f_0 \circ f_5} \\ &\leq \quad \alpha K_1 + \beta K_2 + \alpha K_3 + \beta K_4 + \alpha K_5 + \beta K_0, \end{aligned} \tag{12}$$

where at least one of the inequalities is strict. Adding (11) and (12), we obtain the result.

Theorem 5.2 *For a k-periodic Beverton-Holt equation with minimal period* $k \geq 2$

$$x_{n+1} = \frac{\mu K_n x_n}{K_n + (\mu - 1)x_n}, \quad \mu > 1, \; K_n > 0,$$

there exists a unique globally asymptotically stable k-cycle, $C = \{\xi_0, \xi_1, \ldots, \xi_{k-1}\}$ *and*

$$\operatorname{av}(\xi_n) < \operatorname{av}(K_n).$$

Summary of proof: **Zig-zag induction**

1. Prove it directly for $k = 2$,
2. k odd: True for $(k+1)/2 \Longrightarrow$ True for k,
3. k even: True for $k/2 \Longrightarrow$ True for k.

By judiciously pairing and permuting the maps and using only the formula for the fixed point of 2 maps, (9), it is straightforward to write down the complete proof for any $k > 2$.

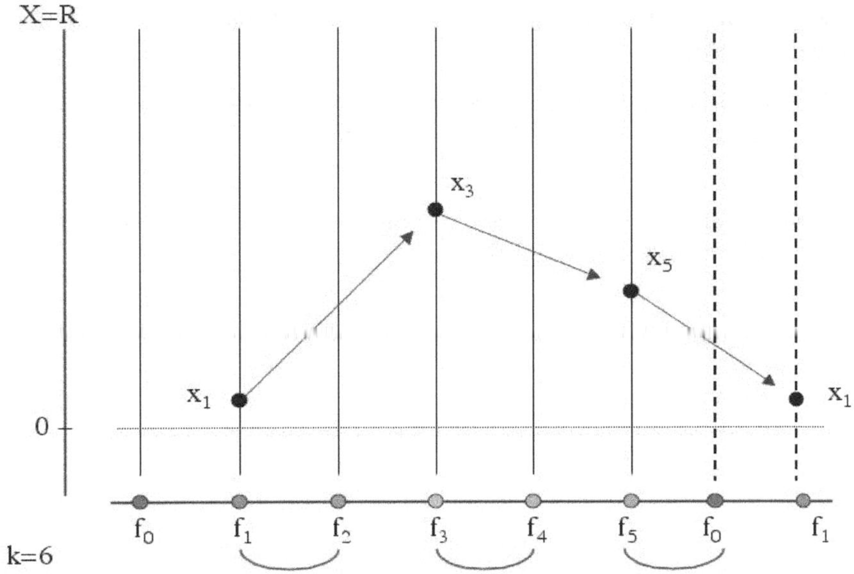

Figure 5: $k = 6$

References

[1] R.J.H. Beverton and S.J. Holt, On the dynamics of exploited fish populations, *Fishery investigations* **19** (1957), 1–533.

[2] C.W. Clark, *The Optimal Management of Renewable Resources*, John Wiley, New York, 1990.

[3] J.M. Cushing and Shandelle M. Henson, Global dynamics of some periodically forced, monotone difference equations, *J. Difference Eq. and Appl.* **7** (2001), 859–872.

[4] J.M. Cushing and Shandelle M. Henson, A periodically forced Beverton-Holt equation, *J. Difference Eq. and Appl.* **8** (2002), 1119–1120.

[5] Saber Elaydi, *An Introduction to Difference Equations*, Undergraduate Texts in Mathematics. Springer, New York, second edition, 1999.

[6] Saber Elaydi and Robert J. Sacker, Global stability of periodic orbits of nonautonomous difference equations, *J Differential Eq.* (in press).

[7] Mark Kot, *Elements of Mathematical Ecology*, Cambridge University Press, United Kingdom, first edition, 2001.

[8] F. Neave, Principles affecting the size of pink and chum salmon populations in British Columbia, *J. Fisheries Research Board of Canada* **9** (1953), 450–491.

[9] Robert J. Sacker and George R. Sell, Lifting properties in skew-product flows with applications to differential equations, *Memoirs Amer Math Soc* **11** (1977).

[10] George R. Sell, *Topological Dynamics and Differential Equations*, Van Nostrand-Reinhold, London, 1971.

Symplectic Factorizations and the Definition of a Focal Point

JULIA ELYSEEVA[1]

Department of Applied Mathematics
Moscow State University of Technology
Moscow, Russia

Abstract We present an approach based on a symplectic factorization of a conjoined basis of symplectic difference systems. It gives us the possibility to formulate important concepts of the oscillation theory of symplectic systems in terms of factors of such symplectic factorization. In particular we connect the number of focal points of a conjoined basis with the index of a symmetric matrix defined by solutions of a transformed Riccati equation.

Keywords Symplectic systems, Riccati difference equation, focal points

AMS Subject Classification 39A10

1 Introduction

We consider the symplectic difference system

$$Y_{i+1} = W_i Y_i \,, i = 0, 1, \ldots, N, \tag{1}$$

and the Riccati difference operator

$$R_W[Q] = C_i - Q_{i+1}A_i + D_iQ_i - Q_{i+1}B_iQ_i,$$

where W_i, Y_i, J are real partitioned matrices with $n \times n$ blocks

$$W_i = \begin{bmatrix} A_i & B_i \\ C_i & D_i \end{bmatrix}, Y_i = \begin{bmatrix} X_i \\ U_i \end{bmatrix}, \ J = \begin{bmatrix} 0 & I \\ -I & 0 \end{bmatrix},$$

and I, 0 are the identity and zero matrices. Note that the symplecity of any matrix W with $n \times n$ blocks A, B, C, D means that $A^TD - C^TB = AD^T - BC^T = D^TA - B^TC = DA^T - CB^T = I$, and the matrices

$$AB^T,\ CD^T,\ C^TA,\ D^TB,\ BA^T,\ DC^T,\ A^TC,\ B^TD$$

[1]Supported by the Federal Goal Program Integration under grant number 43.

are symmetric. In particular, the matrix

$$\Lambda[W] = \operatorname{diag}\left(C^T\ B^T\right) W = W^T \begin{bmatrix} 0 & I \\ 0 & 0 \end{bmatrix} W - \begin{bmatrix} 0 & I \\ 0 & 0 \end{bmatrix} \tag{2}$$

is symmetric (here diag $(C\ B)$ denotes a diagonal block matrix with $n \times n$ blocks). It is well known (see Erbe and Yan [5]) that the Riccati matrix difference equation

$$R_W[Q] = 0,\ i = 0, \ldots, N \tag{3}$$

has a symmetric solution $Q_i^T = Q_i$ iff there exists a conjoined basis of system (1)

$$Y_i^T J Y_i = 0,\ \operatorname{rank} Y_i = n$$

such that the condition

$$\det X_i \neq 0,\ i = 0, \ldots, N+1 \tag{4}$$

holds. If (4) does not hold, we have to consider generalized solutions of Equation (3). According to the definition (see Bohner and Došlý [2, Definition 3]), a conjoined basis of (1) is said to have a focal point in $(i, i+1]$ if the conditions

$$\operatorname{Ker} X_{i+1} \subseteq \operatorname{Ker} X_i, \tag{5}$$

$$X_i X_{i+1}^{\dagger} B_i \geq 0 \tag{6}$$

do not hold (here, $\dagger$ denotes the Moore-Penrose inverse of the matrix A, Ker$A$ denotes the kernel of A, for a symmetric matrix A we write $A \geq 0$ if A is positive semidefinite). If a conjoined basis without focal points in $(0, N+\ 1]$ is considered, then conditions (5), (6) are equivalent to the existence of a symmetric solution of the "implicit Riccati equation"(see Bohner and Došlý [2]) $R_W[Q]\,X_i = 0,\ i = 0, \ldots, N$ such that the condition $\left(D_i^T - B_i^T Q_{i+1}\right) B_i \geqslant 0, i = 0, 1, \ldots, N$ holds. Note that if the condition in (5) does not hold, the matrix in (6) may lose the symmetry. This problem is solved in Kratz [8], where the following concept of the multiplicities of a focal point in $(i, i+1]$ is offered (see Kratz [8, Definition 1]).

Definition 1.1. *A conjoined basis Y_i has a focal point of multiplicity $m_1(i)$ in the point $i+1$ if $m_1(i) = \operatorname{rank}\mathcal{M}_i$, where*

$$\mathcal{M}_i = \left(I - X_{i+1} X_{i+1}^{\dagger}\right) B_i,$$

and this basis has a focal point of multiplicity $m_2(i)$ in the interval $(i, i+1)$ if $m_2(i) = \operatorname{ind}\tilde{D}_i$, $\tilde{D}_i = \mathcal{T}_i^T X_i X_{i+1}^{\dagger} B_i \mathcal{T}_i$, $\mathcal{T}_i = I - \mathcal{M}_i^{\dagger}\mathcal{M}_i$, where indA *is the number of negative eigenvalues of a symmetric matrix A. The number of focal points in $(i, i+1]$ is defined by $m(i) = m_1(i) + m_2(i)$.*

In this work we offer an approach connected with symplectic factorizations of a conjoined basis of (1) and the matrix W_i. We motivate our results by the most elementary special case, when condition (4) holds for $i = 0, 1, \ldots, N+1$. In this case it is easy to verify that there exists a symplectic lower triangular block matrix $L_i = \begin{bmatrix} I & 0 \\ Q_i & I \end{bmatrix}$, such that

$$Y_i = L_i \begin{bmatrix} X_i \\ 0 \end{bmatrix}, \; L_{i+1}^{-1} W_i L_i = H_i = \begin{bmatrix} A_i + B_i Q_i & B_i \\ 0 & (A_i + B_i Q_i)^{-1\,T} \end{bmatrix}, \tag{7}$$

and $\Lambda[H_i] = \operatorname{diag}(0 \; (A_i + B_i Q_i)^{-1} B_i)$, where $\Lambda[H_i]$ is defined by (2) and $Q_i = U_i X_i^{-1}$ is solution of (3). Then, for this particular case $m_1(i) = 0, m_2(i) = m(i) = \operatorname{ind}\Lambda[H_i]$. Moreover, if $Y_i, \tilde{Y}_i$ are two conjoined bases of (1) and condition (4) holds for $\tilde{X}_i$, then $\tilde{Y}_i = \tilde{L}_i \begin{bmatrix} \tilde{X}_i \\ 0 \end{bmatrix}$, the matrices $L_i^{-1}\tilde{L}_i = \begin{bmatrix} I & 0 \\ \tilde{Q}_i - Q_i & I \end{bmatrix}$, $\tilde{Q}_i = \tilde{U}_i \tilde{X}_i^{-1}$ exist for any i and the condition $\Lambda[L_i^{-1}\tilde{L}_i] = \operatorname{diag}(\tilde{Q}_i - Q_i \; 0) \geq 0$ is equivalent to the Riccati inequality $\tilde{Q}_i - Q_i \geq 0$. Note, that $L_{i+1}^{-1} W_i \tilde{L}_i = (L_{i+1}^{-1}\tilde{L}_{i+1})\tilde{L}_{i+1}^{-1} W_i \tilde{L}_i = L_{i+1}^{-1} W_i L_i (L_i^{-1}\tilde{L}_i)$, then we have that the subsequent indentity

$$(L_{i+1}^{-1}\tilde{L}_{i+1})\tilde{H}_i = H_i (L_i^{-1}\tilde{L}_i) \tag{8}$$

holds (here $\tilde{H}_i$ is defined for $\tilde{Y}_i$ by (7)).

Now we pay attention to the symmetric matrix (2). Note that

$$\Lambda[WV] = V^T(\Lambda[W] + \begin{bmatrix} 0 & I \\ 0 & 0 \end{bmatrix})V - \begin{bmatrix} 0 & I \\ 0 & 0 \end{bmatrix} = V^T \Lambda[W] V + \Lambda[V] \tag{9}$$

for any symplectic matrices W, V. Applying Λ to the both sides of (8) and using (9), it is possible to show that $\Lambda[L_{i+1}^{-1}\tilde{L}_{i+1}] + \Lambda[\tilde{H}_i]$ is congruent to $\Lambda[L_i^{-1}\tilde{L}_i] + \Lambda[H_i]$. Taking into account the special structure of the last matrices, we have that $\Lambda[L_{i+1}^{-1}\tilde{L}_{i+1}] \geq 0, \Lambda[\tilde{H}_i] \geq 0$ iff $\Lambda[L_i^{-1}\tilde{L}_i] \geq 0, \Lambda[H_i] \geq 0$. This equivalence is the particular case of the separation results (see Bohner [1, Theorem1]). This example shows that the factorization technique coupled with the technique of operations with (2) based on (9) may be also useful for the case when (4) does not hold.

In this paper we present the first results in this direction. So we introduce the following factorizations:

$$Y_i = L_{j(i)} \mathfrak{N}_{j(i)} \begin{pmatrix} X_{j(i)} \\ 0 \end{pmatrix}, \; \det X_{j(i)} \neq 0, \tag{10}$$

$$L_{j(i+1)}^{-1} W_i L_{j(i)} = \mathfrak{N}_{j(i+1)} H_{j(i)} \mathfrak{N}_{j(i)}^T, \tag{11}$$

$$H_{j(i)} = \begin{bmatrix} \tilde{A}_i + \tilde{B}_i Q_{j(i)} & \tilde{B}_i \\ 0 & (\tilde{A}_i + \tilde{B}_i Q_{j(i)})^{-1\,T} \end{bmatrix}, \tag{12}$$

where Q_j is a solution of the transformed Riccati equation

$$R_{\widetilde{W}}[Q_j] = 0, \widetilde{W}_i = (\mathfrak{N}_{j(i+1)})^T W_i \mathfrak{N}_{j(i)} = \begin{bmatrix} \tilde{A}_i & \tilde{B}_i \\ \tilde{C}_i & \tilde{D}_i \end{bmatrix}, \tag{13}$$

and $L_{j(i)}$ is a symplectic lower triangular block matrix. $\mathfrak{N}_{j(i)}$ are symplectic and orthogonal matrices determined by the values of a function $j = j(i)$, $i = 0, \ldots, N+1$ (it is called an integration path for a conjoined basis Y_i by Elyseeva [3, 4]). The transformations with matrices $\mathfrak{N}_{j(i)}$ are connected with permutations of rows of a conjoined basis in such a manner that condition (4) holds for the transformed basis for any i. The main result is the following theorem.

Theorem 1.2. *Let* $\mathrm{rank}X_i = \mathrm{rank}F_jX_i = \mathrm{rank}F_j$, *where* Y_i *is a conjoined basis of system* (1), F_j, G_j *are* $n\times n$ *matrices,* $F_j + G_j = I$, *and the diagonal of* G_j *is composed of the zeros and ones that constitute the binary representation for the number* $j = j(i)$, $i = 0, \ldots, N+1$. *Let*

$$S_{j(i)} = \mathrm{diag}(F_{j(i)}\ I),\ \Lambda_{j(i)}[W] = S_{j(i)}\Lambda[W]S_{j(i)} \tag{14}$$

for any symplectic matrix W. *Then*

$$m(i) = m_1(i) + m_2(i) = \mathrm{ind}\Lambda_{j(i)}\left[\mathfrak{N}_{j(i+1)}H_{j(i)}\mathfrak{N}^T_{j(i)}\right],\ \mathfrak{N}_j = \begin{bmatrix} F_j & G_j \\ -G_j & F_j \end{bmatrix},$$

where $H_{j(i)}$ *is given by* (12) *and*
$Q_{j(i)} = \left(G_{j(i)}X_i + F_{j(i)}U_i\right)\left(F_{j(i)}X_i - G_{j(i)}U_i\right)^{-1}$ *is the solution of* (13).

Corollary 1.3. $\Lambda_{j(i)}\left[\mathfrak{N}_{j(i+1)}H_{j(i)}\mathfrak{N}^T_{j(i)}\right] \geq 0$ *iff conditions* (5),(6) *hold.*

2 Preliminaries

We introduce the transformations

$$Y_i = \mathfrak{N}_j Y^i_j,\ \mathfrak{N}_j = \begin{bmatrix} F_j & G_j \\ -G_j & F_j \end{bmatrix},\ Y^i_j = \begin{bmatrix} X^i_j \\ U^i_j \end{bmatrix},\ j = j(i). \tag{15}$$

When writing $Y_{j(i)}$, $X_{j(i)}$, $U_{j(i)}$, $Q_{j(i)}$, $L_{j(i)}$, $H_{j(i)}$, we always mean $Y^i_{j(i)}$, $X^i_{j(i)}$, $U^i_{j(i)}$, $Q^i_{j(i)}$, $L^i_{j(i)}$, $H^i_{j(i)}$. For the matrices $Y_{j(i)}$ we have the transformed system

$$Y_{j(i+1)} = \widetilde{W}_i Y_{j(i)},\ \ \widetilde{W}_i = \mathfrak{N}^T_{j(i+1)} W_i\, \mathfrak{N}_{j(i)}. \tag{16}$$

Consider the definition of the matrices $\mathfrak{N}_j$. We say that a matrix $\mathfrak{N} \in \Omega_{\mathfrak{N}}$ if it may be written in form (15) with $n\times n$ diagonal blocks F, G, which obey the conditions:

$$F + G = I,\ FG = 0. \tag{17}$$

It is easy to verify that there exist only 2^n symplectic orthogonal matrices $\mathfrak{N}_j$ defined by (17). The number $j = j\,(i)$ of any $\mathfrak{N}_j$ takes the values from the set $\{0, 1, \ldots, 2^n - 1\}$, and the diagonal of G_j is composed of the zeros and ones that constitute the binary representations of $j = j\,(i)$. In this case, we have $\mathfrak{N}_0 = I$, $\mathfrak{N}_{2^n-1} = J$, $\mathfrak{N}_j = J\mathfrak{N}^T_{2^n-1-j}$.

The treatment of the set $\Omega_{\mathfrak{N}}$ is justified by the following theorem (see Elyseeva [3, Lemma 2.1]) .

Theorem 2.1. *For any conjoined basis of* (1) *there exists* $j = j\,(i)\,,\ i = 0, 1, \ldots, N+1$*:*

$$\det X_{j(i)} \neq 0. \tag{18}$$

In particular $j = j\,(i)\,,\ i = 0, 1, \ldots, N+1$ *can be defined by the conditions*

$$\text{rank} X_i = \text{rank} F_j X_i = \text{rank} F_j. \tag{19}$$

Definition 2.2. *The function* $j = j\,(i)$ *is called an integration path if condition (18) holds, and it is called a special integration path if, in addition, conditions (19) hold.*

Lemma 2.3. *Let* Y_i *be a conjoined basis of (1) with an integration path* $j = j\,(i)$*, and* Q_j *is the solution of* (13)*, then* $j = j\,(i)$ *is a special iff*

$$L_j = \mathfrak{N}_j \begin{bmatrix} I & 0 \\ Q_j & I \end{bmatrix} \mathfrak{N}_j^T = \begin{bmatrix} I + G_j Q_j F_j & -G_j Q_j G_j \\ F_j Q_j F_j & I - F_j Q_j G_j \end{bmatrix} \tag{20}$$

is a symplectic lower triangular block matrix. In this case we see that factorization (10) *holds.*

Proof. It is proved in Elyseeva [4, Lemma 2.4] that $j = j(i)$ is a special path iff $G_j Q_j G_j \equiv 0$. Then matrix L_j given by (20) is lower triangular iff the path $j = j(i)$ is special. By (15), $Y_i = \mathfrak{N}_j Y_j = \mathfrak{N}_j \begin{bmatrix} I & 0 \\ Q_j & I \end{bmatrix} \begin{pmatrix} X_j \\ 0 \end{pmatrix}$, where it is possible use (7) for Y_j. Then (10) does hold. □

Factorization (10) implies that

$$X_i = K_i F_{j(i)} M_i = (K_i F_{j(i)})(F_{j(i)} M_i), \det K_i \neq 0, \det M_i \neq 0, \tag{21}$$

where $K_i = I + G_j Q_j F_j$, $M_i = X_j$. Note that (21) is an analog of the skeleton factorization (see Gantmacher [6]) of X_i, which gives us the possibility to evaluate the orthogonal projections $I - X_{i+1} X_{i+1}^{\dagger}$, $\mathcal{T}_i = I - \mathcal{M}_i^{\dagger} \mathcal{M}_i$ introduced in Definition 1.1 in terms of factors of (21).

Lemma 2.4. *Let* $X = KFM$*, where* $\det K \neq 0$, $\det M \neq 0$ *and* F, G *are defined by* (17)*. Then for any* $n \times n$ *matrix* C *we have*

$$(I - XX^{\dagger})\,C = TGK^{-1}C, \quad C\,(I - X^{\dagger}X) = CM^{-1}GP, \tag{22}$$

where T, P are nonsingular matrices. In patricular,

$$\begin{gathered}\operatorname{Im}C \subseteq \operatorname{Im}X = \operatorname{Im}(KF) \Leftrightarrow \\ GK^{-1}C = 0 \Leftrightarrow K^{-1}C = FK^{-1}C = (KF)^{\dagger} C, \qquad (23) \\ \operatorname{Ker}X = \operatorname{Ker}(FM) \subseteq \operatorname{Ker}C \Leftrightarrow \\ CM^{-1}G = 0 \Leftrightarrow CM^{-1} = CM^{-1}F = C(FM)^{\dagger}. \qquad (24)\end{gathered}$$

Proof. Consider the proof of the first formula in (22), the second one can be proved in a similar way. We have $XX^{\dagger} = (KF)(KF)^{\dagger}$, $C = KK^{-1}C = KFK^{-1}C + KGK^{-1}C$ and $(I - XX^{\dagger})KF = 0$, then $(I - XX^{\dagger})C = \left(I - KF (KF)^{\dagger}\right) KGK^{-1}C = K\left(I - F(KF)^{\dagger} KG\right) GK^{-1}C = TGK^{-1}C$, where $T = K\left(I - F(KF)^{\dagger} KG\right)$. Equivalences (23), (24) are proved in Elyseeva [4]. □

Remark 2.5. *Let $n \times n$ marices $\mathcal{A}$, $\mathcal{B}$ be nonsingular and $\tilde{X} = \mathcal{A}X\mathcal{B}$, then $I - \tilde{X}\tilde{X}^{\dagger} = \tilde{T}T^{-1}(I - XX^{\dagger})\mathcal{A}^{-1}$, $I - \tilde{X}^{\dagger}\tilde{X} = \mathcal{B}^{-1}(I - X^{\dagger}X)P^{-1}\tilde{P}$, where all matrices $\tilde{T}$, $\tilde{P}$, T, P are nonsingular.*

Using Lemma 2.4 and Remark 2.5 it is possible prove that any transformation with a symplectic lower triangular matrix preserves the multiplicities of focal points of a conjoined basis of (1) in sense of Definition 1.1.

Lemma 2.6. *Let Y_i be a conjoined basis of (1) with $m_1(i)$, $m_2(i)$, and $\hat{Y}_k = L_k Y_k$, where $k = i, i+1$ and $L_k = \begin{bmatrix} \mathcal{A}_k & 0 \\ \mathcal{B}_k & (\mathcal{A}_k)^{-1\,T} \end{bmatrix}$ is a symplectic lower triangular block matrix with $n \times n$ blocks. Let $\hat{m}_1(i)$, $\hat{m}_2(i)$ be the multiplicities of focal points of $\hat{Y}_i$. Then $m_1(i) = \hat{m}_1(i)$, $m_2(i) = \hat{m}_2(i)$.*

Proof. Note that $\hat{X}_k = \mathcal{A}_k X_k$, where $\hat{X}_k$, $k = i, i+1$ is the upper block of $\hat{Y}_k$. As $\hat{\mathcal{M}}_i = (I - \hat{X}_{i+1}\hat{X}_{i+1}^{\dagger})\hat{B}_i$ where $\hat{B}_i = \mathcal{A}_{i+1}B_i\mathcal{A}_i^T$ is the block of $\hat{W}_i = L_{i+1}W_iL_i^{-1}$, then, applying Remark 2.5, we have

$$\hat{\mathcal{M}}_i = \hat{T}_i T_i^{-1}(I - X_{i+1}X_{i+1}^{\dagger})B_i\mathcal{A}_i^T = \hat{T}_i T_i^{-1}\mathcal{M}_i\mathcal{A}_i^T, \qquad (25)$$

where $\det \hat{T}_i T_i^{-1} \neq 0$, $\det \mathcal{A}_i \neq 0$. Then $m_1(i) = \operatorname{rank}\mathcal{M}_i = \operatorname{rank}\hat{\mathcal{M}}_i = \hat{m}_1(i)$.

Now we apply Remark 2.5 to (25). We have $\hat{\mathcal{T}}_i = I - \hat{\mathcal{M}}_i{}^{\dagger}\hat{\mathcal{M}}_i = \mathcal{A}_i^{T\,-1}(I - \mathcal{M}_i^{\dagger}\mathcal{M}_i)P_i^{-1}\tilde{P}_i = \mathcal{A}_i^{T\,-1}\mathcal{T}_iP_i^{-1}\tilde{P}_i$, where P_i, $\tilde{P}_i$ are nonsingular matrices. Then $\hat{\mathcal{T}}_i^{\,T}\hat{X}_i\hat{X}_{i+1}^{\dagger}\hat{B}_i\hat{\mathcal{T}}_i$ is congruent to $\mathcal{T}_i{}^T X_i\hat{X}_{i+1}^{\dagger}\mathcal{A}_{i+1}B_i\mathcal{T}_i$. Note that by Kratz [8, p. 141] it holds that $\operatorname{Im}B_i\mathcal{T}_i \subseteq \operatorname{Im}X_{i+1}$, then $\operatorname{Im}\mathcal{A}_{i+1}B_i\mathcal{T}_i \subseteq \operatorname{Im}\mathcal{A}_{i+1}X_{i+1} = \operatorname{Im}\hat{X}_{i+1}$. According to (23) $\hat{X}_{i+1}^{\dagger}\mathcal{A}_{i+1}B_i\mathcal{T}_i = X_{i+1}^{\dagger}B_i\mathcal{T}_i$, and

$$\hat{m}_2(i) = \operatorname{ind}\hat{\mathcal{T}}_i^{\,T}\hat{X}_i\hat{X}_{i+1}^{\dagger}\hat{B}_i\hat{\mathcal{T}}_i = \operatorname{ind}\mathcal{T}_i^T X_iX_{i+1}^{\dagger}B_i\mathcal{T}_i = m_2(i).$$

□

For the proof of Theorem 1.2 we need the following lemma.

Lemma 2.7. $\operatorname{ind}\begin{bmatrix} 0 & X \\ X^T & D \end{bmatrix} = \operatorname{rank} X + \operatorname{ind}(I - X^\dagger X)^T D(I - X^\dagger X)$, *where* $D^T = D$ *and* X, D *are* $n \times n$ *matrices.*

Proof. Let X be factorized as in Lemma 2.4 (note that for any X the factorization of the given type follows from SVD decomposition (see Golub [7])). Then

$$\mathfrak{M} = \begin{bmatrix} 0 & X \\ X^T & D \end{bmatrix} \underset{A}{\sim} \begin{bmatrix} 0 & F \\ F & GM^{-1T}DM^{-1}G \end{bmatrix}, \; A = \begin{bmatrix} K^T & S \\ 0 & M \end{bmatrix},$$

where $S = 0,5M^{-1T}DM^{-1}(I+G)M$ and the notation $B \underset{A}{\sim} C$ means that B is congruent to C, i.e., $B = A^T CA$, $\det A \neq 0$. Then

$$\operatorname{ind}\mathfrak{M} = \operatorname{ind}\begin{bmatrix} 0 & F \\ F & 0 \end{bmatrix} + \operatorname{ind}(GM^{-1T}DM^{-1}G)$$

and

$$\operatorname{ind}\begin{bmatrix} 0 & F \\ F & 0 \end{bmatrix} = \operatorname{rank} F = \operatorname{rank} X,$$
$$\operatorname{ind}(GM^{-1T}DM^{-1}G) = \operatorname{ind}(I - X^\dagger X)^T D(I - X^\dagger X)$$

by (22). □

3 Proof of the main theorem

Lemma 3.1. *Let* $j = j(i)$ *be a special integration path for* Y_i. *Then*

$$m_1(i) = \operatorname{rank}\hat{\mathcal{M}}_i, \; \hat{\mathcal{M}}_i = G_{j(i+1)}P_{j(i)}^T F_{j(i)}, \; P_{j(i)} = \left(\tilde{A}_i + \tilde{B}_i Q_{j(i)}\right)^{-1} \quad (26)$$

$$m_2(i) = \operatorname{ind}\hat{\mathcal{T}}_i^T F_{j(i)} D_{j(i)} F_{j(i)} \hat{\mathcal{T}}_i, \; \hat{\mathcal{T}}_i = I - \hat{\mathcal{M}}_i^\dagger \hat{\mathcal{M}}_i, \; D_{j(i)} = P_{j(i)}\tilde{B}_i \quad (27)$$

where Q_j *is the solution of* (13) *and* m_1, m_2 *are introduced in Definition 1.1.*

Proof. Consider the transformation of Y_i with the matrix L_j given by (20). According to (10), we have $\hat{Y}_i = L_j^{-1} Y_i = \mathfrak{N}_j \begin{pmatrix} X_{j(i)} \\ 0 \end{pmatrix}$, where $\hat{Y}_i$ is a conjoined basis of a transformed system with the matrix defined by (11),(12). As L_j^{-1} is lower triangular, then, by Lemma 2.6, $m_1(i) = \hat{m}_1(i)$, $m_2(i) = \hat{m}_2(i)$, where $\hat{m}_1(i)$, $\hat{m}_2(i)$ are evaluated for $\hat{Y}_i$. As for this case $\hat{X}_i = F_{j(i)} X_{j(i)}$, $I - \hat{X}_{i+1}\hat{X}_{i+1}^\dagger = G_{j(i+1)}$ it is easy to verify that (26),(27) hold. □

Now it is possible to prove the main result connecting the index of (14), where W is given by (11),(12) with the numbers $m_1(i)$, $m_2(i)$. Note that

$$\Lambda_{j(i)}[W] \underset{A}{\sim} \begin{bmatrix} 0 & -\hat{\mathcal{M}}_i \\ -\hat{\mathcal{M}}_i^{\,T} & F_{j(i)}D_{j(i)}F_{j(i)} \end{bmatrix}, \; A = \begin{bmatrix} P_{j(i)}^{-1} & \tilde{B}_i F_{j(i)} \\ 0 & I \end{bmatrix} \mathfrak{N}_{j(i)}^T,$$

where $\hat{\mathcal{M}}_i$, $D_{j(i)}$ are defined in (26),(27). Putting $X = -\hat{\mathcal{M}}_i$, $D = F_{j(i)}D_{j(i)}F_{j(i)}$ by Lemma 2.7 we have that $\operatorname{ind}\Lambda_{j(i)}[\mathfrak{N}_{j(i+1)}H_{j(i)}\mathfrak{N}_{j(i)}^T] = m_1(i) + m_2(i)$.

Remark 3.2. Consider two conjoined bases of (1): Y_i, $\tilde{Y}_i$ and the matrices $L_{j(i)}$, $H_{j(i)}$, $\tilde{L}_{l(i)}$, $\tilde{H}_{l(i)}$ defined by (20), (12) where $j(i)$, $l(i)$ are special paths for Y_i, $\tilde{Y}_i$ respectively. By analogy with the nonsingular case (see Section 1) we have the identity

$$(L_{j(i+1)}^{-1}\tilde{L}_{l(i+1)})\mathfrak{N}_{l(i+1)}\tilde{H}_{l(i)}\mathfrak{N}_{l(i)}^T = \mathfrak{N}_{j(i+1)}H_{j(i)}\mathfrak{N}_{j(i)}^T(L_{j(i)}^{-1}\tilde{L}_{l(i)}). \tag{28}$$

Our purpose is to connect the numbers of focal points of Y_i, $\tilde{Y}_i$ evaluating the index of the operator $\Lambda_{l(i)}$ applied to the both sides of (28). In particular, it is possible to show that

$$\Lambda_{l(i)}[\mathfrak{N}_{j(i)}^T L_{j(i)}^{-1}\tilde{L}_{l(i)}] \geq 0, \quad \Lambda_{j(i)}[\mathfrak{N}_{j(i+1)}H_{j(i)}\mathfrak{N}_{j(i)}^T] \geq 0$$

if and only if

$$\Lambda_{l(i+1)}[\mathfrak{N}_{j(i+1)}^T L_{j(i+1)}^{-1}\tilde{L}_{l(i+1)}] \geq 0, \quad \Lambda_{l(i)}[\mathfrak{N}_{l(i+1)}\tilde{H}_{l(i)}\mathfrak{N}_{l(i)}^T] \geq 0.$$

Note that by Corollary 1.3, the conditions

$$\Lambda_{j(i)}\,[\mathfrak{N}_{j(i+1)}H_{j(i)}\mathfrak{N}_{j(i)}^T] \geq 0, \quad \Lambda_{l(i)}[\mathfrak{N}_{l(i+1)}\tilde{H}_{l(i)}\mathfrak{N}_{l(i)}^T] \geq 0$$

mean that Y_i, $\tilde{Y}_i$ do not have focal points in $(i, i+1]$ while

$$\Lambda_{l(k)}[\mathfrak{N}_{j(k)}^T L_{j(k)}^{-1}\tilde{L}_{l(k)}] \geq 0$$

for $k = i, i+1$ is equivalent to the conditions

$$\operatorname{Im}\tilde{X}_i \subseteq \operatorname{Im}X_i, \; \tilde{X}_i^T(\tilde{Q}_i - Q_i)\tilde{X}_i \geq 0$$

which hold for any symmetric matrices Q_i, $\tilde{Q}_i$ satisfying the equations $X_i^T Q_i X_i = X_i^T U_i$, $\tilde{X}_i^T\tilde{Q}_i\tilde{X}_i = \tilde{X}_i^T\tilde{U}_i$.

References

[1] Bohner, M., Discrete Sturmian Theory, *Math. Inequal. Appl.* **1(3)** (1998), 375–383.

[2] Bohner, M. and Došlý, O., Disconjugacy and transformations for symplectic systems, *Rocky Mountain Journal of Mathematics* **27** (1997), 707–743.

[3] Elyseeva, J., A transformation for symplectic systems and the definition of a focal point, *Comp. Math. Appl.* (in press).

[4] Elyseeva, J., A transformation for the Riccati difference operator, *Proceedings of ICDEA 2001*(2003), 415–422.

[5] Erbe, L. and Yan, P., Disconjugacy for linear Hamiltonian difference systems, *J. Math. Anal. Appl.* **167** (1992), 355–367.

[6] Gantmacher, F., *The Theory of Matrices*, v. 1, Chelsea Publishing Company, New York, 1959.

[7] Golub, G. and Van Loan, C., *Matrix Computations*, The Johns Hopkins University Press, 1996.

[8] Kratz, W., Discrete Oscillation, *J. Difference Equations and Appl.* **9** (2003), 135–147.

Second Smaller Zero of Kneading Determinant for Iterated Maps

SARA FERNANDES[1]

Departamento de Matemática, Universidade de Évora
Évora, Portugal

and

J. SOUSA RAMOS

Departamento de Matemática, Instituto Superior Técnico
Lisbon, Portugal

Abstract The mixing rate on the studies of iterated maps is closely related with the second smaller zero of the kneading determinant. For a certain class of maps (piecewise linear, semiconjugated to cubics with topological entropy equal $\log 2$) we use this zero to classify dynamical systems associated with difference equations.

Keywords Perron-Frobenius operator, second eigenvalue, decay correlations, kneading theory

AMS Subject Classification 37A25, 37A35, 37E05

1 Motivation

In the study of iterated maps there are topological invariants, which are closely related to spectral invariants, as topological entropy is with the first eigenvalue of the transition matrix, well known as the Perron eigenvalue. We can put the question of what does the rest of the spectrum do in the dynamics? It is known that, in certain test function spaces, the absolute value of the second eigenvalue can be used as a measure of mixing.

This work presents an approach to this, combining the results from the kneading theory, transition matrices, and spectral analysis.

Consider $(I, \mathcal{B}, \mu, S)$, where I is the unit interval $I = [0, 1]$, $\mathcal{B}$ are the borelians in I, and μ the Lebesgue measure. We have that $(I, \mathcal{B}, \mu)$ is a probabilistic measure space and $S : I \to I$ will be a nonsingular and measure preserving transformation.

The dynamical system we consider consists of the iterates of the map S :

$$S^0(x) = x, \quad S^n(x) = S^{n-1}(S(x)), \quad n = 1, 2, ...$$

[1]Partially supported by the CIMA-UE.

The authors were partially supported by FCT (Portugal) through the program POCTI.

2 Mixing rate

We can classify various degrees of irregular behaviors: ergodicity, mixing, exactness, etc.

The second eigenvalue is specially related with the mixing property.

Definition 2.1. *The map $S : I \to I$ is called mixing iff*

$$\lim_{n\to\infty} \mu\left(A \cap S^{-n}(B)\right) = \mu(A)\,\mu(B) \qquad \text{for all } A, B \in \mathcal{B},$$

where

$$S^{-n}(B) = \{x \in I : S^{n}(x) \in B\}, \quad n = 1, 2, \ldots.$$

If we consider points x in $A \cap S^{-n}(B)$, when $n \to \infty$, the measure of the set of these points is just $\mu(A)\,\mu(B)$. It means that any set $B \in \mathcal{B}$ under the action of S becomes asymptotically independent of a fixed set $A \in \mathcal{B}$.

Obviously, the mixing property depends on the measure μ, and we will say that the system is (S, μ)-mixing meaning that the map S is mixing with respect to the measure μ.

Let us give another approach to this notion introducing the Perron-Frobenius operator.

We will abstractly define this operator, but, as we will see, in the case of a piecewise linear Markov transformation, the Perron-Frobenius operator will have a very simple representation. For more details see [1], Ch. 4.

Let

$$\lambda(A) = \int_{S^{-1}(A)} f \, d\mu,$$

where $f \in L^1(X, \mu)$ and A is an arbitrary measurable set. Since S is nonsingular, $\mu(A) = 0$ implies $\mu\left(S^{-1}(A)\right) = 0$, which in turn implies that $\lambda(A) = 0$. Hence λ is absolutely continuous with respect to the measure μ. Then, by the Radon-Nikodym theorem, there exists a $\phi \in L^1$, such that for all measurable sets A

$$\lambda(A) = \int_A \phi \, d\mu.$$

ϕ is unique a.e. and depends on S and f.

Definition 2.2. *We define the Perron-Frobenius (P-F) operator $P : L^1 \to L^1$ corresponding to S by*

$$Pf = \phi.$$

3 The correlation function and spectral properties of the Perron-Frobenius operator

Let f and $g : X \to \mathbb{R}$ be C^1 test functions. We define the correlation function of f and g as

$$C_{f,g}(k) = \left| \int_X f \circ S^k \cdot g \, d\mu - \int_X f \, d\mu \cdot \int_X g \, d\mu \right| . \tag{1}$$

Following [4], by putting $f = \chi_A$ and $g = \chi_B$, we see that (1) reduces to the definition of mixing and, by approximating f and g by step functions, it can be shown that $C_{f,g}(k) \to 0$ as $k \to \infty$, for all $f, g \in L^2(X, \mu)$ is equivalent to (S, μ)-mixing.

Definition 3.1. *If there are constants* $c = c(f,g)$ *and* $0 < r_0 < 1$, *such that*

$$C_{f,g}(k) < c(f,g)\, r^k, \qquad \text{for all } r > r_0,$$

we say that we have exponential decay of correlations. We shall call the minimal such r_0, *the rate of decay of correlations for the system* (S, μ) *or the rate of mixing.*

It will depend, obviously, on the test function space that we consider.

We need to impose some conditions on S for the "good behavior" of the P-F operator. So, consider the interval $I = [0, 1]$ and $S : I \to I$ such that the two following conditions hold:

Condition 3.2. *The map* S *is piecewise expanding, i.e., there is a partition* Q$=\{I_i = [a_{i-1}, a_i], i = 1, ..., q\}$ *of* I, *such that* $S|_{I_i}$ *is* C^1 *and* $|S_i'(x)| \geq \alpha > 1$, *for any* i *and for all* $x \in (a_{i-1}, a_i)$.

Condition 3.3. *The function* $g(x) \equiv \frac{1}{|S'(x)|}$ *is a function of bounded variation, where* $S'(x)$ *is the appropriate one-sided derivative at the endpoints of* $\mathcal{Q}$.

For such S, in the space $\mathbf{BV}(I)$ of functions of bounded variation in $L^1(I)$ equipped with the norm $\|f\| = \max\{\|f\|_1, var(f)\}$, the following properties hold for the Perron-Frobenius operator associated with S:

– The operator is Markov.

– The operator is quasi-compact.

– The entire spectrum lies inside the unit disk, moreover $\sigma(P) = \{1\} \cup \Sigma$, where Σ lies inside a circle of radius $\rho < 1$.

– The essential spectrum of the operator lies inside a disk of radius η, where

$$\eta = \lim_{k \to \infty} \left(\sup_x \left(\frac{1}{\left| (S^k)'(x) \right|} \right) \right)^{1/k} . \tag{2}$$

Spectral points of P outside the closed disk of radius η are isolated eigenvalues of P of finite multiplicity.

From these properties follow the exponential decay of correlations (see [9]):

Corollary 3.4. *Given any* $f, g \in \mathbf{BV}(I)$, *there is* $c = c(f,g) > 0$ *such that*

$$\left| \int_X f \circ S^n \cdot g \, d\mu - \int_X f \, d\mu \cdot \int_X g \, d\mu \right| \leq c \rho^n \qquad \text{for all } n \geq 0.$$

4 P-F operator and transition matrices

In the special case, when S is a piecewise linear Markov transformation on the partition $\{I_i\}_{i=1}^n$, we approach P by a $n \times n$ matrix P_{ij} given by

$$P_{ij} = \frac{a_{ij}}{|S_i'|}, \qquad 1 \leq i, j \leq n,$$

where (a_{ij}) is the transition matrix induced by S and the given partition and S_i' is the slope of S in every interval I_i of the partition.

This approximation, being very easy to calculate, gives us a very powerful tool to calculate some eigenvalues, which are important to classify the systems from the rate of decay of correlations. This importance is pointed out in the next lemma.

Lemma 4.1. *(See [2]) Let* S *be a Markov map and* $\mathcal{P}$ *a Markov partition for* S *with cardinality* q. *Denote by* F *the* q*-dimensional space of simple functions on* $\mathcal{P}$, *let* P *be the P-F operator in* $\boldsymbol{BV}(I)$, *induced by* S, *then*

$$\{z \in \sigma(P) : |z| > \eta\} = \{z \in \sigma(P|_F) : |z| > \eta\}.$$

It means that we can find all spectral points outside $\{|z| \leq \eta\}$ by finding eigenvalues of the matrix representing $P|_F$, that is, the transition matrix.

We are now arriving to the pith of the question: the importance of the absolute value of the second eigenvalue of the P-F operator in the estimating the rate of decay of correlations and the simplicity of this calculus in the case of a piecewise linear Markov transformation. Because of Corollary 3.4 the problem is to find ρ, which can be equal to the radius of the essential spectrum η or not. In fact, if we find an eigenvalue of the matrix approximation outside the essential spectrum (the greatest one except 1), it will be ρ, if we do not take η for the rate of decay of correlations.

In the case of constant slope s in every interval of the Markov partition the situation is even better: if the second eigenvalue (in magnitude) of the transition matrix divided by s is greater than $1/s$ (or, what is the same, the second eigenvalue of the transition matrix is greater than 1), we take it for our mixing rate. If the second eigenvalue of the transition matrix is lesser than 1, so we take $1/s$. Precisely we have the following

Theorem 4.2. *(See [3]) Let S be a piecewise linear Markov transformation with constant slope $s > 1$ in every interval of the Markov partition and let $\mathcal{M}$ be the transition matrix. Let λ_2 be the second eigenvalue (in magnitude) of the matrix $\mathcal{M}$. Then, if $|\lambda_2| > 1$, we have exponential decay of correlations with a rate at most $|\lambda_2|/s$. If $|\lambda_2| \leq 1$, we have exponential decay of correlations with a decay rate at most $1/s$.*

5 The kneading theory: bimodal case

Consider $s \in (1,3)$ and the map $S : I \to I$ defined by

$$S(x) = \begin{cases} s\,x & \text{if } \ 0 \leq x < c_1 \\ -s\,x + e & \text{if } \ c_1 \leq x < c_2, \\ s\,x + 1 - s & \text{if } \ c_2 \leq x < 1 \end{cases} \tag{3}$$

where c_1 and c_2 depend on the parameters s and e.

For algebraic s we are in the conditions of Theorem 4.2, because we obtain a partition of the interval I, on which S is Markov.

These maps are the piecewise linear models, semiconjugated to cubic bimodal maps, with the same growth number and the same kneading invariant, introduced by Milnor and Thurston.

From the above the importance of the second eigenvalue of the transition matrix is clear.

In [8] the authors give a characterization of a bimodal family of maps in the interval, using the second topological invariant. In particular they use that to distinguish maps with the same value of topological entropy.

In [3], for certain kinds of trajectories, which correspond to those which have the first eigenvalue (the one corresponding to topological entropy) equal to 2, the general form for the kneading determinant is established. As we will see this result allows us to construct an ordered tree of polynomials, giving rise to a structure based on the second eigenvalue.

We begin with some notations and results from the kneading theory.

Consider the bimodal map defined in (3).

For different values of the parameters s and e we obtain trajectories that can be symbolically expressed by the itineraries of the critical points c_1 and c_2.

If we take a point x in I, we will call the address of x, $A(x)$, one of the symbols L, A, M, B, R according to the following rule:

$$A(x) = \begin{cases} L & \text{if } x < c_1 \\ A & \text{if } x = c_1 \\ M & \text{if } x > c_1 \text{ and } x < c_2. \\ B & \text{if } x = c_2 \\ R & \text{if } x > c_2 \end{cases}$$

The itinerary of $x \in I$ will be then the sequence of symbols $A(x)$, $A(f(x))$, $A(f(f(x)))\ldots$. See [7] for more details.

In [3] we considered only monotone trajectories of the type

$$ARP(BLP)^{\infty} \text{ or } BLP(ARP)^{\infty}, \tag{4}$$

where

$$P = P_1P_2...P_n, \text{with } P_i \in \{L, M, R\}$$

for which we proved the following theorem and corollaries.

Theorem 5.1. *Let $S : I \to I$ be defined by (3) with parameters e and s such that the trajectory of c_1 and c_2 is of type (4). Then, the kneading determinant $D(t)$ satisfies*

$$D(t) = -\frac{d_{\mathcal{M}}(t)}{(-1+t^p)} = \frac{(-1+2t)}{(-1+t^p)}\, p(t),$$

where

$$p(t) = 1 + t + a_2t^2 + ... + a_{n+1}t^{n+1}$$

with $p = n + 2$, n is the length of P and $d_{\mathcal{M}}(t) = \det(I - t\mathcal{M})$, where $\mathcal{M}$ is the transition matrix of the system. The coefficients a_i are given by

$$a_i = \begin{cases} 1 & \text{if the number of } M\text{'s in } \{P_1, ..., P_{i-1}\} \text{ is even} \\ -1 & \text{if the number of } M\text{'s in } \{P_1, ..., P_{i-1}\} \text{ is odd} \end{cases}, \; i = 2, ..., n+1.$$

Remark 5.2. *The zeros of $D(t)$ are the inverse of the transition matrix's eigenvalues.*

Corollary 5.3. *The first eigenvalue of the transition matrix $\mathcal{M}$ is $s = 2$.*

Corollary 5.4. *The second eigenvalue, in magnitude, of the transition matrix $\mathcal{M}$ is the inverse of the smaller zero of $p(t)$.*

6 The tree of polynomials

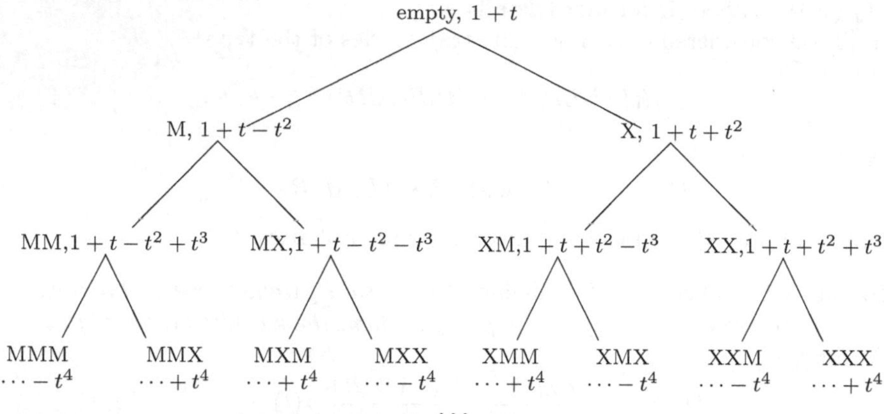

Figure 1: The tree of polynomials $p(t)$ with the correspondent P.

The polynomials obtained by Theorem 5.1 depend only on the M-part of the sequence P. They are disposed in a binary tree as we see in Figure 1. To each branch of this tree corresponds a polynomial $p(t)$, a group of trajectories P, codified with M and X (where X can be replaced by R or L, that it will not make any change in $p(t)$) and the corresponding $|\lambda_2|$ (obtained as the inverse of the absolute value of $p(t)$'s smaller zero). That way, in the same branch P, they happen several situations, corresponding to different choices of the parameter e (s=2). So, for example, corresponding to $P = XMX$, we have the polynomial $p(t) = 1 + t + t^2 - t^3 - t^4$, the second eigenvalue $|\lambda_2| \simeq 1.33539$ and the 8 next possibilities: 4 are admissible $AR\,(RMRBL)^{\infty}$, $BL\,(LMLAR)^{\infty}$, $AR\,(LMLBL)^{\infty}$, $BL\,(RMRAR)^{\infty}$, and 4 are nonadmissible $AR\,(RMLBL)^{\infty}$, $BL\,(LMRAR)^{\infty}$, $AR\,(LMRBL)^{\infty}$, $BL\,(RMLAR)^{\infty}$. From this we see that we cannot establish a trivial correspondence between λ_2 and the second parameter e, as it is done with the first parameter s and λ_1.

Another statement is the existence of an isolated eigenvalue in almost all these maps. In fact, our second eigenvalue is in the interval $[1,2)$ and every eigenvalue in the interval $(1,2)$ corresponds to an isolated eigenvalue of the Perron-Frobenius operator, giving rise to a slower decay of correlations. In Figure 2 we see the distribution of the values $|\lambda_2|$ with the order we choose for the tree: $M < X$. We can see the existence of the values $|\lambda_2| = 1$, namely, in all $P = XX...X$, and we conjecture the following result:

Conjecture 6.1. *Let n be the length of P. Denote by $K(n)$ the number of $|\lambda_2| = 1$ in the level n of the tree. So we have:*

$$\text{if } n = 2m, \text{ then } K(n) = m + 1,$$
$$\text{if } n = 2m + 1, \text{ then } K(n) = 1.$$

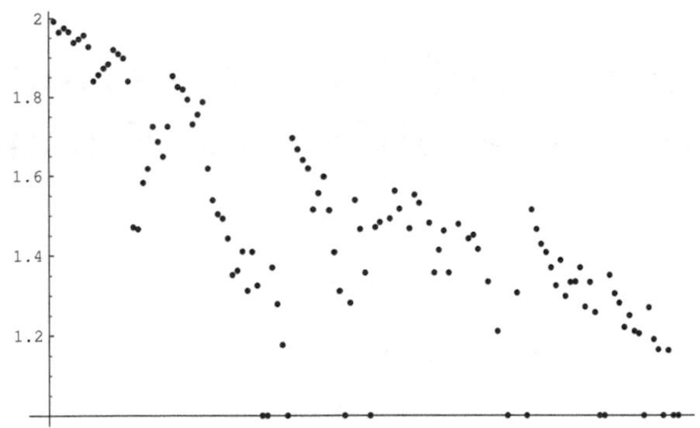

Figure 2: The $|\lambda_2|$ obtained until degree 7 of $p(t)$, ordered like the tree $M < X$.

References

[1] Boyarsky, A., Góra, P., *Laws of Chaos, Birkhauser* (1997).

[2] Dellnitz, M., Froyland, G., Sertl, S., On the isolated spectrum of the Perron-Frobenius operator, *Nonlinearity* **13** (2000), 1171–1188.

[3] Fernandes, S., Sousa Ramos, J., Spectral invariants of iterated maps of the interval, *Grazer Math. Ber.* **346** (2002), 113–122.

[4] Froyland, G., Computer-assisted bounds for the rate of decay of correlations, Comm. Math. Phys., **189** (1997), 237–257.

[5] Lampreia, J.P., Sousa Ramos, J., Symbolic dynamics of bimodal maps, *Portugal. Math.* **54** (1997), 1–18.

[6] Lasota, A., Mackey, M. C., *Chaos, Fractals and Noise*, New York, Springer (1994).

[7] Milnor, J., Thurston, W., *On the Iterated Maps of the Interval, Dynamical Systems: Proc. Univ of Maryland 1986–1987, Lecture Notes in Mathematics* **1342** (Berlin, New York, Springer), 465–563 (1988).

[8] Severino, R., Martins, N., Sousa Ramos, J., Isentropic real cubic maps. Dynamical systems and functional equations (Murcia, 2000). *Internat. J. Bifur. Chaos Appl. Sci. Engrg.* **13** (2003), 1701–1709.

[9] Viana, M., Stochastic dynamics of deterministic systems, *Brazillian Math Coloquium,* IMPA (1997).

Bifurcation of Almost Periodic Solutions in a Difference Equation

YOSHIHIRO HAMAYA

Deptartment of Information Science Okayama University of Science

Okayama, Japan

Abstract This paper is concerned with almost periodic solutions (in short, ap-solutions) originating from an equilibrium state when the parameters of a difference equation are varied. We study the bifurcation of ap-solutions for an ap-difference equation of the form $x_{n+1} = f(n, x_n; \mu)$, using the Green's function for regular ap-operators and Σ-operators.

Keywords Almost-periodic function, bifurcation, difference Shtokalo substitution

AMS Subject Classification 39A10, 39A11

1 Introduction and Notations

In this talk, by constructing the Green's function we study the bifurcation of almost periodic solutions to difference equations constituting the discrete analog to the ordinary differential equations case [4]. We will extend those results to the difference equations [cf. 2]. Our focus is on the creation of almost periodic oscillations from an equilibrium state of the difference equation.

Let R^m denote the Euclidean m-space and $|\cdot|$ the Euclidean norm. We introduce an almost periodic function $f(n, x) : Z \times D^* \to R^m$ where D^* is an open set in R^m.

Definition 1.1. *$f(n, x)$ is said to be almost periodic in n uniformly with respect to $x \in D^*$ (in short ap function) if for any $\epsilon > 0$ and any compact set K in D^* there exists a positive integer $L^*(\epsilon, K)$ such that any interval of length $L^*(\epsilon, K)$ contains an integer τ for which*

$$|f(n + \tau, x) - f(n, x)| \leq \epsilon \tag{1}$$

for all $n \in Z$ and all $x \in K$. Such a number τ is called an ϵ-translation number of $f(n, x)$.

Discrete almost periodic functions have many properties in common with continuous almost periodic functions $f(t,x) \in C(R \times D^*, R^m)$ [cf. 3,5]. To see this we refer to [1].

By $B(R^m)$ we denote the space of bounded functions on Z with the norm $\|x_n\|_{B(R^m)} = \sup_{-\infty<n<\infty} |x_n|$. A matrix $A(n)$ with almost periodic elements $a_{ij}(n)$ $(i,j = 1,\cdots,m)$ is called an *ap matrix*. The norm of an ap matrix is defined by

$$\|A(n)\|_{B(R^m)} = \sup_{-\infty<n<\infty} |A(n)|, \quad \text{where } |A| = \max_{x\in R^m, |x|=1} |Ax|.$$

Let $A_i(n)$ $(i = 1,\cdots,p)$ be square matrices of order m with ap elements. We consider the difference expression

$$Dx_n = x_{n+p} + A_1(n)x_{n+p-1} + \cdots + A_p(n)x_n. \tag{2}$$

This defines a difference operator D, an *ap-operator* as we say, which can be considered in various function spaces. An ap operator is said to be *regular* if the equation $Dx_n = f(n)$ has a unique solution $x_n \in B(R^m)$, for some ap function $f(n) \in B(R^m)$.

Furthermore, we consider an equation

$$\tilde{D}u_n \equiv u_{n+1} + Q(n)u_n = \tilde{f}(n, u_n), \tag{3}$$

where $Q(n)$ is a matrix. The trivial solution of Equation (3) is said to be *stable* if for any $\epsilon > 0$ there exists a $\delta > 0$ such that any solution u_n of Equation (3) with $|u_0| < \delta$ satisfies the inequality $|u_n| < \epsilon$ $(0 \le n < \infty)$. If there exist $\delta_0, q \in (0,1)$ and $M_0 > 0$ such that if $|u_0| < \delta_0$ then

$$|u_n| \le M_0 q^n |u_0| \text{ for } n \ge 0,$$

the trivial solution of Equation (3) is said to be *exponentially stable*. If the trivial solution is not stable, it is said to be *unstable*.

2 Bifurcation of almost periodic solutions

We consider the regular operator

$$Dx_n = x_{n+1} + A(n)x_n$$

in $B(R^m)$ and denote its Green's function by $G(n,l)$. Let the vector function $f(n;\epsilon)$ be almost periodic with respect to n for any fixed ϵ. Then the homogeneous linear equation $Dx_n = f(n;\epsilon)$ has a unique ap solution $x_n(\epsilon)$ for any fixed $\epsilon \in (0,\epsilon_0)$ for some $\epsilon_0 > 0$. Let $A_1(n)$ be an ap matrix with mean value $A_0 := \lim_{T\to\infty} \frac{1}{2T}\sum_{l=-T}^{T} A_1(l)$. We consider the equation

$$x_{n+1} = \epsilon A_1(n)x_n + \epsilon B(n;\epsilon)x_n \quad (0 < \epsilon < \epsilon_0), \tag{4}$$

where

$$\lim_{\epsilon \to 0} \|B(n;\epsilon)\|_{B(R^m)} = 0.$$

Later we shall need the following result to Equation (4) which, for differential equations, is due to Shtokalo [cf. 4].

Theorem 2.1. *If $a_{j_0} < 0$, then the ap operators*

$$D(\epsilon)x_n = x_{n+1} - (\epsilon A_1(n) + \epsilon^2 A_2(n) + \cdots + \epsilon^k A_k(n) + \epsilon^{k+1} C(n;\epsilon))x_n \quad (5)$$

are regular and stable for small positive ϵ. On the other hand, if $a_{j_0} > 0$, then the ap operators (5) are regular and unstable for small positive ϵ.

3 Main theorem

We suppose that for every value of μ the system

$$x_{n+1} = f(n, x_n; \mu) \quad (6)$$

whose right-hand side is almost periodic in n admits the trivial solution ($f(n, 0; \mu) \equiv 0$). A value μ_0 of the parameter μ is called a *bifurcation point* if for any $\eta > 0$ there exists a $\mu \in (\mu_0 - \eta, \mu_0 + \eta)$ such that system (6) has a nontrivial ap-solution satisfying $\|x_n(\mu)\|_{B(R^m)} < \eta$. We suppose the right-hand side of Equation (6) is sufficiently smooth with respect to the space variables. We consider the ap operators

$$D(\mu)x_n = x_{n+1} - f_x(n, 0; \mu)x_n. \quad (7)$$

Let the operator $D(\mu_0)$ be regular for some μ_0 and let the ap matrices $f_x(n, 0; \mu)$ converge regularly to $f_x(n, 0; \mu_0)$ as $\mu \to \mu_0$. Then Equation (6) has no small nontrivial ap-solutions for μ close to μ_0, as we may see by applying the contracting mapping principle to the operator

$$\Pi(\mu)x_n = \sum_{l=-\infty}^{\infty} G_0(n, l; \mu)\{f(l, x_l; \mu) - f_x(l, 0; \mu)x_l\},$$

where $G_0(n, l; \mu)$ is the Green's function of the operator (7). It follows that bifurcation points can occur only at values of μ for which the ap operator (7) is not regular. Let μ_0 be such a value. Set $\mu = \mu_0 + \epsilon$. Then Equation (6) becomes

$$x_{n+1} = g(n, x_n; \epsilon). \quad (8)$$

This reduces the problem of finding small nontrivial ap-solutions of Equation (8) for small ϵ. We suppose that Equation (8) has an ap-solution $x_n^*(\epsilon)$ continuous in ϵ and converging to the trivial solution as $\epsilon \to 0$. We consider the family of ap operators

$$D(\epsilon)x_n = x_{n+1} - g_x(n, x_n^*(\epsilon); \epsilon)x_n. \quad (9)$$

These operators are, in general, regular for small ϵ. Thus, the ap operators

$$D_1(\epsilon)x_n = x_{n+1} - g_x(n, x_n^{(0)}(\epsilon); \epsilon)x_n$$

are also regular, provided the ap functions $x_n^{(0)}(\epsilon)$ are sufficiently "close" to $x_n^*(\epsilon)$. When the ap functions $x_n^{(0)}(\epsilon)$ are known and the ap-solutions $x_n^*(\epsilon)$ are unknown, one may pass from difference equation (8) to the equation

$$x_n = \sum_{l=-\infty}^{\infty} G(n, l; \epsilon)\{g(l, x_l; \epsilon) - g_x(l, x_l^{(0)}(\epsilon); \epsilon)x_l\}, \tag{10}$$

and then try to use successive approximations. It may be used in proving the existence of ap-solutions $x_n^*(\epsilon)$. In this case the ap functions $x_n^{(0)}(\epsilon)$ must be chosen in such a way that Equation (10) has solutions for small ϵ. In the cases considered below, Equation (8), as a rule, has several families of ap-solutions; each family is to be determined using a specific ap function $x_n^{(0)}(\epsilon)$.

Let us assume Equation (8) is of the form

$$\begin{aligned} x_{n+1} &= \epsilon A_1(n)x_n + \cdots + \epsilon^k A_k(n)x_n + \epsilon^{k+1}A(n;\epsilon)x_n \\ &\quad + \epsilon F(n, x_n; \epsilon) + \epsilon\omega(n, x_n; \epsilon). \end{aligned} \tag{11}$$

Here $A_1(n), \cdots, A_k(n)$ are $m \times m$ matrices whose elements are trigonometric polynomials, and $A(n;\epsilon)$ is a matrix almost periodic in n, continuous in ϵ uniformly with respect to $n \in Z$. $F(n, x; \epsilon)$ is a form of degree p $(p \geq 2)$ in the space variables; its coefficients are trigonometric polynomials, continuous in ϵ uniformly with respect to n. The remainder term $\omega(n, x; \epsilon)$ contains terms that vanish with x to order greater than p : $\omega(n, 0; \epsilon) \equiv 0$ and $|\omega(n, x; \epsilon) - \omega(n, y; \epsilon)| \leq q(r)|x - y| \quad (|x|, |y| \leq r; n \in Z)$, where $\lim_{r\to 0} r^{-p+1}q(r) = 0$. $\omega(n, x; \epsilon)$ is almost periodic in n and continuous in ϵ uniformly with respect to $n \in Z$. To simplify matters, we shall restrict ourselves to the case that ϵ is positive. By substituting Shtokalo's equation into Equation (11), we obtain an equation of the form

$$\begin{aligned} y_{n+1} &= \epsilon B_1 y_n + \cdots + \epsilon^k B_k y_n + \epsilon^{k+1}B(n;\epsilon)y_n \\ &\quad + \epsilon F(n, y_n; \epsilon) + \epsilon^2 F_1(n, y_n; \epsilon) + \epsilon\omega_1(n, y_n; \epsilon) \end{aligned} \tag{12}$$

where $B_1, B_2, \cdots, B_k$ are constant matrices. The matrix $B(n;\epsilon)$ has the same properties as $A(n;\epsilon)$, and $F_1(n, y; \epsilon)$ is a form of degree p with the same properties as $F(n, y; \epsilon)$, and the remainder term $\omega_1(n, y; \epsilon)$ has the same properties as $\omega(n, y; \epsilon)$. It is evident from Shtokalo's equation that finding the ap-solutions of Equation (11) for small ϵ is equivalent to finding the ap-solutions of Equation (12). We define an operator J on the set of all trigonometric polynomials

$$f(t) = a_0 + \sum_{i=1}^{l}(a_i \sin \lambda_i t + b_i \cos \lambda_i t),$$

where $\lambda_i \neq 0$, by $J(f(n)) = f_t(n)$. Next we apply the Bogolyubov substitution

$$y = z + \epsilon J(F(n, z; \epsilon)) \tag{13}$$

to (12), where $J(F(n, z; \epsilon))$ is the form obtained by applying the operator J to the coefficients of $F(n, y; \epsilon)$. Obviously we get

$$\begin{aligned} & z_{n+1} + \epsilon F(n, z_n; \epsilon) - \epsilon \bar{F}(z_n; \epsilon) + \epsilon J(F_z(n, z_n; \epsilon)) z_{n+1} \\ & - \ \epsilon B_1 z_n + \cdots + \epsilon^k B_k z_n + \epsilon^{k+1} B(n; \epsilon) z_n \\ & \quad + \epsilon F(n, z_n; \epsilon) + \epsilon^2 F_2(n, z_n; \epsilon) + \epsilon \omega_2(n, z_n : \epsilon), \end{aligned}$$

where $F_2(n, z; \epsilon)$ and $\omega_2(n, z; \epsilon)$ have the same properties as $F_1(n, z; \epsilon)$ and $\omega_1(n, z; \epsilon)$, respectively, and

$$\bar{F}(z; \epsilon) = \lim_{T \to \infty} \frac{1}{T} \sum_{l=0}^{T} F(l, z; \epsilon) \quad (z \in R^m).$$

From the last equation, we have

$$\begin{aligned} z_{n+1} &= \epsilon B_1 z_n + \cdots + \epsilon^k B_k z_n + \epsilon^{k+1} B(n; \epsilon) z_n + \\ & \quad + \epsilon \bar{F}(z_n; \epsilon) + \epsilon^2 F_3(n, z_n; \epsilon) + \epsilon \omega_3(n, z_n; \epsilon). \end{aligned} \tag{14}$$

Let us now introduce a new time variable t. Then Equation (14) assumes the form

$$\begin{aligned} z_{t+1} &= B_1 z_t + \cdots + \epsilon^{k-1} B_k z_t + \bar{F}(z_t; \epsilon) \\ & \quad + \epsilon^k B([\tfrac{t}{\epsilon}]; \epsilon) z_t + \epsilon F_3([\tfrac{t}{\epsilon}], z_t; \epsilon) + \omega_3([\tfrac{t}{\epsilon}], z_t; \epsilon), \end{aligned} \tag{15}$$

where $[\tau]$ denotes the greatest integer $\leq \tau$. As a result of several substitutions, we have reduced the problem of finding small ap-solutions of Equation (11) to the same problem for Equation (9). The advantage of the latter is that the "principal" terms on its right do not depend on time. Of course, Equation (15) is meaningful only for $\epsilon \neq 0$. For $\epsilon = 0$, a natural choice for the equation is

$$z_{n+1} = B_1 z_n + \bar{F}(z_n; 0) + \bar{\omega}_3(z_n; 0), \tag{16}$$

where $\bar{\omega}_3(z; \epsilon) = \lim_{T \to \infty} \frac{1}{T} \sum_{l=0}^{T} \omega_3(l, z; \epsilon) \quad (z \in R^m)$. Equation (16) is known as the *averaged equation of the first approximation.* Equation (15) has small nontrivial ap solutions for small ϵ only if the ap operator

$$D z_n = z_{n+1} - B_1 z_n$$

is not regular or, equivalently, if the matrix B_1 has eigenvalues on the unit circle. For the sake of simplicity we only consider the case where zero is a simple eigenvalue of B_1 and all other eigenvalues lie inside the unit circle. Let

$$\lambda(\epsilon) = a_1 \epsilon + a_2 \epsilon^2 + \cdots + a_j \epsilon^j + \cdots \tag{17}$$

be the eigenvalue of the matrix

$$B(\epsilon) = B_1 + \epsilon B_2 + \cdots + \epsilon^{k-1} B_k.$$

Let a_{j_0} be the first nonvanishing coefficient in (17). We assume that

$$j_0 \leq k - 1. \tag{18}$$

We denote by e_0 and g_0, respectively, eigenvectors of the matrices B_1 and B_1^* corresponding to the zero eigenvalue, normalized so that $(e_0, g_0) = 1$. We set $\alpha_0 = (\bar{F}(e_0, 0), g_0)$ and assume henceforth that

$$\alpha_0 \neq 0. \tag{19}$$

Let $V(r)$ denote the ball $\|x_n\| \leq r$ in $B(R^m)$. Now, we have the main result of this talk.

Theorem 3.1. *Let conditions (18) and (19) be satisfied. Then there exist* $\epsilon_0, r_0 > 0$ *such that the following is true:*

1. *Equation (16) has no nontrivial ap-solutions in the ball* $V(r_0)$.
2. *If p is even then for* $0 < \epsilon < \epsilon_0$ *Equation (15) has a unique ap-solution in* $V(r_0)$. *This solution is exponentially stable if* $a_{j_0} > 0$ *and unstable if* $a_{j_0} < 0$.
3. *If p is odd and* $\alpha_0 a_{j_0} < 0$ *then for* $0 < \epsilon < \epsilon_0$ *Equation (15) has exactly two nontrivial ap-solutions in* $V(r_0)$. *They are exponentially stable if* $a_{j_0} > 0$ *and unstable if* $a_{j_0} < 0$.
4. *If p is odd and* $\alpha_0 a_{j_0} > 0$ *then for* $0 < \epsilon < \epsilon_0$ *Equation (15) has no nontrivial ap solutions in* $V(r_0)$.

Outline of Proof. We can prove the existence of the ap-solutions of equation (15) by the contraction mapping principle [2]. Let us begin with part 1. Equation (16), upon the substitution $Z_t = U(\epsilon)w_t$ (with $\epsilon = 0$), takes the form

$$w_{t+1} = C_0(0)w_t + \Phi_0(w_t; 0) + \bar{\Omega}(w_t; 0), \tag{20}$$

where

$$\bar{\Omega}(w; 0) = \lim_{T\to\infty} \frac{1}{T} \sum_{n=0}^{T} \Omega(n, w; 0).$$

For any $\rho > 0$ one can find $r = r(\rho) > 0$ such that the small nontrivial ap-solutions of Equation (20) lie entirely either in cones $K(\rho)$ or in $-K(\rho)$. Therefore, part 1 will be proved if we have shown that, for small ρ and r, no nontrivial ap-solution of Equation (20) lies entirely in $K(\rho, r)$ or $-K(\rho, r)$. We suppose the contrary is true. To fix ideas, we assume that for small ρ and r there is at least one nontrivial ap-solution w_t lying entirely in $K(\rho, r)$. We consider the scalar almost periodic function $v_t = (w_t, e_1)$ $(-\infty < t < \infty)$ where $e_1 = \{1, 0, 0, \cdots, 0\}$. Clearly, $v_{t+1} = (\Phi_0(w_t; 0), e_1) + (\bar{\Omega}(w_t; 0), e_1)$. We write w_t in the form $w_t = \zeta_t e_1 + x_t$ where $(x_t, e_1) = 0$ and ζ is the first component of the vector w. Then $v_{t+1} = \alpha_0 \zeta_t^p + (\Phi_0(\zeta_t e_1 + x_t; 0) - \Phi_0(\zeta_t e_1; 0), e_1) + (\bar{\Omega}(w_t; 0), e_1)$. Since $\alpha_0 \neq 0$, it follows that, for sufficiently small ρ and r, $\operatorname{sign}(v_{t+1}) = \operatorname{sign}(\alpha_0)$ for all $t \in Z$. Therefore v_t is strictly monotone, contradicting the fact that it is almost periodic.

We omit the proofs of parts 2, 3, and 4 in this outline. □

Example 3.2. We consider the equation

$$\Delta^2 \theta_n + 2\alpha\epsilon\Delta\theta_n + (\epsilon^2 k^2 - \epsilon p(n+2)) \sin\theta_n = 0 \tag{21}$$

describing the oscillations of a pendulum with a vibrating point of suspension. Here, α and ϵ are positive parameters, Δ and Δ^2 are difference operators and $p(n)$ is a trigonometric polynomial with zero mean. For small ϵ, the lower equilibrium position ($\theta_n = 0$) is always stable. The upper equilibrium position ($\theta_n = \pi$) may be stable or unstable, depending on $p(n)$, as shown by Bogolyubov and Kapitsa for differential equations. We are interested in the case that ap-solutions of Equation (21) originate from the upper equilibrium position. In order to use Theorem 3.1 we need to transform (21) to the form of (11). To do this, we replace the unknown function θ_n by two unknown functions $x_{(1)n}$ and $x_{(2)n}$ via Bogolyubov's substitution. Now, we replace Equation (21) by the system of two equations

$$\begin{aligned}
x_{(1)n+1} &= \frac{\epsilon x_{(2)n}}{1 - \epsilon p(n) \cos x_{(1)n}}, \\
x_{(2)n+1} &= p(n+2)\{\sin x_{(1)n} - \sin(x_{(1)n} - \epsilon p(n) \sin x_{(1)n})\} \\
&\quad + \frac{\epsilon p(n+1) x_{(2)n} \cos x_{(1)n}}{1 - \epsilon p(n) \cos x_{(1)n}} - 2\alpha\epsilon x_{(2)n} \\
&\quad + 2\alpha\epsilon p(n+1) \sin x_{(1)n} + \epsilon k^2 \sin(x_{(1)n} - \epsilon p(n) \sin x_{(1)n}).
\end{aligned} \tag{22}$$

Expanding the right-hand sides in powers of ϵ and introducing vector notation we obtain the equation

$$x_{n+1} = \epsilon A_1(n) x_n + \epsilon^2 A_2(n) x_n + \cdots + \epsilon F(n, x_n) + \cdots, \tag{23}$$

where

$$\begin{aligned}
A_1(n) &= \begin{pmatrix} 0 & 1 \\ p(n)p(n+2) + 2\alpha p(n+1) + k^2 & p(n+1) - 2\alpha \end{pmatrix}, \\
A_2(n) &= \begin{pmatrix} 0 & p(n) \\ -p(n)k^2 & p(n+1)p(n) \end{pmatrix}, \\
F(n, x) &= \{0, -\frac{1}{6}[k^2 + 4p(n)p(n+2) + 2\alpha p(n+1)] x_{(1)n}^3 \\
&\quad - \frac{1}{2} p(n+1) x_{(1)n}^2 x_{(2)n}\}^T.
\end{aligned}$$

Using the substitutions of Bogolyubov and Shtokalo as well as (13) we get from Equation (23) to the equation

$$z_{n+1} = \epsilon B_1 z_n + \epsilon^2 B_2 z_n + \cdots + \epsilon \bar{F}(z_n) + \cdots,$$

where

$$B_1 = \begin{pmatrix} 0 & 1 \\ k^2 - M[(p(n+1))^2] & -2\alpha \end{pmatrix},$$

$$B_2 = \begin{pmatrix} 0 & 0 \\ 2M[p(n)(p(n+1))^2] & 0 \end{pmatrix},$$

$$\bar{F}(z_n) = \{0, \frac{2}{3}M[(p(n+1))^2]z_{(1)n}^3 - \frac{1}{6}k^2 z_{(1)n}^3\}^T,$$

and $M[\cdot] = Mean[\cdot]$. Clearly, when $M[(p(n+1))^2] > k^2$, the upper equilibrium position is stable. If $M[(p(n+1))^2] < k^2$, then the upper equilibrium position is unstable. In the latter situation, for sufficiently small ϵ, there are no almost periodic oscillations in the neighborhood of the upper equilibrium position.

Finally, we are interested in what happens when

$$M[(p(n+1))^2] = k^2. \tag{24}$$

In this case, in order to examine the stability of the upper equilibrium position, we must construct the sequence of matrices

$$B_1 + \epsilon B_2,\ B_1 + \epsilon B_2 + \epsilon^2 B_3,\ B_1 + \epsilon B_2 + \epsilon^2 B_3 + \epsilon^3 B_4, \cdots, \tag{25}$$

continuing until one of them is nonsingular for small nonzero ϵ. Thus, for example, if $M[(p(n+1))^2] \neq 0$, this will be the matrix $B_1 + \epsilon B_2$. The stability of the trivial solution depends on the first nonsingular matrix in (25). For example, if

$$M[p(n)(p(n+1))^2] < 0$$

then the upper equilibrium position is stable, but if

$$M[p(n)(p(n+1))^2] > 0$$

then it is unstable. Therefore, we shall assume that one of the matrices (25) is nonsingular for small nonzero ϵ.

Theorem 3.1. implies the following theorem.

Theorem 3.3. *Let condition (24) hold for Equation (22). Then we get:*

1. *If the upper equilibrium position of the pendulum is unstable there are no ap oscillations in its neighborhood for small positive ϵ.*

2. *If the upper equilibrium position is stable then in each of its neighborhoods there are, for small ϵ, exactly two ap oscillations, both unstable.*

For any other neighborhood $V(r)$ there exists an $\epsilon_0 = \epsilon_0(r) > 0$ such that for $0 < \epsilon < \epsilon_0$ Equation (22) has solutions with initial values in $V(r)$, and as n increases these solutions leave $V(r_0)$. This means that, if (24) holds, the upper equilibrium position is practically always unstable.

References

[1] C. Corduneanu, Almost periodic discrete processes, *Libertas Mathematica*, **2** (1982), 159–169.

[2] Y. Hamaya, Bifurcation of almost periodic solutions in difference equations, J. Difference Equ. Appl., **10** (2004), 257-297.

[3] Y. Hino, T. Naito, N. V. Minh and J. S. Shin, *Almost Periodic Solutions of Differential Equations in Banach Space, Stability and Control: Theory, Methods and Applications*, Vol. 15, Taylor and Francis, London and New York, 2002.

[4] M. A. Krasnosel'skii, V. Sh. Burd, and Yu. S. Kolesov, *Nonlinear Almost Periodic Oscillations,* John Wiley and Sons, New York, 1973.

[5] T. Yoshizawa, *Stability Theory and the Existence of Periodic Solutions and Almost Periodic Solutions,* Applied Math. Sciences 14, Springer-Verlag, New York, 1975.

The Harmonic Oscillator – An Extension Via Measure Chains

STEFAN HILGER
Katholische Universität Eichstätt
Eichstätt, Germany

Key words Harmonic oscillator, discretization, measure chains, time scales

AMS Subject Classification 39A70, 34L40

1 Introduction

In this note we present an idea of how to discretize the harmonic oscillator eigenvalue equation

$$-f''(x) + x^2 f(x) = \lambda\, f(x), \qquad x \in \mathbb{R}, \tag{1}$$

which means that we will replace the underlying continuous set $\mathbb{R}$ by the discrete subset $h\mathbb{Z}$. The parameter $h > 0$ allows us to monitor the influence of the discrete structure, we will see that the limit process $h \searrow 0$ will result in the continuous theory.

An essential observation in this process of discretization is, that any standard method from numerical computation may provide approximative solutions of (1), but in general it will destroy the beautiful algebraic background structure, the so-called ladder formalism of creation and annihilation operators.

When playing around with Equation (1) again and again, one can find out how to perform the "right" structural discretization. The resulting discrete counterpart of (1) is a sequence of equations:

$$\begin{aligned} &-(1+\frac{nh^2}{2})^2 \cdot \frac{f(x+2h) - 2f(x) + f(x-2h)}{4h^2} \\ &\qquad + \frac{(x+h)^2 f(x+2h) + 2x^2 f(x) + (x-h)^2 f(x-2h)}{4} \\ &\qquad\qquad = \lambda f(x), \quad x \in h\mathbb{Z}, \quad n \in \mathbb{N}_0. \end{aligned} \tag{2}$$

When analyzing the ladder formalism of these discretized harmonic oscillator equations, it turns out, that it is not only preserved, it even becomes more complicated and fascinating.

We remark that the above considerations are stimulated by the spirit of measure chains, a language incorporating continuous and discrete analysis. The calculus on measure chains was initiated in [5] and is now gaining more and more interest, see, e.g., [2, 3].

In [1], continued in [4], also discussed in [8, 9, 10], one can find other approaches to the discrete harmonic oscillator. The authors start with discrete analogs of the solutions of the harmonic oscillator Equation (1). These are the binomial distribution on a finite discrete interval combined with the so-called Kravchuk (orthogonal) polynomials. Then they derive corresponding operator equations and algebraic identities.

In this paper we go in the opposite direction; we start with the ladder formalism for the classical equation (1), find out about its perseveration under discretization, then end up with the "right" equations and solutions. It seems that the approach presented here is much more lean and natural.

The authors of [7] or [12] analyze the discrete harmonic oscillator by the methods of operator algebra. This access is completely different from the one presented here.

In this paper we only present the basic facts about the discretization process. After pointing out the ideas leading to the discrete equations (2) we will compute eigenvalues and eigenfunctions. The well-known linear structure of the spectrum of the continuous harmonic oscillator operator will turn out to be perturbed by an additional quadratic term controlled by the discretization parameter h.

In a much more elaborate paper to be submitted soon we will generalize the results of this note in various directions:

- The harmonic oscillator equation on the dual group of $h\mathbb{Z}$, which is the circle $\frac{1}{h}\,\mathbb{S}$ with radius $\frac{1}{h}$ will be studied. This will provide the framework for Fourier transforming the discrete harmonic oscillator. The resulting equations on the circle are the so-called Mathieu equations (see [11]).

- A connection to the (discrete) heat equation will be established.

- We will look for connections to Heisenberg's uncertainty relation (see also [6]).

- We will drop the restriction to solutions of finite support, which allows an easy description of the eigenspaces of (2) in this note.

- The discussion of solutions of (2) for negative n will be included. This is of interest on its own. It will turn out that the role of $\ell^2(h\mathbb{Z})$ as the appropriate space for solutions has to be taken over by a scale of certain Hilbert spaces.

- We will present the orthogonal polynomials arising from (2) in a more abstract operator language.

2 The classical theory of the harmonic oscillator equation

In order to prepare the discretization we recall the classical algebraic theory for the harmonic oscillator. We consider smooth functions $f : \mathbb{R} \to \mathbb{C}$. There are two operators, the *differential operator* or *moment operator* D defined by

$$Df(x) = f'(x) \tag{3}$$

and its Fourier counterpart, the *multiplication operator* (or *location operator*)

$$\widehat{D}f(x) = ix \cdot f(x). \tag{4}$$

The two operators fulfill the well-known Heisenberg commutator relation

$$[D, \widehat{D}] := D\widehat{D} - \widehat{D}D = iI,$$

where I is the identity operator. We define the *creation operator*

$$\overset{+}{A} = -D - i\widehat{D} \tag{5}$$

and its formal adjoint, the *annihilation operator*

$$A = D - i\widehat{D}. \tag{6}$$

We also call these operators *Gauss operators.* We are very much interested in the difference and average of the two products built by A and $\overset{+}{A}$:

$$A\overset{+}{A} = -D^2 - \widehat{D}^2 + i[\widehat{D}, D] = -D^2 - \widehat{D}^2 + I \tag{7}$$

$$\overset{+}{A}A = -D^2 - \widehat{D}^2 - i[\widehat{D}, D] = -D^2 - \widehat{D}^2 - I \tag{8}$$

$$[A, \overset{+}{A}] := A\overset{+}{A} - \overset{+}{A}A = 2I \tag{9}$$

$$H := \frac{A\overset{+}{A} + \overset{+}{A}A}{2} = -D^2 - \widehat{D}^2 = A\overset{+}{A} - I = \overset{+}{A}A + I. \tag{10}$$

The average in the last line is just the *harmonic oscillator operator*:

$$Hf(x) = -f''(x) + x^2 f(x).$$

An important fact now is that in (10) the identity operator appears, which commutes — of course — with the operators A and $\overset{+}{A}$. This makes the ladder formalism work, as we will see in the following lemma. By $\text{Eig}(H, \lambda) := \ker(H - \lambda I)$ we denote the eigenspace (including 0) of H for the eigenvalue $\lambda \in \mathbb{C}$. In case that λ is not an eigenvalue this space may be trivial.

Lemma 2.1 (Ladder lemma). *The following statements hold:*

- *We have* $\ker A \subseteq \mathrm{Eig}(H, 1)$.
- *In the diagram with a given fixed number* $\nu \in \mathbb{C}$

$$\mathrm{Eig}(H, \nu - 1) \underset{A}{\overset{\overset{+}{A}}{\rightleftarrows}} \mathrm{Eig}(H, \nu + 1)$$

  the operators $\overset{+}{A}$ *and* A *are well-defined.*

- *If* $\nu \neq 0$ *then the operators* $\overset{+}{A}$ *and* $\frac{1}{\nu}A$ *in the diagram are inverse of each other, the two eigenspaces are isomorphic.*

Proof. The first statement follows from the identity $H = \overset{+}{A}A + I$ above. For $f \in \mathrm{Eig}(H, \nu - 1)$ we have

$$\begin{aligned} H(\overset{+}{A}f) &= (\overset{+}{A}A + I)\overset{+}{A}f = \overset{+}{A}(A\overset{+}{A} + I)f \\ &= \overset{+}{A}(H + 2I)f = (\nu + 1)\overset{+}{A}f; \end{aligned}$$

hence $\overset{+}{A}f \in \mathrm{Eig}(H, \nu + 1)$. Also

$$A\overset{+}{A}f = (H + I)f = \nu f.$$

Dually for $f \in \mathrm{Eig}(H, \nu + 1)$

$$H(Af) = (\nu - 1)Af \qquad \text{and} \qquad \overset{+}{A}Af = \nu f.$$

By patching together the above isomorphisms the harmonic oscillator ladder is created. It contains the spaces $\mathcal{E}_j := \mathrm{Eig}(H, 2j + 1) = \ker(H - (2j + 1)I)$, $j \in \mathbb{N}_0$, and the operators A and $\overset{+}{A}$ operating between them as isomorphisms:

$$\ker A \quad \hookrightarrow \quad \mathcal{E}_0 \underset{A}{\overset{\overset{+}{A}}{\rightleftarrows}} \mathcal{E}_1 \underset{A}{\overset{\overset{+}{A}}{\rightleftarrows}} \cdots \underset{A}{\overset{\overset{+}{A}}{\rightleftarrows}} \mathcal{E}_j \underset{A}{\overset{\overset{+}{A}}{\rightleftarrows}} \cdots \tag{11}$$

The kernel of the annihilation operator A is given by the solutions of

$$Af = f'(x) + xf(x) = 0.$$

These are the Gauss bell-shaped functions:

$$f(x) = C \cdot \exp(-\frac{x^2}{2}).$$

If one defines Hermite polynomials q_j and Hermite functions f_j recursively by

$$\begin{aligned} f_0(x) &= \exp(-\frac{x^2}{2}), \quad q_0 \equiv 1, \\ f_{j+1}(x) &= q_{j+1}(x) \cdot e^{-\frac{x^2}{2}} := \overset{+}{A}f_j(x) = (-D - i\widehat{D})[q_j(x) \cdot e^{-\frac{x^2}{2}}] \end{aligned}$$

or equivalently

$$q_{j+1}(x) = 2x\, q_j(x) - q_j'(x), \tag{12}$$

then it follows with the ladder lemma that

$$f_j \in \operatorname{Eig}(H, 2j+1), \quad j \in \mathbb{N}_0.$$

□

Thus, we have constructed eigenfunctions for the eigenvalues $\lambda \in 2\mathbb{N}_0 + 1$. Within a functional analytic framework we can show that there are no other eigenfunctions in $\mathcal{L}^2(\mathbb{R})$. This means

$$\operatorname{Spec} H = 2\mathbb{N}_0 + 1 \qquad \text{and} \qquad \dim \operatorname{Eig}(H, 2j+1) = 1, \quad j \in \mathbb{N}_0,$$

if H is considered as an unbounded operator in $\mathcal{L}^2(\mathbb{R})$.

3 Discretization of the harmonic oscillator: first attempt

We now try to transfer the above considerations to the discrete case. This means that we consider functions $f : h\mathbb{Z} \to \mathbb{C}$. As a first attempt we simply replace the differential operator by the *central difference operator*:

$$Df(x) := \frac{f(x+h) - f(x-h)}{2h}. \tag{13}$$

Our first crucial observation is that the Heisenberg commutator relation takes on the form

$$[D, \widehat{D}] = iM, \tag{14}$$

where M is the so-called *mix operator*

$$Mf(x) := \frac{f(x+h) + f(x-h)}{2}. \tag{15}$$

A first hint that this operator may play a crucial role in the whole process also comes from an observation within calculus on measure chains, where the generalized Leibnitz product rule is one of the cornerstones. For the special measure chain $\mathbb{T} = h\mathbb{Z}$ its symmetrized version reads as

$$D(f \cdot g) = Mf \cdot Dg + Df \cdot Mg.$$

It is easy to verify further commutator relations

$$[\widehat{D}, M] = -ih^2 D \qquad \text{and} \qquad [D, M] = 0. \tag{16}$$

We define the creation and annihilation operators according to (5) and (6) and recalculate the commutator and average

$$\begin{aligned} A\overset{+}{A} &= -D^2 - \widehat{D}^2 + i[\widehat{D}, D] = -D^2 - \widehat{D}^2 + M \\ \overset{+}{A}A &= -D^2 - \widehat{D}^2 - i[\widehat{D}, D] = -D^2 - \widehat{D}^2 - M \end{aligned}$$

$$[A, \overset{+}{A}] = A\overset{+}{A} - \overset{+}{A}A = 2M \tag{17}$$

$$H = \frac{A\overset{+}{A} + \overset{+}{A}A}{2} = -D^2 - \widehat{D}^2 = A\overset{+}{A} - M = \overset{+}{A}A + M. \tag{18}$$

We have encountered a serious problem: In the relations (17), (18) the M operator appears instead of the identity operator I. But this operator does not commute with the operators A and $\overset{+}{A}$, the ladder formalism fails.

4 Discretization of the harmonic oscillator: second attempt

Some basic considerations from another area of discrete mathematical physics, especially from [6], suggest replacing the operator $\widehat{D}$ given in (4) by the following modified multiplication operator

$$\widehat{C} := \frac{M\widehat{D} + \widehat{D}M}{2}. \tag{19}$$

Its action on a function f is given by

$$\widehat{C}f(x) = \frac{i \cdot (x + \frac{h}{2})f(x+h) + i \cdot (x - \frac{h}{2})f(x-h)}{2}. \tag{20}$$

Since D and M commute, it turns out that

$$\begin{aligned} [D, \widehat{C}] &= D\widehat{C} - \widehat{C}D = \frac{DM\widehat{D} + D\widehat{D}M - M\widehat{D}D - \widehat{D}MD}{2} \\ &= \frac{M[D, \widehat{D}] + [D, \widehat{D}]M}{2} = iM^2. \end{aligned}$$

After redefining the Gauss operators as

$$\overset{+}{G} := D - i\widehat{C}, \qquad G := -D - i\widehat{C}$$

we test again the fundamental identities (7)–(10) and find that

$$\begin{aligned} GG^{+} &= -D^2 - \widehat{C}^2 + i[\widehat{C}, D] = -D^2 - \widehat{C}^2 + M^2 \\ G^{+}G &= -D^2 - \widehat{C}^2 - i[\widehat{C}, D] = -D^2 - \widehat{C}^2 - M^2 \end{aligned}$$

$$\begin{aligned} [G, G^{+}] &= GG^{+} - G^{+}G = 2M^2 && (21) \\ H &= \frac{GG^{+} + G^{+}G}{2} \\ &= -D^2 - \widehat{C}^2 \overset{(*)}{=} GG^{+} - M^2 = G^{+}G + M^2. && (22) \end{aligned}$$

M^2 is again an operator that does not commute with G and G^{+}. So it seems at first glance that the ladder formalism still does not work.

5 Discretization of the harmonic oscillator: third attempt

It is a kind of wonder that there is a remedy for the serious problem just described. We compute the difference between the operator M^2 and the identity operator

$$M^2 - I = h^2 D^2. \tag{23}$$

This is the key relation in our theory. We are able now to replace $M^2 = h^2D^2 + I$ in (22). Since we would like then to bring the term h^2D^2 to the left side of equation $(*)$ in (22), we have to keep the D^2 part of the operator H variable. This means that we have to consider appropriate sequences of operators G, G^{+} and H instead of constant ones.

In order to introduce the necessary generalizations we restrict our scope to a special case that is characterized by the following condition on the step size h:

$$\frac{2}{h^2} \in \mathbb{N}. \tag{24}$$

As we will see below this condition allows eigenfunctions with compact (finite) support. A much more elaborate discussion of more general eigenfunctions will be presented in the paper that we already mentioned in the introduction.

We define the integer

$$n_0 := 2 - \frac{2}{h^2} \leq 1.$$

For $n \geq n_0 - 1$ let

$$\alpha_n := 1 + n\frac{h^2}{2} > 0 \tag{25}$$

and then define

$$G_n^+ \;:=\; -\alpha_n D - i\widehat{C} \quad \text{and} \quad G_n := \alpha_n D - i\widehat{C}. \tag{26}$$

Within this new framework we resume the computation in (7) and find

$$\begin{aligned}
G_n G_n^+ &= (\alpha_n D - i\widehat{C}) \cdot (-\alpha_n D - i\widehat{C}) = -\alpha_n^2 D^2 - \widehat{C}^2 + i\alpha_n[\widehat{C}, D] \\
&= -\alpha_n^2 D^2 - \widehat{C}^2 + \alpha_n M^2 = -\alpha_n^2 D^2 - \widehat{C}^2 + \alpha_n(h^2 D^2 + I) \\
&= -\alpha_n(\alpha_n - h^2)D^2 - \widehat{C}^2 + \alpha_n I = -\alpha_n\alpha_{n-2}D^2 - \widehat{C}^2 + \alpha_n I.
\end{aligned}$$

Similarly

$$G_n^+ G_n \;=\; -\alpha_n\alpha_{n+2}D^2 - \widehat{C}^2 - \alpha_n I.$$

Then the other identities (9) and (10) read as

$$G_{n+1}G_{n+1}^+ - G_{n-1}^+ G_{n-1} = 2\alpha_n I \tag{27}$$

$$H_n := -\alpha_{n-1}\alpha_{n+1}D^2 - \widehat{C}^2 = \tag{28}$$

$$G_{n+1}G_{n+1}^+ - \alpha_{n+1}I = G_{n-1}^+ G_{n-1} + \alpha_{n-1}I, \tag{29}$$

where $n \geq n_0$. The identity operator is back again, the ladder formalism is fixed, and we can proceed developing the theory.

In order to rewrite the operator H_n we compute with (16)

$$\begin{aligned}
\widehat{C}^2 &= \left(\frac{M\widehat{D} + \widehat{D}M}{2}\right)^2 \\
&= \frac{M\widehat{D}(2M\widehat{D} - ih^2 D) + \widehat{D}M(2\widehat{D}M + ih^2 D)}{4} \\
&= \frac{(M\widehat{D})^2}{2} + \frac{(\widehat{D}M)^2}{2} + \frac{ih^2(\widehat{D}M - M\widehat{D})D}{4} \\
&= \frac{(M\widehat{D})^2}{2} + \frac{(\widehat{D}M)^2}{2} + \frac{h^4}{4}D^2
\end{aligned}$$

and

$$\alpha_{n-1}\alpha_{n+1} \;=\; \left[1 + (n-1)\frac{h^2}{2}\right] \cdot \left[1 + (n+1)\frac{h^2}{2}\right] = \alpha_n^2 - \frac{h^4}{4}.$$

Hence with (28)

$$\begin{aligned}
H_n &= -\alpha_{n-1}\alpha_{n+1}D^2 - \widehat{C}^2 \\
&= -(\alpha_n^2 - \frac{h^4}{4})D^2 - \left[\frac{(M\widehat{D})^2}{2} + \frac{(\widehat{D}M)^2}{2} + \frac{h^4}{4}D^2\right] \\
&= -\alpha_n^2 D^2 - \frac{(M\widehat{D})^2 + (\widehat{D}M)^2}{2}.
\end{aligned}$$

When applied to a function f we arrive at the expression

$$
\begin{aligned}
[H_n f](x) \quad &= \quad -\alpha_n^2 \frac{f(x+2h) - 2f(x) + f(x-2h)}{4h^2} \\
&\quad + \frac{(x+h)^2 f(x+2h) + 2x^2 f(x) + (x-h)^2 f(x-2h)}{4},
\end{aligned} \tag{30}
$$

which is already given in (2).

6 The discretized ladder lemma

Now we are able to present the following.

Lemma 6.1 (Ladder lemma, discrete version). *For $n \geq n_0$ the following statements hold:*

- $\ker G_n \subseteq \mathrm{Eig}(H_{n+1}, \alpha_n)$.
- *In the diagram with a given fixed number $\nu \in \mathbb{C}$*

$$
\mathrm{Eig}(H_{n-1}, \nu - \alpha_n) \underset{G_n}{\overset{G_n^+}{\rightleftarrows}} \mathrm{Eig}(H_{n+1}, \nu + \alpha_n)
$$

  the operators G_n^+ and G_n are well-defined.

- *If $\nu \neq 0$ then the operators G_n^+ and $\frac{1}{\nu} G_n$ in the diagram are inverse of each other, the two eigenspaces are isomorphic.*

Proof. The proof is a simple generalization of the proof of the continuous version. The first statement follows from the identity $H_{n+1} = G_n^+ G_n + \alpha_n I$ (see (29) above). For $f \in \mathrm{Eig}(H_{n-1}, \nu - \alpha_n)$ we have

$$
\begin{aligned}
H_{n+1}(G_n^+ f) = (G_n^+ G_n + \alpha_n I)\, G_n^+ f = G_n^+ \, (G_n G_n^+ + \alpha_n I) f \\
= G_n^+ (H_{n-1} + 2\alpha_n I) f = (\nu + \alpha_n)\, G_n^+ f;
\end{aligned}
$$

hence $G_n^+ f \in \mathrm{Eig}(H_{n+1}, \nu + \alpha_n)$. Also

$$
G_n G_n^+ f = (H_{n-1} + \alpha_n I) f = \nu f.
$$

Dually for $f \in \mathrm{Eig}(H_{n+1}, \nu + \alpha_n)$

$$
H_{n-1}(G_n f) = (\nu - \alpha_n) G_n f \qquad \text{and} \qquad G_n^+ G_n f = \nu f.
$$

□

7 The discrete harmonic oscillator ladder

In order to describe the new harmonic oscillator ladder we introduce a triangle index set depending on n_0 as

$$\mathcal{T} := \Big\{(n,j) \;\Big|\; n \in n_0 + 2\mathbb{N}_0,\ j = 0,1,2,\ldots,\frac{n-n_0}{2}\Big\}.$$

One can depict this set in the following diagram:

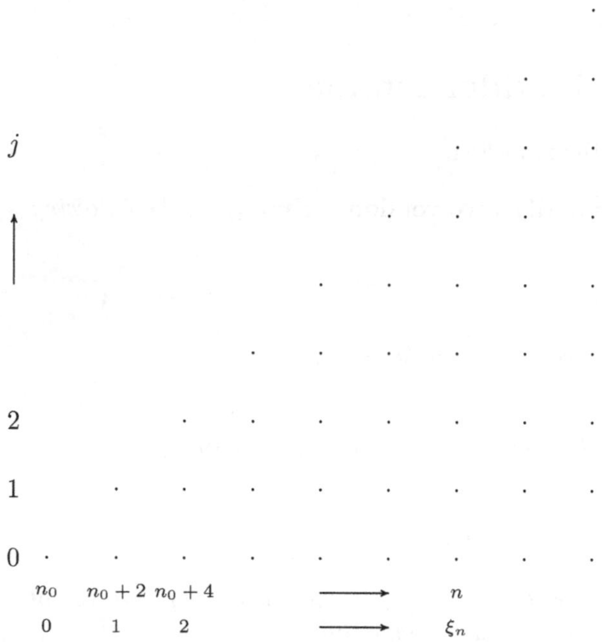

j

2

1

0

n_0 $\quad n_0+2$ $\quad n_0+4$ $\longrightarrow$ n

0 $\quad$ 1 $\quad$ 2 $\longrightarrow$ ξ_n

Here we also define

$$\xi_n := \frac{n-n_0}{2} = \frac{1}{h^2} - 1 + \frac{n}{2} = \frac{\alpha_{n-2}}{h^2} \in \mathbb{N}_0.$$

The point $(n,j) \in \mathcal{T}$ is supposed to indicate the jth eigenvalue λ_{nj} and eigenspace $\mathcal{E}_{nj} := \mathrm{Eig}(H_n, \lambda_{nj}) := \ker(H_n - \lambda_{nj} I)$ of the operator H_n. The isomorphism of the ladder lemma operates in south-west $\leftrightarrow$ north-east direction

$$\mathrm{Eig}(H_n, \lambda_{nj}) \ \underset{G_{n+1}}{\overset{G^+_{n+1}}{\rightleftarrows}}\ \mathrm{Eig}(H_{n+2}, \lambda_{n+2,j+1}).$$

8 The eigenvalues of the discrete harmonic oscillator ladder

We define for $(n,j) \in \mathcal{T}$ the real numbers

$$\lambda_{nj} \;:=\; (2j+1)[1 + (n-1)\frac{h^2}{2}] - j^2 h^2. \tag{31}$$

They fulfill the relations

$$\begin{aligned} 0 < \alpha_{n-1} &= \lambda_{n0} < \lambda_{n1} < \cdots < \lambda_{n,\xi_n} \\ \lambda_{n+2,j+1} - \lambda_{nj} &= 2\,\alpha_{n+1}; \end{aligned}$$

hence by the discrete ladder lemma 6.1 they are eigenvalues of the H_n operators. The chain of inequalities can be seen by differentiating λ_{nj} with respect to j.

In the diagram on the last page of this note we have plotted the system of eigenvalues λ_{nj} for

$$h = 1 \quad \Longleftrightarrow \quad n_0 = 0 \quad \Longleftrightarrow \quad \xi_n = \alpha_{n-2} = \frac{n}{2}, \tag{32}$$

which implies that

$$\lambda_{nj} \quad = \quad (j + \frac{1}{2})(n+1) - j^2.$$

One can see that for fixed j the function $n \mapsto \lambda_{nj}$ is linear increasing. In the diagram these functions are drawn as solid straight lines. Note that by (31) these lines turn into horizontal lines $\lambda_{nj} = 2j + 1$ for $h \to 0$.

Counting the eigenvalues by k from above for fixed n yields the function

$$\widetilde{\lambda}_{nk} := \lambda_{n,\xi_n - k}, \quad k = 0, 1, \ldots \xi_n.$$

For k fixed these are quadratic functions in n. Plugging in (32) again we find

$$\widetilde{\lambda}_{nk} = \lambda_{n,\frac{n}{2}-k} = \frac{n^2}{4} + n - k(k+1) + \frac{1}{2}.$$

These functions are drawn as a dotted parabola in the diagram.

9 The eigenfunctions of the discrete harmonic oscillator ladder

It remains to compute eigenfunctions f_{nj} of H_n for $(n, j) \in \mathcal{T}$. As we already mentioned, in this paper we are only interested in eigenfunctions with finite support. In order to determine the function f_{n0} in the first eigenspace $\mathcal{E}_{n0}$ of the ladder (the ground state) we have to solve the equation

$$G_{n-1} f = (\alpha_{n-1} D - i\widehat{C}) f = 0 \tag{33}$$

for $n \geq n_0$. Using (20) we write out this equation:

$$\left(1 + (n-1)\frac{h^2}{2}\right) \frac{f(x+h) - f(x-h)}{2h} + \frac{(x + \frac{h}{2}) f(x+h) + (x - \frac{h}{2}) f(x-h)}{2} = 0.$$

It is equivalent to

$$\left(\xi_n + 1 + \frac{x}{h}\right) f(x+h) - \left(\xi_n + 1 - \frac{x}{h}\right) f(x-h) = 0. \tag{34}$$

The space of solutions with finite support is one-dimensional. The normalized solution f_{n0} is given by a centered binomial distribution

$$f_{n0}(x) = \frac{2^{-\xi_n}}{h} \cdot \binom{\xi_n}{\frac{\xi_n}{2} + \frac{x}{2h}} \cdot \chi_{\xi_n}(x), \tag{35}$$

where χ_{ξ_n} is the characteristic function of the set

$$\left\{ -\xi_n h, (-\xi_n + 2)h, \ldots\ldots, (\xi_n - 2)h, \xi_n h \right\} \subseteq h\mathbb{Z}, \tag{36}$$

the support of f_{n0} containing $\xi_n + 1$ elements.

For $n = n_0$ we get $\xi_{n_0} = 0$, the corresponding eigenfunction is given by the Kronecker–Delta function on $h\mathbb{Z}$:

$$f_{n_0,0}(x) = \frac{1}{h} \cdot \delta_0(x).$$

Now we define recursively the *Kravchuk polynomials* q_{nj}, $(n, j) \in \mathcal{T}$, by

$$q_{n0}(x) \equiv 1$$

and

$$\begin{aligned} q_{n+2,j+1}(x) &= \left[2xM - \left(\alpha_n + \frac{h^2}{\alpha_n} x^2 \right) D \right] q_{nj}(x) & (37) \\ &= \left[2xM - (h^2 \xi_{n+2} + \frac{x^2}{\xi_{n+2}}) D \right] q_{nj}(x). & (38) \end{aligned}$$

From the definitions of M in (15) and D in (13) one can see that q_{nj} has degree j, and its parity is given by

$$q_{nj}(x) = (-1)^j q_{nj}(-x).$$

For $h \to 0$ the dependence on n in (37) disappears, since $\alpha_n \to 1$ for all n. For $h = 0$ we recover the defining recurrence relation (12) for the Hermite polynomials. We compute the Kravchuk polynomials for $j = 1, 2, 3$:

$$\begin{aligned} q_{n1}(x) &= 2x, \\ q_{n2}(x) &= \left[2xM - \left(\alpha_n + \frac{h^2}{\alpha_n} x^2 \right) D \right] (2x) = 4x^2 - \left(\alpha_n + \frac{h^2}{\alpha_n} x^2 \right) 2 \\ &= 4x^2 \cdot (1 - \frac{\frac{h^2}{2}}{\alpha_n}) - 2\alpha_n = 4 \frac{\alpha_{n-1}}{\alpha_n} x^2 - 2\alpha_n. \end{aligned}$$

Now we define the *Kravchuk functions*:

$$f_{nj} := q_{nj} \cdot f_{n0}, \quad (n,j) \in \mathcal{T}. \tag{39}$$

In order to show that these functions are precisely the eigenfunctions of H_n we first compute the action of G^+_{n+1} analogously to (33) → (34):

$$\begin{aligned}
(G^+_{n+1} f)(x) &= (-\alpha_{n+1} D - i\widehat{C}) f(x) \\
&= \frac{h}{2}\Big[(\frac{x}{h} - \xi_{n+2}) f(x+h) + (\frac{x}{h} + \xi_{n+2}) f(x-h) \Big].
\end{aligned}$$

A further preparation is the following functional equation, which follows from (35):

$$\begin{aligned}
&\frac{-\frac{x}{h} + \xi_{n+2}}{\xi_{n+2}} \cdot f_{n+2,0}(x) \\
&= \frac{-\frac{x}{2h} + \frac{\xi_{n+2}}{2}}{\xi_{n+2}} \cdot 2 \cdot \frac{2^{-\xi_{n+2}}}{h} \cdot \chi_{\xi_{n+2}}(x) \cdot \frac{\xi_{n+2}!}{(\frac{\xi_{n+2}}{2} - \frac{x}{2h})!(\frac{\xi_{n+2}}{2} + \frac{x}{2h})!} \\
&= \frac{2^{-\xi_n}}{h} \cdot \chi_{\xi_{n+2}}(x) \cdot \frac{\xi_n!}{(\frac{\xi_{n+2}}{2} - \frac{x}{2h} - 1)!(\frac{\xi_{n+2}}{2} + \frac{x}{2h})!} \\
&= \frac{2^{-\xi_n}}{h} \cdot \chi_{\xi_n}(x+h) \cdot \frac{\xi_n!}{(\frac{\xi_n}{2} - \frac{x+h}{2h})!(\frac{\xi_n}{2} + \frac{x+h}{2h})!} = f_{n0}(x+h),
\end{aligned}$$

dually

$$\frac{\frac{x}{h} + \xi_{n+2}}{\xi_{n+2}} f_{n+2,0}(x) = f_{n0}(x-h).$$

Now we are able to show by the discrete ladder lemma 6.1 that the functions f_{nj} are the eigenfunctions of H_n also for $j \geq 1$:

$$\begin{aligned}
&[G^+_{n+1} f_{nj}](x) = [G^+_{n+1}(q_{nj} \cdot f_{n0})](x) \\
&= \frac{h}{2}\Big[(\frac{x}{h} - \xi_{n+2}) \cdot q_{nj}(x+h) \cdot f_{n0}(x+h) \\
&\qquad + (\frac{x}{h} + \xi_{n+2}) \cdot q_{nj}(x-h) \cdot f_{n0}(x-h) \Big] \\
&= \frac{-h}{2\xi_{n+2}} \Big[(\frac{x}{h} - \xi_{n+2})^2 q_{nj}(x+h) \\
&\qquad - (\frac{x}{h} + \xi_{n+2})^2 q_{nj}(x-h) \Big] \cdot f_{n+2,0}(x) \\
&= \frac{-h}{2\xi_{n+2}} \Big[\left(\frac{x^2}{h^2} + \xi_{n+2}^2 \right) 2h\, Dq_{nj}(x)
\end{aligned}$$

$$-2\xi_{n+2}\frac{x}{h}\cdot 2\cdot Mq_{nj}(x)\Big]\cdot f_{n+2,0}(x)$$

$$= \left[2xMq_{nj}(x) - (h^2\xi_{n+2} + \frac{x^2}{\xi_{n+2}})Dq_{nj}(x)\right]\cdot f_{n+2,0}(x)$$

$$= q_{n+2,j+1}(x)\cdot f_{n+2,0}(x) = f_{n+2,j+1}(x).$$

We have found $\xi_n + 1$ eigenfunctions f_{nj} of H_n for different eigenvalues λ_{nj}. Since they are multiples of f_{n0}, this system of eigenfunctions forms an orthogonal basis for the $(\xi_n + 1)$–dimensional space of functions with finite support given by (36).

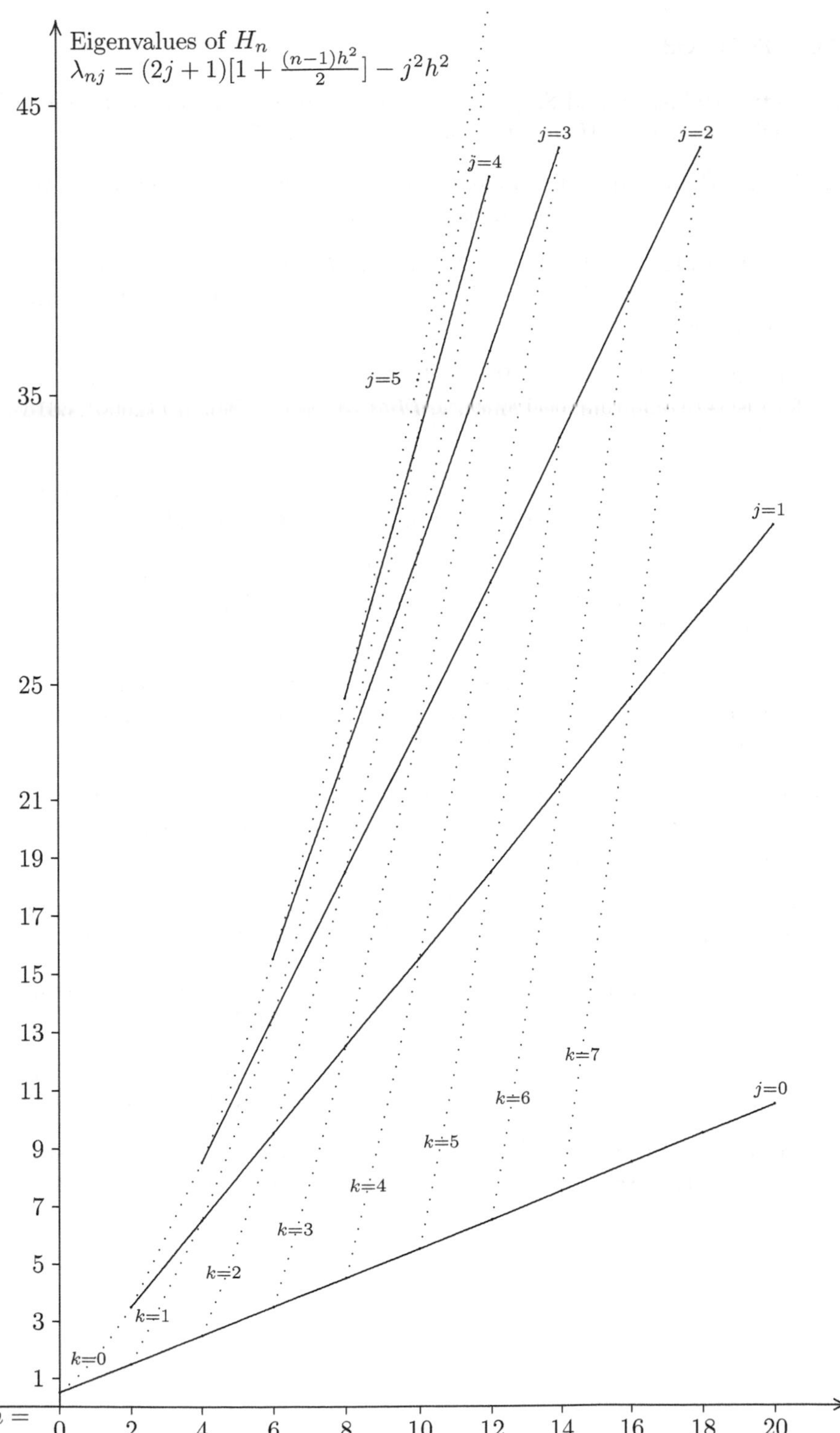
Eigenvalues of H_n
$\lambda_{nj} = (2j+1)[1 + \frac{(n-1)h^2}{2}] - j^2h^2$
45
35
25
21
19
17
15
13
11
9
7
5
3
1
j=5
j=4
j=3
j=2
j=1
j=0
k=0
k=1
k=2
k=3
k=4
k=5
k=6
k=7
n =
0
2
4
6
8
10
12
14
16
18
20

References

[1] Natig Atakishiev and Sergej Suslov, Difference analog of the harmonic oscillator, *Theor. Math. Phys.* **85** (1991), 1055–1062.

[2] Martin Bohner and Allan Peterson, editors, *Advances in Dynamic Equations on Time Scales*, Birkhäuser, Boston, 2003.

[3] Martin Bohner and Allan C. Peterson, *Dynamic Equations on Time Scales — An Introduction with Applications*, Birkhäuser, Boston, Basel, Berlin, 2001.

[4] Tugrul Hakioglu and Kurt Bernardo Wolf, The canonical Kravchuk basis for discrete quantum mechanics, *J. Phys. A, Math. Gen.* **33** (2000), 3313–3323.

[5] Stefan Hilger, Analysis on measure chains — a unified approach to continuous and discrete calculus, *Results Math.* **18** (1990), 18–56.

[6] Stefan Hilger, An application of calculus on measure chains to fourier theory and heisenberg's uncertainty principle, *J. Differ. Equations Appl.* **8** (2002), 897–936.

[7] Decio Levi, Piergiulio Tempesta, and Pavel Winternitz, Umbral calculus, difference equations and the discrete Schrödinger equation, Los Alamos preprint arXives nlinSI/0305047.

[8] Miguel Lorente, Creation and annihilation operators for orthogonal polynomials of continuous and discrete variables, *Electronic Transactions on Numerical Analysis* **9** (1999), 102–111.

[9] Miguel Lorente, Continuous vs. discrete models for the quantum harmonic oscillator and the hydrogen atom, *Phys. Lett. A* **285** (2001), 119–126.

[10] Miguel Lorente, Integrable systems on the lattice and orthogonal polynomials of discrete variable, *J. Comput. Appl. Math.* **153** (2003), 321–330.

[11] M.A. Rego-Monteiro, The quantum harmonic oscillator on a circle and a deformed Heisenberg algebra, *Eur. Phys. J. C* **21** (2001) 749–756.

[12] Alexander Turbiner, Canonical discretization I. Discrete faces of (an)harmonic oscillator, *Int. J. Mod. Phys. A* **16** (2001), 1579–1603.

Solution of Dirichlet Problems with Discrete Double-Layer Potentials

ANGELA HOMMEL

Chair of Applied Mathematics, University of Weimar
Weimar, Germany

Abstract As the main tool in the theory of difference potentials we split the difference potential on the boundary into a discrete single and double-layer potential. For a long time it was not possible to prove solvability conditions for Dirichlet problems based on discrete double-layer potentials. In this paper we show how we can overcome the difficulties. We prove that for arbitrarily chosen boundary values a unique solution of the interior and exterior discrete Dirichlet problems in the plane exists and describe this solution with the help of the discrete double-layer potential.

Keywords finite difference operator, difference potential, double-layer potential

AMS Subject Classification 39A70, 39A12

1 Introduction

More than 30 years ago Ryabenkij [4] studied difference potentials and related methods to solve Dirichlet and Neumann problems. The aim of this theory is the solution of linear systems of equations on the boundary such that the equivalence to the original problem is preserved. The definition of the discrete potentials is based on a famous theorem on boundary projections for right invertible operators (see [5]). In [2] and [3] it was possible to split the difference potential on the boundary into a discrete single and double-layer potential. Similar to the continuous case we can prove solvability conditions for Neumann problems based on the discrete single-layer potential. The discrete double-layer potential has a special structure in the form of an additional summand. Therefore the adjoint equation to the Neumann problem is different from the equation we have to study in relation to the double-layer potential.
In the following we show how we can overcome these difficulties. It can be proved that the interior and exterior discrete Dirichlet problem in the plane has a unique solution that can be described with the help of the discrete double-layer potential.

2 Difference potentials on the boundary

Let $\mathbb{R}^2$ be the two-dimensional Euclidean space. An equidistant lattice with the mesh width $h > 0$ is defined by $\mathbb{R}_h^2 = \{mh = (m_1 h, m_2 h) : m_1, m_2 \in \mathbb{Z}\}$. We consider a bounded domain $G \subset \mathbb{R}^2$ with a piecewise smooth boundary Γ and denote the discrete domain by $G_h = G \cap \mathbb{R}_h^2$. In addition we introduce the set $K = \{(0,0),\ (-1,0),\ (1,0),\ (0,-1),\ (0,1)\}$ and define the five-point star $N_m = \{m+k : \ k \in K\}$ for all $mh \in G_h$. The union $\bigcup_{mh \in G_h} N_m$ is denoted by N. At all points $r = (r_1, r_2) \in N$ the set $K_r = \{k \in K : \ (r+k)h \notin G_h\}$ is analyzed. We look at the discrete domain G_h with the double-layer boundary $\gamma_h = \{rh : r \in N \text{ and } K_r \neq \emptyset\}$. In detail all points rh with $k = (0,0) \in K_r$ are mesh points of the outer layer γ_h^-. The points $rh \in \gamma_h \setminus \gamma_h^-$ belong to the inner boundary layer γ_h^+. The same notation can be used if we look at an exterior domain. To distinguish interior and exterior problems we use the index a for exterior problems.
We consider now the difference equation

$$-\Delta_h u_h(mh) = \sum_{k \in K} a_k u_h(mh - kh) = f_h(mh) \qquad \forall\, mh \in G_h$$

with the coefficients $a_k = \begin{cases} -1/h^2 & \text{for } k \in K, k \neq (0,0) \\ 4/h^2 & \text{for } k = (0,0). \end{cases}$
Each solution of the equation

$$-\Delta_h E_h(mh) = \begin{cases} 1/h^2 & \text{if } mh = (0,0), \\ 0 & \text{if } mh \neq (0,0), \end{cases}$$

which does not grow more than $\ln |mh|$ at infinity is called *discrete fundamental solution.* For the details see [1], [6], [7], and [2].
In analogy to the integral representation for functions in C^2 Ryabenkij [4] proved that

$$\sum_{rh \in \gamma_h} \Big(\sum_{k \in K_r} E_h(lh - (r+k)h) a_k h^2 \Big) u_h(rh)$$
$$- \sum_{mh \in G_h} E_h(lh - mh) \Delta_h u_h(mh) h^2 = \begin{cases} u_h(lh) & l \in N \\ 0 & l \notin N. \end{cases}$$

In [2] the difference potential on the boundary was split into a discrete single and double-layer potential: We denote by $u_R(rh) = u_h(rh)$ the boundary values and by $u_A(rh) = h^{-1} \sum_{k \in K \setminus K_r} (u_h(rh) - u_h((r+k)h)$ the discrete normal derivatives and obtain

$$\sum_{rh \in \gamma_h} \Big(\sum_{k \in K_r} E_h(lh - (r+k)h) a_k h^2 \Big) u_h(rh) = (P_h^E u_A)(lh) - (P_h^D u_R)(lh),$$

$\forall l \in N$ with the discrete single-layer potential

$$(P_h^E u_A)(lh) = \sum_{rh \in \gamma_h^-} u_A(rh) E_h(lh - rh)h,$$

and the double-layer potential

$$(P_h^D u_R)(lh) = \sum_{rh \in \gamma_h^-} \sum_{k \in K \setminus K_r} \frac{E_h(lh-rh) - E_h(lh-(r+k)h)}{h} u_R(rh)h - \kappa(lh) u_R(lh),$$

with $\kappa(lh) = 1$ for all $lh \in \gamma_h^-$, and zero at all other mesh points.

3 The solution of interior Dirichlet problems

We use the discrete double-layer potential to solve the discrete problem

$$\begin{aligned} -\Delta_h u_h(mh) &= 0 & \forall mh \in G_h \qquad & (\mathbf{D}_i) \\ u_h(rh) &= \varphi_h(rh) & \forall rh \in \gamma_h^- . & \end{aligned}$$

Theorem 3.1. *If the system*

$$\varphi_h(lh) = \sum_{rh \in \gamma_h^-} \sum_{k \in K \setminus K_r} (E_h(lh-rh) - E_h(lh-(r+k)h)) u_R(rh) - u_R(lh) \quad (1)$$

is solvable $\forall lh \in \gamma_h^-$ *then the discrete double-layer potential* $(P_h^D u_R)(mh)$ *is a solution of the problem* $(\mathbf{D}_i)$ *for all* $mh \in G_h \cup \gamma_h^-$.

Proof. If (1) is solvable then the boundary condition is automatically fulfilled. In order to prove $-\Delta_h (P_h^D u_R)(mh) = 0$ for all $mh \in G_h \setminus \gamma_h^+$ we simply use the definition of the discrete fundamental solution. For all $mh \in \gamma_h^+$ the additional summand comes inside. We have

$$\begin{aligned} & -\Delta_h (P_h^D u_R)(mh) \\ = & \sum_{rh \in \gamma_h^-} \sum_{k \in K \setminus K_r} (-\Delta_h E_h(mh - rh) + \Delta_h E_h(mh - (r+k)h) u_R(rh) \\ & + h^{-2} \sum_{k \in K_m} u_R((m+k)h) \\ = & -h^{-2} \sum_{-k \in K_m} u_R((m + (-k))h) + h^{-2} \sum_{k \in K_m} u_R((m+k)h) = 0. \end{aligned}$$

□

We now study the solvability of system (1). For the proof of the next theorem we use the first Green formula in exterior domains: Let w_h and u_h be two *discrete harmonic functions in the exterior domain* G_h^a, which means functions that fulfill the properties $-\Delta_h w_h(mh) = -\Delta_h u_h(mh) = 0$ for all $mh \in G_h^a$ and $|w_h(mh)| < C$ as well as $|u_h(mh)| < C$ for $|mh| \to \infty$. C is an arbitrary constant. We split the boundary layer γ_h^{-a} into the parts $\gamma_{hi}^{-a} = \{rh \in \gamma_h^{-a} : (r+k_i)h \in G_h^a\}$, $i = 1, \dots, 4$ with $k_1 = (-1,0)$, $k_2 = (0,-1)$, $k_3 = (1,0)$, $k_4 = (0,1)$ and obtain

$$\begin{aligned}
&\sum_{mh \in G_h^a} w_h(mh)\Delta u_h(mh)h^2 \\
=\; & -\sum_{mh \in G_h^a}\sum_{i=1}^{2}\Big(\frac{w_h(mh) - w_h((m+k_i)h)}{h}\Big)\Big(\frac{u_h(mh) - u_h((m+k_i)h)}{h}\Big)h^2 \\
&+ \sum_{rh \in \gamma_h^{-a}}\sum_{k \in K \setminus K_r^a} w_h(rh)\Big(\frac{u_h(rh) - u_h((r+k)h)}{h}\Big)h \\
&-\sum_{i=1}^{2}\sum_{rh \in \gamma_{hi}^{-a}}\Big(\frac{w_h(rh) - w_h((r+k_i)h)}{h}\Big)\Big(\frac{u_h(rh) - u_h((r+k_i)h)}{h}\Big)h^2.
\end{aligned}$$

For the proof of this Green formula we refer to [2, Theorem 2.6].

Theorem 3.2. *System (1) has a unique solution for arbitrarily chosen boundary values* $\varphi_h(lh)$.

Proof. The assertion follows from the Fredholm alternative if the homogeneous system (1) has only the trivial solution. We consider the adjoint system

$$0 = \sum_{rh \in \gamma_h^-}\sum_{k \in K \setminus K_l} (E_h(lh - rh) - E_h((l+k)h - rh))v_h(rh) - v_h(lh) \qquad (2)$$

for all $lh \in \gamma_h^-$. Using the Gauss formula

$$\sum_{lh \in \gamma_h^-}\sum_{k \in K \setminus K_l} (E_h(rh - lh) - E_h(rh - (l+k)h) = 0 \quad \forall rh \in \gamma_h^-$$

(see [2]) and the symmetry of the fundamental solution it follows from (2)

$$\sum_{lh \in \gamma_h^-} v_h(lh)h = 0. \qquad (3)$$

We extend $v_h(lh)$ and fix the values in the outer corners.
Let γ_h^A be the set of all outer corners with respect to G_h. In addition to the domain G_h we consider the exterior domain $G_h^a = \{mh \in \mathbb{R}_h^2 : mh \notin (G_h \cup \gamma_h^- \cup \gamma_h^A)\}$ with $\gamma_h^{-a} \subset (\gamma_h^- \cup \gamma_h^A)$. We define

$$\tilde{v}_h(lh) = \begin{cases} v_h(lh) & \forall lh \in \gamma_h^- \\ 0 & \forall lh \in (\gamma_h^{-a} \setminus \gamma_h^-). \end{cases}$$

Furthermore we use the notation $M_1 = K \setminus K_l$, $M_2 = K_l \setminus [K \setminus K_l^a]$ and $M_3 = K \setminus K_l^a$, where the last set is related to points in the exterior domain. Based on the definition of the discrete Laplacian we can write at each mesh point $lh \in (\gamma_h^- \cup \gamma_h^{-a})$

$$\begin{aligned}\tilde{v}_h(lh) &= \sum_{rh\in(\gamma_h^-\cup\gamma_h^{-a})} h^2(-\Delta_h E_h(lh - rh))\tilde{v}_h(rh) \qquad (4)\\ &= \sum_{rh\in(\gamma_h^-\cup\gamma_h^{-a})} \sum_{k\in M_1\cup M_2\cup M_3} (E_h(lh - rh) - E_h((l+k)h - rh))\tilde{v}_h(rh).\end{aligned}$$

For all $lh \in \gamma_h^-$ we conclude in relation with (2) and (4)

$$\sum_{rh\in(\gamma_h^-\cup\gamma_h^{-a})} \sum_{k\in M_2\cup M_3} (E_h(lh - rh) - E_h((l+k)h - rh))\tilde{v}_h(rh) = 0.$$

At the points $lh \in (\gamma_h^{-a} \setminus \gamma_h^-)$ the set M_1 is empty and we obtain from (4)

$$\sum_{rh\in(\gamma_h^-\cup\gamma_h^{-a})} \sum_{k\in M_2\cup M_3} (E_h(lh - rh) - E_h((l+k)h - rh))\tilde{v}_h(rh) = \tilde{v}_h(lh) = 0.$$

We now consider the discrete single-layer potential with the density $\tilde{v}_h(lh)$. From the last two equations it follows for all $lh \in (\gamma_h^{-a} \cup \gamma_h^-)$ that

$$h^{-1} \sum_{k\in M_2\cup M_3} (P_h^E \tilde{v}_h)(lh) - (P_h^E \tilde{v}_h)((l+k)h) = 0. \qquad (5)$$

We show that this system has only the constant solution $(P_h^E \tilde{v}_h)(lh) = C$ $\forall\, lh \in (\gamma_h^- \cup \gamma_h^{-a})$. At first we investigate the system

$$h^{-1} \sum_{k\in M_2} (P_h^E \tilde{v}_h)(lh) - (P_h^E \tilde{v}_h)((l+k)h) = f_h(lh) \qquad (6)$$

for all $lh \in (\gamma_h^- \cup \gamma_h^{-a})$. Let n be the number of all mesh points in $\gamma_h^- \cup \gamma_h^{-a}$. The system (6) can be written in the form

$$A\, P_h^E \tilde{v}_h = h\, f_h, \qquad (7)$$

where $P_h^E \tilde{v}_h$ and f_h are vectors with n elements and the matrix $A = (a_{ij})_{i=1,\ldots,n}^{j=1,\ldots,n}$ has the structure

$$\begin{aligned} &a_{ii} = 2,\ i = 1,\ldots,n, && a_{ii+1} = -1,\ i = 1,\ldots,n-1,\\ &a_{1n} = a_{n1} = -1, && a_{ii-1} = -1,\ i = 2,\ldots,n\end{aligned}$$

and all other elements a_{ij} are zero. From (5) and (6) it follows that

$$h^{-1} \sum_{k\in M_3} (P_h^E \tilde{v}_h)(lh) - (P_h^E \tilde{v}_h)((l+k)h) = -f_h(lh) \qquad \forall\, lh \in \gamma_h^{-a}. \qquad (8)$$

At the mesh points $lh \in \gamma_h^- \setminus \gamma_h^{-a}$ the set M_3 is empty and we conclude from (5) and (6) $f_h(lh) = 0$.

Based on the property $\sum\limits_{lh \in (\gamma_h^- \cup \gamma_h^{-a})} \tilde{v}_h(lh)h = \sum\limits_{lh \in \gamma_h^-} v_h(lh)h = 0$ we can prove that the single-layer potential $(P_h^E \tilde{v}_h)(mh)$ is a discrete harmonic function in the exterior domain G_h^a. (For the proof of the behavior at infinity we refer to [2], proof of Theorem 2.11.) Using the first Green's formula we obtain in relation with the system (8)

$$\begin{aligned} 0 &= \sum_{mh \in G_h^a} \Delta_h (P_h^E \tilde{v}_h)(mh) h^2 \\ &= \sum_{lh \in \gamma_h^{-a}} \sum_{k \in M_3} \frac{(P_h^E \tilde{v}_h)(lh) - (P_h^E \tilde{v}_h)((l+k)h)}{h} h = - \sum_{lh \in \gamma_h^{-a}} f_h(lh)h. \end{aligned}$$

Consequently we look for a solution of the system (7) for all $lh \in (\gamma_h^- \cup \gamma_h^{-a})$ under the conditions $f_h(lh) = 0 \quad \forall lh \in \gamma_h^- \setminus \gamma_h^{-a}$ and $\sum\limits_{lh \in \gamma_h^{-a}} f_h(lh)h = 0$.

Using again the first Green's formula in exterior domains and the equations (7) and (8) it follows that

$$\begin{aligned} 0 &\leq \sum_{lh \in \gamma_h^{-a}} \sum_{k \in M_3} (P_h^E \tilde{v}_h)(lh) \left(\frac{(P_h^E \tilde{v}_h)(lh) - (P_h^E \tilde{v}_h)((l+k)h)}{h} \right) h \\ &= - \sum_{lh \in (\gamma_h^- \cup \gamma_h^{-a})} (P_h^E \tilde{v}_h)(lh)\, f_h(lh)h \\ &= - \sum_{lh \in (\gamma_h^- \cup \gamma_h^{-a})} (P_h^E \tilde{v}_h)(lh)(A\, P_h^E \tilde{v}_h)(lh). \end{aligned}$$

Let us enumerate all mesh points of the set $(\gamma_h^- \cup \gamma_h^{-a})$ from $l_1 h$ to $l_n h$. An easy calculation shows that

$$\begin{aligned} &\sum_{lh \in (\gamma_h^- \cup \gamma_h^{-a})} (P_h^E \tilde{v}_h)(lh)(A\, P_h^E \tilde{v}_h)(lh) \\ = \;& [(P_h^E \tilde{v}_h)(l_1 h) - (P_h^E \tilde{v}_h)(l_2 h)]^2 + [(P_h^E \tilde{v}_h)(l_2 h) - (P_h^E \tilde{v}_h)(l_3 h)]^2 \\ &+ \cdots + [(P_h^E \tilde{v}_h)(l_n h) - (P_h^E \tilde{v}_h)(l_1 h)]^2 \geq 0. \end{aligned} \tag{9}$$

Consequently the equation $\sum\limits_{lh \in (\gamma_h^- \cup \gamma_h^{-a})} (P_h^E \tilde{v}_h)(lh)(A\, P_h^E \tilde{v}_h)(lh) = 0$ is fulfilled and from (9) it follows that

$$(P_h^E \tilde{v}_h)(lh) = C \quad \forall lh \in (\gamma_h^- \cup \gamma_h^{-a}).$$

By this way we proved that the discrete single-layer potential $(P_h^E \tilde{v}_h)(mh)$ is a discrete harmonic function in G_h^a with $(P_h^E \tilde{v}_h)(lh) = C$ for all $lh \in \gamma_h^{-a}$.

From the uniqueness theorem for exterior Dirichlet problems (see [Theorem 2.9][2]) we obtain

$$(P_h^E \tilde{v}_h)(mh) = C \ \forall mh \in G_h^a.$$

On the other side the discrete single-layer potential is constant at all points $lh \in \gamma_h^-$ and we can conclude from the uniqueness theorem for interior Dirichlet problems (see [2], Theorem 2.7)

$$(P_h^E \tilde{v}_h)(mh) = C \ \forall mh \in G_h.$$

Consequently we proved $(P_h^E \tilde{v})(mh) = C \ \forall mh \in \mathbb{R}_h^2$ and from the relation

$$0 = -\Delta_h (P_h^E \tilde{v}_h)(lh) = \sum_{rh \in (\gamma_h^- \cup \gamma_h^{-a})} \Delta_h E_h(lh - rh) \tilde{v}_h(rh) h - \tilde{v}_h(lh) h^{-1},$$

which holds for all $lh \in (\gamma_h^- \cup \gamma_h^{-a})$ it follows $\tilde{v}_h(lh) = 0$.
We showed that the system (5) has only the solution $(P_h^E \tilde{v}_h)(lh) = C \ \forall lh \in (\gamma_h^- \cup \gamma_h^{-a})$ and from $\tilde{v}_h(lh) = 0$ we obtain $C = 0$. From (4) we conclude that for all $lh \in (\gamma_h^- \cup \gamma_h^{-a})$ the system

$$0 = \sum_{rh \in (\gamma_h^- \cup \gamma_h^{-a})} \sum_{k \in M_1} (E_h(lh - rh) - E_h((l+k)h - rh)) \tilde{v}_h(rh) - \tilde{v}_h(lh)$$

has also only the trivial solution. We remark that the set M_1 is empty at all points $lh \in (\gamma_h^{-a} \setminus \gamma_h^-)$ in which per definition $\tilde{v}_h(lh) = 0$. Therefore the system (2) has for all $lh \in \gamma_h^-$ only the trivial solution. Based on the Fredholm theorem we can also conclude that the homogeneous system (1) has only the trivial solution. Using the Fredholm alternative the proof is complete $\square$

4 Theorems for exterior Dirichlet problems

Similar to interior problems the discrete double-layer potential has the structure

$$\begin{aligned}(P_h^{D,a} u_R)(lh) &= \sum_{rh \in \gamma_h^{-a}} \sum_{k \in K \setminus K_r^a} \frac{E_h(lh - rh) - E_h(lh - (r+k)h)}{h} u_R(rh) h \\ &\quad - \kappa(lh)\, u_R(lh)\end{aligned}$$

with $\kappa(lh) = 1$ for all $lh \in \gamma_h^{-a}$ and zero otherwise. This potential is a discrete harmonic function that tends to zero at infinity. In the two-dimensional case it is not necessary for the solution of exterior Dirichlet problems to tend to zero at infinity. Therefore we expand the above potential with the help of an additive constant. For problems in exterior domains the following theorems can be proved:

Theorem 4.1. *If the system*

$$\begin{aligned}\varphi_h^a(lh) &= \sum_{rh\in\gamma_h^{-a}} \sum_{k\in K\setminus K_r^a} (E_h(lh-rh)-E_h(lh-(r+k)h))u_R(rh) - u_R(lh) \\ &\quad + \sum_{rh\in\gamma_h^{-a}} u_R(rh)h \qquad (10)\end{aligned}$$

is solvable for all $lh \in \gamma_h^{\ a}$ *then the potential* $(P_h^{D,a}u_R)(mh) + \sum_{rh\in\gamma_h^{-a}} u_R(rh)h$ *is a solution of the problem*

$$\begin{aligned}-\Delta_h u_h(mh) &= 0 \quad \forall mh \in G_h^a \\ |u_h(mh)| &\leq C \quad \text{if } |\, mh| \to \infty \\ u_h(rh) &= \varphi_h^a(rh) \quad \forall rh \in \gamma_h^{-a}.\end{aligned}$$

Theorem 4.2. *The system (10) has a unique solution for arbitrarily chosen boundary values* $\varphi_h^a(lh)$.

References

[1] Boor, C., Höllig K., and Riemenschneider, S., Fundamental solutions for multivariate difference equations, *Amer. J. Math.* **111** (1989), 403–415.

[2] Hommel, A., Fundamental solutions of partial difference operators and the solution of discrete boundary value problems based on difference potentials, Dissertation, Bauhaus University Weimar (1998) (German), digital version: www.db-thueringen.de, document 618.

[3] Hommel, A., On finite difference potentials, in *Proceedings of the Sixth Internation Conference on Difference Equations*, S. Elaydi, G. Ladas and B. Aulbach (eds.), 469–476, CRC, Boca Raton, FL, 2004.

[4] Ryabenkij, V.S., The method of difference potentials for some problems in continuum mechanics, *Nauka, Moskau* (1987) (Russian).

[5] Seeley, R.T., Singular integrals and boundary value problems, *Amer. J. Math.* **88** (1966),781–809.

[6] Sobolev, S.L., About a difference equation, *Doklady Akad. Nauk SSSR* **87** (1952), 341–343 (Russian).

[7] Van der Pol, B., The finite-difference analogy of the periodic wave equation and the potential equation, Appendix IV in *Prohability and Related Topics in Physical Sciences*, Marc Kac, Interscience Publishers, New York (1957)

Moments of Solutions of Linear Difference Equations

KLARA JANGLAJEW[1]
Institute of Mathematics, University of Bialystok
Bialystok, Poland
and
KIM VALEEV
Department of Mathematics
Kiev National University of Economics
Kiev, Ukraine

Abstract In this paper we obtain moments of solutions for a linear equation with coefficients dependent on two successive values of a Markov chain.

Keywords Markov chain, intial moments, linear difference equations

AMS Subject Classification 39A99

1 Introduction

Difference equations for moments apparently have been introduced in [6] by Khrisanov. Several important papers on the subject appeared afterward (see [7] and the references given there).

An asymptotic method for the construction of moment equations for a system of linear difference equations with random coefficients and with a small parameter has been proposed in [3]. For yet another method of the construction of moments equations we refer the reader to [2], where a system of linear difference equations with coefficients dependent on a Markov chain has been studied.

Moment equations may be used for the investigation of the stability of random solutions. In [5] we gave conditions for the stability of moments of solutions of a system of linear difference equations with coefficients dependent on a Markov chain.

Our main objective in this paper is to extend the study of moment equations of linear nonhomogeneous difference equations with coefficients dependent on two successive values of a Markov chain.
This type of difference equation has been used in a mathematical model of passenger transport [4].

[1]Supported by the Institute of Mathematics, University of Bialystok, under BST-140.

2 Preliminaries

Consider a Markov chain $\zeta_n(n = 0, 1, \dots)$ taking values $\theta_1, \dots, \theta_q$ with probabilities

$$p_k(n) = P\{\zeta_n = \theta_k\} \qquad (k = 1, ..., q). \tag{1}$$

We assume that the probabilities $p_k(n)$ $(k = 1, ..., q)$ satisfy a system of difference equations

$$p_k(n+1) = \sum_{j=1}^{q} \pi_{kj}(n) p_j(n) \qquad (k = 1, ..., q), \tag{2}$$

where coefficients satisfy the well-known conditions [1]

$$\pi_{kj}(n) \geq 0, \qquad \sum_{k=1}^{q} \pi_{kj}(n) = 1.$$

Let the coefficients of the difference equation

$$x_{n+1} = a(\zeta_{n+1}, \zeta_n)x_n + b(\zeta_{n+1}, \zeta_n) \quad (n = 0, 1, 2, ...) \tag{3}$$

be random and depend on ζ_n, ζ_{n+1}, $a(\theta_k, \theta_j) \neq 0, (k, j = 1, ..., q)$.

Let us introduce the density distribution $f(n, x, \zeta)$ of the stochastic process (x_n, ζ_n).

Since x_n is continuously distributed, and ζ_n is a finite-valued random variable,

$$f(n, x, \zeta) = \sum_{k=1}^{q} f_k(n, x)\delta(\zeta - \theta_k), \tag{4}$$

where $\delta(\zeta)$ is the Dirac delta function. We will write it simply as x, ζ when no confusion can arise. The stochastic value x_n has a density distribution

$$f(n, x) = \sum_{k=1}^{q} f_k(n, x) \tag{5}$$

The functions $f_k(n, x)$, $(k = 1, \dots, q)$ are called the particular density distributions [7]. They may be defined by the equalities

$$P\{x_n < \alpha, \zeta_n = \theta_k\} = \int_{-\infty}^{\alpha} f_k(n, x)dx \qquad (k = 1, ...q).$$

It is assumed that the Markov chain ζ_n does not depend on values that take variables x_n.

We now obtain equations connecting particular density distributions. For simplicity of notation, we write a_{ks} instead of $a(\theta_k, \theta_s)$ and b_{ks} instead of

$b(\theta_k, \theta_s)$. We have

$$\begin{aligned}\int_{-\infty}^{\alpha} f_k(n+1,x)dx &= P\{x_{n+1} < \alpha,\ \zeta_{n+1} = \theta_k\} \\ &= \sum_{s=1}^{q} P\{x_{n+1} < \alpha,\ \zeta_{n+1} = \theta_k,\ \zeta_n = \theta_s\} \\ &= \sum_{s=1}^{q} P\{\zeta_{n+1} = \theta_k |\ x_{n+1} < \alpha,\ \zeta_n = \theta_s\},\end{aligned}$$

and the last expression equals

$$\sum_{s=1}^{q} P\{\zeta_{n+1} = \theta_k |\ \zeta_n = \theta_s\} \cdot P\{a_{ks}x_n + b_{ks} < \alpha,\ \zeta_n = \theta_s\}.$$

Applying the equalities

$$P\{\zeta_{n+1} = \theta_k | \zeta_n = \theta_s\} = \pi_{ks},$$

$$P\{x_n < \frac{\alpha - b_{ks}}{a_{ks}},\ \zeta_n = \theta_s\} = \int_{-\infty}^{\frac{\alpha - b_{ks}}{a_{ks}}} f_s(n,x)dx,$$

we get the system of equations for particular density distributions

$$\int_{-\infty}^{\alpha} f_k(n+1,x)dx = \sum_{s=1}^{q} \pi_{ks} \int_{-\infty}^{\frac{\alpha - b_{ks}}{a_{ks}}} f_s(n,x)dx,\ (k = 1, ..., q)$$

By differentiating the last equations under the condition $a_{ks} > 0$ $(k, s = 1, \dots, q)$ we obtain

$$f_k(n+1,\alpha) = \sum_{s=1}^{q} \pi_{ks} f_s(n, \frac{\alpha - b_{ks}}{a_{ks}}) \frac{1}{a_{ks}} \tag{6}$$

If the coefficients a_{ks} $(k, s = 1, \dots, q)$ can take negative values, we get for particular density distributions the system

$$f_k(n+1,x) = \sum_{s=1}^{q} \pi_{ks} f_s(n, \frac{x - b_{ks}}{a_{ks}}) \frac{1}{|a_{ks}|} \tag{7}$$

3 Moment equations

We will derive the initial moments of order p $(p = 0, 1, 2, \dots)$ to a random solution x_n

$$m_p(n) \equiv \int_{-\infty}^{\infty} x^p f(n,x)dx \tag{8}$$

To compute the $m_p(n)$ we use the particular moments

$$m_{p,k}(n) = \int_{-\infty}^{\infty} x^p f_k(n,x)dx, \qquad (p = 0,1,2,\dots;\ k = 1,\dots q)$$

$$m_p(n) = \sum_{k=1}^{q} m_{pk}(n). \tag{9}$$

Multiplying System (7) by x^p and integrating over the interval $(-\infty,\infty)$ we obtain the system of difference equations

$$m_{p,k}(n+1) = \sum_{s=1}^{q} \pi_{ks} \int_{-\infty}^{\infty} x^p f_s(n, \frac{x-b_{ks}}{a_{ks}}) \frac{dx}{|a_{ks}|}$$

Using the change of the integration variable $\frac{x-b_{ks}}{a_{ks}} = y$ we get

$$\begin{aligned} m_{p,k}(n+1) &= \sum_{s=1}^{q} \pi_{ks} \int_{-\infty}^{\infty} (a_{ks}y + b_{ks})^p f_s(n,y)dy \\ &= \sum_{s=1}^{q} \pi_{ks} \sum_{l=0}^{p} C_p^l a_{ks}^{p-l} b_{ks}^l \int_{-\infty}^{\infty} y^{p-l} f_s(n,y)dy \end{aligned}$$

or

$$m_{p,k}(n+1) = \sum_{s=1}^{q} \pi_{ks} \sum_{l=0}^{p} C_p^l a_{ks}^{p-l} b_{ks}^l m_{p-l,s}(n) \tag{10}$$

$$(k = 1,\dots,q;\ p = 0,1,\dots;\ n = 0,1,\dots)$$

Summarizing,we have the following theorem.

Theorem 3.1. *If the coefficients of the linear difference equation (3) depend on two successive values of the Markov chain ζ_n defined by (2), then the initial moments (8) of a random solution x_n are defined by (9), where the particular moments $m_{p,k}$ satisfy (10).*

For $p = 0,1,2$ we get systems of difference equations

$$m_{0,k}(n+1) = \sum_{s=1}^{q} \pi_{ks} m_{0,s}(n), \quad m_{0,s}(n) \equiv p_s(n),$$

$$m_{1,k}(n+1) = \sum_{s=1}^{q} \pi_{ks}(a_{ks} m_{1,s}(n) + b_{ks} m_{0,s}(n)),$$

$$m_{2,k}(n+1) = \sum_{s=1}^{q} \pi_{ks}(a_{ks}^2 m_{2,s}(n) + 2a_{ks}b_{ks}m_{1,s}(n) + b_{ks}^2 m_{0,s}(n)),$$

$$(k = 1,\dots,q).$$

4 Generalization

Analogously, it is possible to consider a system of m linear difference equations with coefficients dependent on two successive values of a Markov chain ζ_n

$$X_{n+1} = A(\zeta_{n+1}, \zeta_n)X_n + B(\zeta_{n+1}, \zeta_n). \tag{11}$$

We introduce the vectors of the first moments

$$M(n) = E(X_n), \qquad M(n) = \sum_{k=1}^{q} M_k(n)$$

and the matrices of the second moments

$$D(n) = E(X_n, X_n^*), \qquad D(n) = \sum_{k=1}^{q} D_k(n).$$

Here $*$ indicates the transpose of a vector.
For particular moments we obtain systems of difference equations

$$\begin{aligned} M_k(n+1) &= \sum_{s=1}^{q} \pi_{ks}(A_{ks}M_s(n) + B_{ks}P_s(n)) \\ D_k(n+1) &= \sum_{s=1}^{q} \pi_{ks}(A_{ks}P_s(n)A_{ks}^* + A_{ks}M_s(n)B_{ks}^* \\ &\quad + B_{ks}M_k^*(n)A_{ks}^* + B_{ks}B_{ks}^*P_k(n), \end{aligned}$$

where $A_{ks} \equiv A(\theta_k, \theta_s)$, $B_{ks} \equiv B(\theta_k, \theta_s)$, $(k, s = 1, \ldots, q)$.

Remark 4.1. *The particular case in which the coefficients of System (11) depend only on ζ_n has been investigated in [7].*

References

[1] Elaydi, S., *An Introduction to Difference Equations*, Second Edition, Springer, New York, 1999.

[2] Janglajew, K., Moment equations for stochastic difference equations, in *iProceedings of the Sixth International Conference on Difference Equations* (Eds. Elaydi, E., Ladas, G., Aulbach, B.), 479-484, CRC, Boca Raton, FL, 2004.

[3] Janglajew, K. and Lavrenyk, O., On the asymptotic method for construction of moment equations, in *Control and Self-Organization in Nonlinear Systems (Bialystok, 2000)*, Technical University of Bialystok Press, 2000, 57–62.

[4] Janglajew, K. and Lavrenyk, O., A mathematical model of passenger transport, *Miscellanea Algebraicae*, WZiA AS, Kielce, **2** (2001), 55–62.

[5] Janglajew, K. and Valeev, K., On the stability of solutions of a system of difference equations with random coefficients, *Miscellanea Methodological*, WSP, Kielce, **3** (1998), 35–41 (in Polish).

[6] Khrisanov, S., Difference equations with Markov's coefficients, *Probability Theory and Mathematical Statistic*, Kiev, **25** (1981), 149–153 (in Russian).

[7] Valeev, K., Karelova, O., and Gorelov, V., *The Optimization of Linear Systems with Random Coefficients*, Moscow, 1996 (in Russian).

Time-Variant Consensus Formation in Higher Dimensions

ULRICH KRAUSE
Department of Mathematics, University of Bremen
Bremen, Germany
E-mail: krause@math.uni-bremen.de

Abstract The classical result by De Groot on consensus formation for one-dimensional opinions and fixed coefficients of confidence is extended to a discrete dynamical system with opinions of any finite dimension and a confidence pattern that may depend on time.

Keywords Discrete dynamical system, opinions, consensus
AMS Subject Classification Primary 39A11; Secondary 92H30

1 Introduction

This paper extends the classical result of De Groot [2] on consensus formation among experts to time-variant interaction of the experts and to multi-dimensional assessment by the experts. A first attempt in this direction, which also treats bounded confidence but which deals with one dimension only, was made in Krause [6].

Section 2 gives a short presentation of the classical model of consensus formation.

Section 3 presents the main result on time-variant consensus formation in higher dimensions.

Section 4 illustrates the main theorem by an example where each agent interacts only with his next neighbors.

2 Classical consensus formation

Consider a group of experts who have to make a joint assessment of a certain magnitude, say, the overall wheat production on earth in the year 2020. Each of the experts has his own assessment but is open to learn from the experience and methods of his colleagues. The learning process may lead to revisions and the question arises if this iterative process of opinion formation will tend to a consensus among the experts concerning the value of the magnitude.

Let $x_i(t)$ be the assessment made by expert, individual, **agent** $i \in \{1, \ldots, n\}$ at time $t \in \mathbb{N}_0 = \{0, 1, 2, \ldots\}$. In the classical model it is assumed that $x_i(t) \in \mathbb{R}_+$, $\mathbb{R}_+$ being the set of nonnegative real numbers.

Furthermore, it is assumed that agent i forms his assessment for the next period $t+1$ by taking a weighted arithmetic mean over the assessments of all agents — including himself — made in period t. Thus, a discrete dynamical system is defined by

$$x_i(t+1) = a_{i1}x_1(t) + \cdots + a_{in}x_n(t) \tag{1}$$

for each $i \in \{1, \ldots, n\}$, each $t \in \mathbb{N}_0$ and numbers $a_{ij} \geq 0$ with $\sum_{j=1}^{n} a_{ij} = 1$. Number a_{ij} can be interpreted as the extent to which agent i trusts agent j. The matrix $A = (a_{ij})_{1 \leq i,j \leq n}$ is a row stochastic matrix and with $x(t)$ the column vector of the $x_i(t)$, system (1) may be written as

$$x(t+1) = Ax(t), \ t \in \mathbb{N}_0, x(0) \in \mathbb{R}^n_+ \text{ given.} \tag{2}$$

This system has the **consensus property** if for every initial pattern $x(0)$ there is a **consensus** $c = c(x(0)) \in \mathbb{R}_+$, which depends on $x(0)$ such that

$$\lim_{t \to \infty} x_i(t) = c \text{ for } \mathbf{all} \ i \in \{1, \ldots, n\}.$$

The consensus property may be looked at as a property of global **joint** stability. Since the system (2) is linear and autonomous it is not difficult to specify conditions under which the system shows the consensus property. For this the methods of the Perron–Frobenius theory or Markov chain theory are available. For example the **Basic Limit Theorem** for regular Markov chains yields that

$$\lim_{t \to \infty} A^t = B = (b_{ij}) \text{ with } 0 \leq b_{ij} = b_{kj} \text{ for all } i, j, k \text{ and } \sum_{j=1}^{n} b_{ij} = 1 \tag{3}$$

for A regular, i.e., some power A^k is strictly positive (see Luenberger [9]). Under these assumptions, therefore,

$$\lim_{t \to \infty} x_i(t) = \sum_{j=1}^{n} b_{ij} x_j(0) \tag{4}$$

and the consensus property holds with consensus $c(x(0)) = \sum_{j=1}^{n} b_{ij} x_j(0)$, which value is the same for all $i \in \{1, \ldots, n\}$. (See De Groot [2].)

The above is also true if the regularity of A is replaced by the following assumption:

There exists k such that A^k has a strictly positive column (*).

Obviously, each regular matrix satisfies (*) but the reverse does not hold as can be seen from example $A = \begin{bmatrix} 1 & 0 \\ \frac{1}{2} & \frac{1}{2} \end{bmatrix}$.

3 A theorem on global joint stability for time-variant systems in higher dimensions

Consider the following extension of the classical system (1)

$$x^i(t+1) = a_{i1}(t)x^1(t) + \cdots + a_{in}(t)x^n(t), \ t \in \mathbb{N}_0, \tag{5}$$

where for each $i \in \{1, \ldots, n\}$

$$x^i(t) \in \mathbb{R}^d_+ \text{ with components } x^i_j(t), \ 1 \leq j \leq d$$

$$a_{ij}(t) \geq 0 \text{ with } \sum_{j=1}^{n} a_{ij}(t) = 1.$$

System (5) is said to possess **global joint stability** if for every given initial conditions $x^1(0), \ldots, x^n(0) \in \mathbb{R}^d_+$ there exists c in the convex hull $\text{conv}\{x^1(0), \ldots, x^n(0)\}$ such that

$$\lim_{t \to \infty} x^i(t) = c \text{ for } \textbf{all } 1 \leq i \leq n, \tag{6}$$

where, in general, $c = c(x^1(0), \ldots, x^n(0))$ will depend on the initial conditions. In what follows let $A(t)$ be the matrix of the $a_{ij}(t)$ and for $s, t \in \mathbb{N}_0, \ s < t$, let $B(t,s) = (b_{ij}(t,s))$ be the **accumulation matrix**

$$A(t-1)A(t-2)\ldots A(s).$$

Theorem 3.1. *Suppose there exists a sequence $0 = t_0 < t_1 < \cdots$ in $\mathbb{N}_0$ and a sequence $\delta_1, \delta_2, \ldots$ in $[0,1]$ with $\sum_{m=1}^{\infty} \delta_m = \infty$ such that for all $m \in \mathbb{N}$, all $i, j \in \{1, \ldots, n\}$*

$$\sum_{k=1}^{n} \min\{b_{ik}(t_m, t_{m-1}), \ b_{jk}(t_m, t_{m-1})\} \geq \delta_m. \tag{7}$$

Then for the system given by (5) there holds global joint stability and the following sensitivity property:

$$\| \ c(x^1(0), \ldots, x^n(0)) - c(y^1(0), \ldots, y^n(0)) \ \| \leq \max_{1 \leq i,j \leq n} \| \ x^i(0) - y^j(0) \ \|$$

for any norm $\| \cdot \|$ on $\mathbb{R}^d$.

Proof. For $s < t$ it holds by definition of the accumulation matrix that $x^i(t) = \sum_{j=1}^{n} b_{ij}(t,s)x^j(s)$. Let $M(t) = \text{conv}\{x^1(t), \ldots, x^n(t)\}$ for $t \in \mathbb{N}_0$.

1. In the first step we show that the diameter $\Delta M(t) = \sup\{\|u(t)-v(t)\| \mid u(t), v(t) \in M(t)\}$ is shrinking for t increasing, namely for $s < t$,

$$\Delta M(t) \leq (1 - r(t,s))\Delta M(s) \tag{8}$$

provided that $\sum_{k=1}^{n} \min\{b_{ik}(t,s),\ b_{jk}(t,s)\} \geq r(t,s)$ for all i, j. To prove (8), let $u(t) = \sum_{j=1}^{n} \alpha_j x^j(t),\ v(t) = \sum_{j=1}^{n} \beta_j x^j(t)$ from $M(t)$ with $0 \leq \alpha_j, \beta_j$ and $\sum_{j=1}^{n} \alpha_j = \sum_{j=1}^{n} \beta_j = 1$. It follows that

$$\| u(t) - v(t) \| \leq \max_{1 \leq i,j \leq n} \| x^i(t) - x^j(t) \| . \tag{9}$$

By definition

$$\| x^i(t) - x^j(t) \| = \| \sum_{k=1}^{n} b_{ik}(t,s) x^k(s) - \sum_{k=1}^{n} b_{jk}(t,s) x^k(s) \| .$$

Let

$$\alpha_{ik} = b_{ik}(t,s) - \min\{b_{ik}(t,s), b_{jk}(t,s)\} \geq 0$$
$$\beta_{jk} = b_{jk}(t,s) - \min\{b_{ik}(t,s), b_{jk}(t,s)\} \geq 0.$$

Obviously,

$$\sum_k \alpha_{ik} = \sum_k \beta_{jk} = 1 - \sum_k \min\{b_{ik}(t,s), b_{jk}(t,s)\} \leq 1 - r(t,s).$$

From
$\| x^i(t) - x^j(t) \| = \| \sum_k \alpha_{ik} x^k(s) - \sum_k \beta_{jk} x^k(s) \|$ it follows, similarly as the proof for (9), that

$$\| x^i(t) - x^j(t) \| \leq (1 - r(t,s)) \max_{1 \leq h,k \leq n} \| x^h(s) - x^k(s) \| .$$

Together with (9) we obtain

$$\| u(t) - v(t) \| \leq (1 - r(t,s))\Delta M(s).$$

This proves inequality (8).

2. Next we apply (8) to $s = t_{m-1},\ t = t_m$ to get by assumption (7)

$$\Delta M(t_m) \leq (1 - \delta_m)\Delta M(t_{m-1}). \tag{10}$$

By the mean value theorem $1 - \delta_m \leq e^{-\delta_m}$ and by iterating (10) we obtain

$$\Delta M(t_m) \leq e^{-(\delta_1 + \cdots + \delta_m)} \Delta M(0).$$

From this and assumption $\sum\limits_{m=1}^{\infty} \delta_m = \infty$ we get

$$\lim_{m\to\infty} \Delta M(t_m) = 0.$$

Furthermore, $x^i(t+1) \in M(t)$ for all i implies that $M(t+1) \subset M(t)$, and, hence, $\lim\limits_{t\to\infty} \Delta M(t) = 0$. Since $M(t)$ is compact there exists $c \in \bigcap\limits_{t=0}^{\infty} M(t)$ for which we must have that $\| c - x^i(t) \| \leq \Delta M(t)$ for all i and all t. Therefore, $\lim\limits_{t\to\infty} x^i(t) = c$ for all i. This proves the global joint stability. For different initial conditions one has that $c(x^1(0),\ldots,x^n(0))$ and $c(y^1(0),\ldots,y^n(0))$ are convex combinations of the respective initial conditions. This yields the sensitivity property by an argument similar to the one leading to inequality (9).

□

Remark 3.2. The next remarks complement the main result of this section:

1. The inequality (8) in part 1 of the proof for the theorem can be considered a higher-dimensional extension of Seneta [10, Theorem 3.1].

2. For the theorem in one dimension see also Chatterjee and Seneta [1] and Krause [6], and in higher dimensions Krause [7].

3. For further results in the area of opinion dynamics see Dittmer [3], Hegselmann and Krause [4] and [5], and Lorenz [8].

4 An example: Next neighbor interaction

Consider a structure of communication among the agents where in each time period every agent takes only himself and his immediate neighbors into account. Thus, for $t \in \mathbb{N}_0$ one has $a_{ij}(t) > 0$ for $2 \leq i \leq n-1$ and $i-1 \leq j \leq i+1$, and $a_{11}(t) > 0,\ a_{12}(t) > 0,\ a_{n,n-1}(t) > 0,\ a_{n,n}(t) > 0$.

Suppose further that for the smallest strictly positive entry of $A(t)$ one has that

$$\mu(A(t)) \geq t^{-\frac{1}{n-1}} \tag{11}$$

for all $t \geq t^*$, where $t^* \in \mathbb{N}$.

Choose for the theorem of the previous section $t_m = (n-1)m$ and $\delta_m = t_m^{-1}$. It is not difficult to show (see Dittmer [3]) that for the accumulation matrix B one has that

$$\begin{aligned} b_{ij}(t_m, t_{m-1} &\geq (\min\{\mu(A(t)) \mid t_{m-1} \leq t \leq t_m\})^{n-1} \\ &\geq \tfrac{1}{t_m} = \delta_m \text{ for } t_{m-1} \geq t^*. \end{aligned}$$

Obviously, $\sum_{m=1}^{\infty} \delta_m = \infty$ and condition (7) of the theorem is satisfied. Thus, global joint stability holds for this example. Consider the special case of two agents, that is $n = 2$. The above condition (11)means that $\mu(A(t)) \geq \frac{1}{t}$ for t big enough. The theorem does not apply if for the above communication structure one has only that $\mu(A(t)) \geq \frac{1}{t^2}$ for t big enough. Actually, in this case the system is not globally joint stable as can be seen from the example

$$A(t) = \begin{bmatrix} 1 & 0 \\ \frac{1}{t^2} & 1 - \frac{1}{t^2} \end{bmatrix} \quad \text{for} \quad t \geq 2.$$

Actually, it is not difficult to see that

$$A(t) \cdots A(2) = \begin{bmatrix} 1 & 0 \\ 1 - \prod_{i=2}^{t} (1 - \frac{1}{i^2}) & \prod_{i=2}^{t} (1 - \frac{1}{i^2}) \end{bmatrix}.$$

and, hence $\lim_{t \to \infty} B(t, 2) = \begin{bmatrix} 1 & 0 \\ \frac{1}{2} & \frac{1}{2} \end{bmatrix}$.

This shows that there will be no global joint stability except for the case that $x_1(0) = x_2(0)$. In other words, there will be never a consensus formation except for the trivial case that there was a consensus already in the beginning. This may be interpreted as a "too-fast hardening of positions" in that the weight of trust of $\frac{1}{t^2}$, which agent 2 gives to 1, decreases too fast to zero. Roughly speaking, the conditions of the theorem in the previous section just mean that the admitted hardening of positions should not be too fast.

References

[1] Chatterjee, S. and Seneta, E., Toward consensus: some convergence theorems on reperated averaging. *J. Applied Probability*, **14** (1977), 89–97.

[2] De Groot, M.H., Reaching a consensus. *J. Amer. Statist.*, **69** (1974) 118–121.

[3] Dittmer, J.C., Consensus formation under bounded confidence. *Nonlinear Analysis*, **47** (2001) 4615–4621.

[4] Hegselmann, R. and Krause, U., Opinion dynamics and bounded confidence: models, analysis and simulation. *J. Artificial Societies and Social Simulation*, **5(3)** (2002). http://jasss.soc.surrey.ac.uk/5/3/2.html.

[5] Hegselmann, R. and Krause, U., Collective dynamics of interacting agents when driven by PAM. Paper presented at the conference *Complexity 2003*, Aix-en-Provence, 2003.http://ideas.repec.org/s/sce/cplx03.html.

[6] Krause, U., A discrete nonlinear and non–autonomous model of consensus formation. In S. Elaydi, G. Ladas, J. Popenda, and J. Rakowski, editors, *Communications in Difference Equations*, 227–236. Gordon and Breach Science Publ., Amsterdam, 2000.

[7] Krause, U., Positive particle interaction. In L. Benvenuti, A. De Santis, L. Farina, editors, *Positive Systems*, 199–206. Springer, Berlin, 2003.

[8] Lorenz, J., Multidimensional opinion dynamics when confidence changes. Paper presented at the conference *Complexity 2003*, Aix-en-Provence, 2003.

[9] Luenberger, D.G., *Introduction to Dynamic Systems. Theory, Models, and Applications.* Wiley & Sons, New York, 1979.

[10] Seneta, E., *Non-negative Matrices and Markov Chains*, 2nd edition. Springer, New York, 1980.

Asymptotic Properties of Solutions of the Difference Equation $\Delta x(t) = -ax(t) + bx(\tau(t))$

PETR KUNDRÁT[1]

Institute of Mathematics, Faculty of Mechanical Engineering
Brno University of Technology
Brno, Czech Republic

Abstract In this paper, we derive the asymptotic bounds of all solutions of the delay difference equation

$$\Delta x(t) = -ax(t) + bx(\tau(t)), \quad t \in [t_0, \infty)$$

with real constants $a > 0$, $b \neq 0$. This equation is obtained via the discretization of a delay differential equation and we show the resemblance in the asymptotic bounds of both equations.

Keywords Difference equation, delayed argument, asymptotic behavior

AMS Subject Classification 39A11, 34K25

1 Introduction

In this paper, we discuss the asymptotic properties of the difference equation

$$\Delta x(t) = -ax(t) + bx(\tau(t)), \quad t \in I := [t_0, \infty), \tag{1}$$

where $a > 0$, $b \neq 0$ are real constants and $\tau(t)$ is a continuous delayed argument fulfilling some additional requirements.

Equation (1) is a discretization of the delay differential equation

$$\dot{z}(t) = -az(t) + bz(\tau(t)), \quad t \in I. \tag{2}$$

One of the important theoretical questions about this discretization is whether the solution $x(t)$ of (1) can approach the long-time dynamical behavior of solution $z(t)$ of (2). In this paper we pay attention to the case when the lag $t - \tau(t)$ is unbounded as $t \to \infty$. Particularly, if $\tau(t) = \lambda t, 0 < \lambda < 1$, then

[1]Supported by grant # 201/01/0079 of the Czech Grant Agency.

(2) becomes the so-called pantograph equation describing a motion of the pantograph head along a troley wire (see Ockendon and Tayler [10]). The pantograph equation and its modifications have been the subject of many qualitative investigations (see, e.g., Iserles [4], Liu [6], or Makay and Terjéki [8]). Similar questions have also been solved for the discrete pantograph equation (e.g., in Iserles [5], Liu [7], Péics [11]), where the stability problems have been discussed. Our aim is to show that specific asymptotic bounds of solutions valid in the continuous case also remain preserved (under certain restrictions) in the corresponding discrete case. Note that similar questions have also been discussed for other types of differential equations, e.g., the questions on the existence of the asymptotic equilibrium of nonlinear delay differential and difference equations have been discussed in Györi and Pituk [2]. The related asymptotics of differential equations and its discretizations is described also for equations without deviations (see, e.g., Mickens [9]).

2 Preliminaries

Throughout this paper we assume (except for Proposition 2.1, where some additional requirements are involved) that $\tau(t)$ is an increasing continuous function on I such that $\tau(t+h)-\tau(t)$ is a nonincreasing function for any real $0 < h \leq 1$ on I. Then we can put $t_{-1} = \tau(t_0)$ and $t_k = \tau^{-k}(t_0)$, $k = 1, 2, \ldots$, where τ^{-k} means the kth iteration of the inverse τ^{-1}. If we set $I_m := [t_{m-1}, t_m]$ for all $m = 0, 1, 2, \ldots$, then τ is mapping I_{m+1} onto I_m.

Now we recall the following asymptotic result on the delay differential equation (2) first derived in Heard [3, Theorem 3.1].

Proposition 2.1. *Let $a > 0$, $b \neq 0$ be scalars, $\tau \in C^2([t_0, \infty))$ be such that $\dot{\tau}$ is positive and decreasing on $[t_0, \infty)$ and $\lambda = \dot{\tau}(t_0) < 1$. Then for any solution z of (2) there exists a continuous periodic function g of period $\log \lambda^{-1}$ such that*

$$z(t) = (\varphi(t))^\alpha g(\log \varphi(t)) + O\left((\varphi(t))^{\alpha_r - 1}\right) \qquad as \quad t \to \infty, \tag{1}$$

where φ is a solution of

$$\varphi(\tau(t)) = \lambda \varphi(t), \qquad t \in I \tag{2}$$

given by

$$\varphi(t) = \lambda^{-n} \varphi_0(\tau^n(t)), \qquad t_{n-1} \leq t \leq t_n,\ n = 0, 1, 2, \ldots,$$

$\alpha = \log(b/a)/\log \lambda^{-1}$ and $\alpha_r = \operatorname{Re} \alpha$.

Note that formula (1) has been proved under weaker assumptions on τ in Čermák [1]. It is easy to verify that this formula particularly yields that $z(t) = O(\rho(t))$ as $t \to \infty$, where $\rho(t)$ is a solution of the functional equation

$$0 = -a\rho(t) + |b|\rho(\tau(t)), \qquad a > 0, \quad b \neq 0, \quad t \in I. \tag{3}$$

Our goal is to derive that this estimate is also valid for the corresponding discrete case.

First we show that under the above assumptions on τ there exists a solution $\rho(t)$ of (3) having certain properties.

Proposition 2.2. *Consider Equation (3), where $a > 0, b \neq 0$ are reals and let all the assumptions on τ be fulfilled.*

(i) *If $|b|/a \geq 1$, then there exists a positive continuous nondecreasing solution $\rho(t)$ of Equation (3).*

(ii) *If $|b|/a < 1$, then there exists a positive continuous decreasing solution $\rho(t)$ of Equation (3) such that $\rho(t+h) - \rho(t)$ is nondecreasing on I for any real $0 < h \leq 1$.*

Proof. Using the step method we can easily verify that there exists a positive continuous solution $\rho(t)$ of (3), which is nondecreasing or decreasing according to $|b|/a \geq 1$ or $|b|/a < 1$, respectively. Indeed, let $\hat{t} > t_0$ and let $\rho_0(t)$ be a continuous function defined on $[\tau(\hat{t}), \hat{t}]$ and having the required properties (i.e., it is a positive nondecreasing or positive decreasing function). If $\rho_0(\hat{t}) = (|b|/a)\rho_0(\tau(\hat{t}))$, then the formula

$$\rho(t) = (|b|/a)^n \rho_0(\tau^n(t)), \quad \tau^{-n+1}(\hat{t}) \leq t \leq \tau^{-n}(\hat{t}), \quad n = 0, 1, 2, \ldots$$

defines a unique solution of (3) with the required property on the interval $[\tau(\hat{t}), \infty)$ such that $\rho(t) = \rho_0(t)$ for any $t \in [\tau(\hat{t}), \hat{t}]$.

Further, we assume that $|b|/a < 1$ and show that the function $\rho(t+h) - \rho(t)$ is nondecreasing on I for all $0 < h \leq 1$. We choose the decreasing function ρ defined on the initial interval I_0 such that $\rho(t_0) = (|b|/a)\rho(t_{-1})$ and let $\rho(t+h) - \rho(t)$ be nondecreasing on I_0. Further let $t^*, t^{**} \in I_1$, $t^* < t^{**}$. If we denote $h^* := \tau(t^* + h) - \tau(t^*)$, $h^{**} := \tau(t^{**} + h) - \tau(t^{**})$, then $h^* \geq h^{**}$ and we can write

$$\begin{aligned}
&\rho(t^* + h) - \rho(t^*) = \\
&= \frac{|b|}{a}\Big(\rho(\tau(t^* + h)) - \rho(\tau(t^*))\Big) = \frac{|b|}{a}\Big(\rho(\tau(t^*) + h^*) - \rho(\tau(t^*))\Big) \\
&\leq \frac{|b|}{a}\Big(\rho(\tau(t^{**}) + h^*) - \rho(\tau(t^{**}))\Big) \leq \frac{|b|}{a}\Big(\rho(\tau(t^{**}) + h^{**}) - \rho(\tau(t^{**}))\Big) \\
&= \frac{|b|}{a}\Big(\rho(\tau(t^{**} + h)) - \rho(\tau(t^{**}))\Big) = \rho(t^{**} + h) - \rho(t^{**})
\end{aligned}$$

by use of the assumptions of proposition. Thus $\rho(t+h) - \rho(t)$ is nondecreasing on I_1 and repeating this procedure for intervals $I_2, I_3, \ldots$, we obtain that the function $\rho(t+h) - \rho(t)$ is nondecreasing on I. □

3 Main result

Theorem 3.1. *Let $x(t)$ be a solution of Equation (1), where $0 < a < 1$, $b \neq 0$ are reals and let all the assumptions put on τ be fulfilled. Let $\rho(t)$ be a positive continuous solution of Equation (3) with the properties guaranteed by Proposition 2.2.*

(i) *If $|b|/a \geq 1$, then $x(t) = O(\rho(t))$ as $t \to \infty$.*

(ii) *If $|b|/a < 1$ and moreover $\sum_{k=1}^{\infty} \frac{-\Delta\rho(t_k - 1)}{\rho(t_{k+1})} < \infty$, then $x(t) = O(\rho(t))$ as $t \to \infty$.*

Proof. First we rewrite the difference equation (1) as

$$x(t+1) = \tilde{a}x(t) + bx(\tau(t)), \quad t \in I, \tag{4}$$

where $\tilde{a} := 1 - a$, i.e., $0 < \tilde{a} < 1$.

We introduce the substitution $y(t) = x(t)/\rho(t)$ in (4) to obtain

$$\rho(t+1)y(t+1) = \tilde{a}\rho(t)y(t) + b\rho(\tau(t))y(\tau(t)), \quad t \in I \tag{5}$$

and show that every solution $y(t)$ of (5) is bounded as $t \to \infty$. Multiplying the previous equality by $1/\tilde{a}^{t+1}$ we get

$$\frac{\rho(t+1)y(t+1)}{\tilde{a}^{t+1}} = \frac{\rho(t)y(t)}{\tilde{a}^t} + \frac{b}{\tilde{a}^{t+1}}\rho(\tau(t))y(\tau(t)),$$

i.e.,

$$\Delta\left(\frac{\rho(t)y(t)}{\tilde{a}^t}\right) = \frac{b}{\tilde{a}^{t+1}}\rho(\tau(t))y(\tau(t)). \tag{6}$$

Now we take any $\bar{t} \in I_{m+1}$, $m = 1, 2, \ldots$. We define nonnegative integers $k_m(\bar{t}) := [\bar{t} - t_m]$, where $[\,]$ is an integer part and denote $\bar{t}_m := \bar{t} - k_m(\bar{t}) - 1$. Summing Equation (6) from $\bar{t}_m$ to $\bar{t} - 1$, we get

$$y(\bar{t}) = \frac{\rho(\bar{t}_m)\tilde{a}^{\bar{t}}}{\rho(\bar{t})\tilde{a}^{\bar{t}_m}}y(\bar{t}_m) + \frac{\tilde{a}^{\bar{t}}}{\rho(\bar{t})}\sum_{s=\bar{t}_m}^{\bar{t}-1}\frac{b}{\tilde{a}^{s+1}}\rho(\tau(s))y(\tau(s)).$$

Let us denote $M_m := \sup\left\{|y(t)|, \quad t \in \bigcup_{j=0}^{m} I_j\right\}$. In accordance with (3) we obtain

$$|y(\bar{t})| \leq \frac{\rho(\bar{t}_m)\tilde{a}^{\bar{t}}}{\rho(\bar{t})\tilde{a}^{\bar{t}_m}}M_m + \frac{\tilde{a}^{\bar{t}}}{\rho(\bar{t})}\sum_{s=\bar{t}_m}^{\bar{t}-1}\frac{(1-\tilde{a})\rho(s)}{\tilde{a}^{s+1}}M_m.$$

Using the relation $\frac{(1-\tilde{a})}{\tilde{a}^{s+1}} = \Delta\left(\frac{1}{\tilde{a}}\right)^s$ we get

$$|y(\bar{t})| \leq \frac{\rho(\bar{t}_m)\tilde{a}^{\bar{t}}}{\rho(\bar{t})\tilde{a}^{\bar{t}_m}}M_m + \frac{\tilde{a}^{\bar{t}}}{\rho(\bar{t})}\sum_{s=\bar{t}_m}^{\bar{t}-1}\left(\Delta\left(\frac{1}{\tilde{a}}\right)^s\right)\rho(s)M_m$$

and summing by parts we finally have

$$\begin{aligned}|y(\bar{t})| &\leq M_m \left\{ \frac{\rho(\bar{t}_m)\tilde{a}^{\bar{t}}}{\rho(\bar{t})\tilde{a}^{\bar{t}_m}} + \frac{\tilde{a}^{\bar{t}}}{\rho(\bar{t})} \left(\frac{\rho(\bar{t})}{\tilde{a}^{\bar{t}}} - \frac{\rho(\bar{t}_m)}{\tilde{a}^{\bar{t}_m}} - \sum_{s=\bar{t}_m}^{\bar{t}-1} \left(\frac{1}{\tilde{a}}\right)^{s+1} \Delta\rho(s) \right) \right\} \\ &= M_m \left\{ 1 - \frac{\tilde{a}^{\bar{t}}}{\rho(\bar{t})} \sum_{s=\bar{t}_m}^{\bar{t}-1} \left(\frac{1}{\tilde{a}}\right)^{s+1} \Delta\rho(s) \right\}. \end{aligned} \tag{7}$$

The common part of the proof ends here and we continue for the cases (i) and (ii) separately.

ad (i): If $|b|/a \geq 1$, then in accordance with Proposition 2.2 we choose a nondecreasing function $\rho(t)$ on I. Then $\Delta\rho(t)$ is nonnegative on I, hence $|y(\bar{t})| \leq M_m$. Since $\bar{t} \in I_{m+1}$ was arbitrary, we have $M_{m+1} \leq M_m$, i.e., M_m is bounded as $m \to \infty$. Hence the function $y(t)$ is bounded and the statement (i) is proved.

ad (ii): If $|b|/a < 1$, then in accordance with Proposition 2.2 we choose a decreasing function $\rho(t)$ on I such that $\Delta\rho(t)$ is nondecreasing on I. Then from (7) we have

$$\begin{aligned}|y(\bar{t})| &\leq M_m \left\{ 1 + \frac{\rho(\bar{t}_m) - \rho(\bar{t}_m + 1)}{\rho(\bar{t})} \sum_{s=\bar{t}_m}^{\bar{t}-1} \left(\frac{\tilde{a}^{\bar{t}}}{\tilde{a}^{s+1}} \right) \right\} \\ &\leq M_m \left\{ 1 + \frac{\rho(\bar{t}_m) - \rho(\bar{t}_m + 1)}{\rho(\bar{t})} \alpha \right\} \leq M_m \left\{ 1 + \alpha \frac{-\Delta\rho(t_m - 1)}{\rho(t_{m+1})} \right\}, \end{aligned}$$

where $\alpha := 1/(1 - \tilde{a})$. The repeated application of this procedure yields

$$|y(\bar{t})| \leq M_1 \prod_{j=1}^{m} \left(1 + \alpha \frac{-\Delta\rho(t_j - 1)}{\rho(t_{j+1})} \right),$$

i.e.,

$$M_{m+1} \leq M_1 \prod_{j=1}^{m} \left(1 + \alpha \frac{-\Delta\rho(t_j - 1)}{\rho(t_{j+1})} \right).$$

By the assumption, the product converges as $m \to \infty$; hence M_m is bounded as $m \to \infty$. The theorem is proved. □

4 Examples and remarks

In this section we illustrate the conclusions of Theorem 3.1 for some particular cases of Equation (1).

Example 4.1. *Consider the discrete pantograph equation*

$$\Delta x(t) = -ax(t) + bx(\lambda t), \qquad t \geq t_0 > 0, \tag{8}$$

where $0 < a < 1$, $b \neq 0$, $0 < \lambda < 1$ *are real constants. It is easy to verify that the solution of the appropriate equation (3) is*

$$\rho(t) = t^r, \qquad r = \frac{\log \frac{|b|}{a}}{\log \lambda^{-1}}.$$

Further, $t_k = \lambda^{-k} t_0$, $k = -1, 0, 1, \ldots$ *Then we get*

$$\begin{aligned} \frac{-\Delta\rho(t_k)}{\rho(t_{k+1})} &= \frac{(\lambda^{-k}t_0 - 1)^r - \lambda^{-kr}t_0^r}{\lambda^{-r(k+1)}t_0^r} = \frac{\left(\frac{\lambda^{-k}t_0 - 1}{\lambda^{-k}t_0}\right)^r - 1}{\lambda^{-r}} \\ &= \frac{\left(1 - \frac{1}{\lambda^{-k}t_0}\right)^r - 1}{\lambda^{-r}} = O(\lambda^k) \qquad \text{as } k \to \infty \end{aligned}$$

by use of the binomial formula. Hence all the assumptions of Theorem 3.1 are fulfilled and the estimate

$$x(t) = O(t^r), \quad r = \frac{\log \frac{|b|}{a}}{\log \lambda^{-1}} \qquad \text{as } t \to \infty$$

holds for any solution x *of (8) regardless of the values* $0 < a < 1$, $b \neq 0$.

Remark 4.2. The discrete pantograph equation has been investigated also in Liu [7]. There has been used the convergent numerical discretization

$$x_{n+1} = (1 - ah)x_n + bhx_{[\lambda n]}, \qquad n = 0, 1, \ldots$$

of the pantograph equation, where $x(0) = 1$, x_n is an approximation of $x(t)$ at $t = nh$, $n = 1, 2, \ldots$, h is the stepsize. The author discusses some special numerical problems concerning the relationship between asymptotic behavior of the pantograph equation and the above mentioned discretization.

Example 4.3. Consider the difference equation

$$\Delta x(t) = -ax(t) + bx(t^\gamma), \qquad t \geq t_0 > 1, \tag{9}$$

where $0 < a < 1$, $b \neq 0$, $0 < \gamma < 1$ are real constants. It is easy to verify that the solution of the appropriate equation (3) is

$$\rho(t) = (\log t)^r, \qquad r = \frac{\log \frac{|b|}{a}}{\log \gamma^{-1}}.$$

Further, $t_k = t_0^{1/\gamma^k}$, $k = -1, 0, 1, \dots$ and similarly as in Example 4.1 we get

$$\begin{aligned}
\frac{-\Delta\rho(t_k)}{\rho(t_{k+1})} &= \frac{(\log(t_0^{1/\gamma^k} - 1))^r - (\log(t_0^{1/\gamma^k}))^r}{(\log(t_0^{1/\gamma^{k+1}}))^r} \\
&= \frac{(\log(t_0^{1/\gamma^k} - 1))^r - (\frac{1}{\gamma})^{kr}(\log t_0)^r}{(\frac{1}{\gamma})^{(k+1)r}(\log t_0)^r} \\
&= \frac{\left(\frac{\log(t_0^{1/\gamma^k} - 1)}{\log t_0^{1/\gamma^k}}\right)^r - 1}{\left(\frac{1}{\gamma}\right)^r} \le \frac{\left(\frac{t_0^{1/\gamma^k} - 1}{t_0^{1/\gamma^k}}\right)^r - 1}{\left(\frac{1}{\gamma}\right)^r} \\
&= \gamma^r\left(\left(1 - \frac{1}{t_0^{1/\gamma^k}}\right)^r - 1\right) = O(t_0^{-k/\gamma}) \qquad \text{as } k \to \infty.
\end{aligned}$$

Hence the estimate

$$x(t) = O((\log t)^r), \quad r = \frac{\log \frac{|b|}{a}}{\log \gamma^{-1}} \qquad \text{as } t \to \infty$$

holds for any solution x of (9) regardless of the values $0 < a < 1$, $b \neq 0$.

Example 4.4. Consider the difference equation with a constant delay

$$\Delta x(t) = -\frac{1}{2}x(t) + \left(\frac{1}{e^2} - \frac{1}{2e}\right)x(t-1), \qquad t \ge t_0. \tag{10}$$

It is easy to verify that the solution of the appropriate equation (3) is

$$\rho(t) = \exp(rt), \qquad r = \log\left(\frac{1}{e} - \frac{2}{e^2}\right) < -1.$$

However, Equation (10) admits the solution $x(t) = \exp(-t)$, i.e., the estimate $x(t) = O(\rho(t))$ as $t \to \infty$ from Theorem 3.1 is not valid for this solution. This is a result of the fact that the series $\sum_{k=1}^{\infty} \frac{-\Delta\rho(t_k+1)}{\rho(t_{k+1})}$ mentioned in the assumptions of Theorem 3.1 diverges in this case.

Remark 4.5. The main result is valid also for equations of the type

$$x(t+h) - x(t) = -ahx(t) + bhx(\tau(t)), \qquad t \in I, \tag{11}$$

where $a > 0, b \neq 0, 0 < h \le 1$, which is the simplest numerical discretization of (2). Assume that all the assumptions imposed on τ are fulfilled. Then putting $h < 1/a$ we can ensure the validity of the estimate occurring in Theorem 3.1. In other words, Equation (2) and its discretization (11) admit the same asymptotic estimate of all solutions provided the stepsize h is chosen sufficiently small.

References

[1] Čermák, J., The asymptotic of solutions for a class of delay differential equations, *Rocky Mountain J. Math.* (in press).

[2] Györi, I. and Pituk, M., Comparison theorems and asymptotic equilibrium for delay differential and difference equations, *Dynam. Systems Appl.* **5** (1996), 277–302.

[3] Heard, M.L., A change of variables for functional differential equations, *J. Differential Equations* **18** (1975), 1–10.

[4] Iserles, A., On the generalized pantograph functional–differential equation, *Euro. J. Appl. Math.* **4** (1993), 1–38.

[5] Iserles, A., Exact and discretized stability of the pantograph equation, *Appl. Numer. Math.*, **24** (1997), 295–308.

[6] Liu, Y., Asymptotic behaviour of functional–differential equations with proportional time delays, *Euro. J. Appl. Math.* **7** (1996), 11–30.

[7] Liu, Y., Numerical investigation of the pantograph equation, *Appl. Numer. Math.* **24** (1997), 309–317.

[8] Makay, G. and Terjéki, J., On the asymptotic behaviour of the pantograph equations, *Electron. J. Qual. Theory Differ. Equ.***2** (1998), 1–12.

[9] Mickens, R.E., Asymptotic solutions to a discrete Airy equation, *J. Difference Equ. Appl.* **7** (2001), 851–858.

[10] Ockendon, J.R. and Tayler, A.B., The dynamics of a current collection system for an electric locomotive, *Proc. Roy. Soc. London Ser. A* **322** (1971), 447–468.

[11] Péics, H., On the asymptotic behaviour of difference equations with continuous arguments, *Ser A. Math. Anal.* **9** (2002), 257–273.

Oscillation Theorems for a Class of Fourth Order Nonlinear Difference Equations

MALGORZATA MIGDA, ANNA MUSIELAK, and EWA SCHMEIDEL
Institute of Mathematics
Poznań University of Technology
Poznań, Poland

Abstract We consider a class of fourth order nonlinear difference equations. Necessary and sufficient conditions for oscillation of each solution of considered equation are the main results of this paper.

Keywords Nonlinear difference equation, oscillatory solution, nonoscillatory solution, fourth order

AMS Subject Classification 39A10

1 Introduction

Consider the difference equation

$$\Delta(a_n\Delta(b_n\Delta(c_n\Delta y_n))) + f(n, y_{\delta_n}) = 0, \quad n \in \mathcal{N}, \tag{E}$$

where $\mathcal{N} = \{0, 1, 2, ...\}$, Δ is the forward difference operator defined by $\Delta y_n = y_{n+1} - y_n$, (a_n), (b_n) and (c_n) are sequences of positive real numbers, δ is an integer-valued function defined on $\mathcal{N}$ such that

$$\lim_{n\to\infty} \delta_n = \infty.$$

Function $f : \mathcal{N} \times \mathcal{R} \to \mathcal{R}$. By a solution of Equation (E) we mean a sequence (y_n), which is defined for $n \geq \delta_n$ and satisfies Equation (E) for n sufficiently large. We consider only such solutions that are nontrivial for all large n. A solution of Equation (E) is called oscillatory if its terms are not eventually positive or eventually negative. Otherwise it is called nonoscillatory. Equation (E) is called oscillatory if each solution (y_n) of this equation is oscillatory. Equations of the form (E) are conveniently classified according to the nonlinearity of $f(k,y)$ with respect to y. Equation (E) is said to be *superlinear* if, for each fixed integer k, $\frac{f(k,y)}{y}$ is nondecreasing in y for $y>0$ and nonincreasing in y for $y<0$. Equation (E) is called *strongly superlinear* if there is a number $\sigma > 1$ such that, for each fixed integer k, $\frac{f(k,y)}{|y|^{\sigma} sgn\ y}$ is nondecreasing in y for $y>0$ and nonincreasing in y for $y<0$. Equation (E) is called *sublinear* if, for each fixed integer k, $\frac{f(k,y)}{y}$ is nonincreasing in y for $y>0$ and nondecreasing in y for $y<0$. It is called *strongly sublinear* if there is a positive number $\eta < 1$ such that, for each fixed integer k, $\frac{f(k,y)}{|y|^{\eta} sgn\ y}$ is nonincreasing in y for $y>0$ and nondecreasing in y for $y<0$. Clearly, if Equation (E) is superlinear, then $f(\cdot, y)$ is nondecreasing on $(0,\infty)$.

In the last few years there has been an increasing interest in the study of oscillatory and asymptotic behavior of solutions of difference equations. Compared to second-order difference equations, the study of higher-order equations, and in particular fourth order equations (see for example [2]-[4], [6]-[18]) has recived considerably less attention. An important special case of fourth order difference equations is the discrete version of the Schrödinger equation.

The purpose of this paper is to establish some necessary and sufficient conditions for Equation (E) to be oscillatory.

Throughout the rest of our investigations, one or several of the following assumptions will be imposed:

$(H1)$ $\sum_{i=1}^{\infty} \frac{1}{a_i} = \sum_{i=1}^{\infty} \frac{1}{b_i} = \infty.$

$(H2)$ $yf(n,y) > 0$ for all $y \neq 0$ and $n \in \mathcal{N}$.

$(H3)$ Function $f(n,y)$ is continuous in the second argument, for each fixed $n \in \mathcal{N}$.

$(H4)$ Function $f(n,y)$ is decreasing in y, for each $n \in \mathcal{N}$.

$(H5)$ Sequence (a_n) is a nondecreasing sequence .

$(H6)$ Sequence (a_n) fulfill condition $a_1 \geq 1$.

$(H7)$ There exist constants K_1 and K_2, such that $0 < K_1 \leq c_n \leq K_2$, for each $n \in \mathcal{N}$.

It is clear that, from $(H7)$, we get $\sum_{i=1}^{\infty} \frac{1}{c_i} = \infty$.

We start with the following known lemma.

Lemma 1.1. *Assume that (H1), (H2) and (H7) hold. Let (y_n) be an eventually positive solution of Equation (E). Then exactly one of the following statements holds:*

$$\begin{array}{lllll} (i) & y_n > 0, & \Delta y_n > 0, & \Delta(c_n\Delta y_n) > 0, & \Delta(b_n\Delta(c_n\Delta y_n)) > 0 \\ (ii) & y_n > 0, & \Delta y_n > 0, & \Delta(c_n\Delta y_n) < 0, & \Delta(b_n\Delta(c_n\Delta y_n)) > 0 \end{array}$$

for all sufficiently large n.

Proof. The prove you can find in [7]. □

Now we introduce the notation

$$R_{n,N} = \sum_{i=N}^{n-1} \frac{1}{c_i} \sum_{j=N}^{i-1} \frac{j-N}{b_j},$$

$$Q_{n,N} = \sum_{k=N}^{n-1} \frac{1}{c_k} \sum_{j=N}^{k-1} \frac{1}{b_j} \sum_{i=N}^{j-1} \frac{1}{a_i}.$$

Notice that

$$Q_{n,N} \leq R_{n,N}, \tag{1}$$

and that $Q_{n,N}$ can be written in the form

$$Q_{n,N} = \sum_{i=N}^{n-1} \frac{1}{a_i} \sum_{j=i+1}^{n-1} \frac{1}{b_j} \sum_{k=j+1}^{n-1} \frac{1}{c_k}.$$

2 Oscillation theorems

In this section we are concerned with oscillation theorems for Equation (E). But first, we present the following lemmas.

Lemma 2.1. *Assume that Equation (E) is strongly sublinear. If (u_n), (v_n), (w_n) and (ν_n) are positive sequences such that for $n > N$, we have $0 < \mu_n \leq 1$, $u_n \leq kw_n$ for some positive number k, and*

$$u_n \geq \mu_n w_n \sum_{i=n}^{\infty} \nu_i f(i, u_i),$$

then

$$\sum_{i=N}^{\infty} \nu_i \mu_n f(i, kw_i) < \infty.$$

Proof. See proof of Lemma 2.4 in [18]. □

Lemma 2.2. *Assume that conditions $(H1)$, $(H2)$, and $(H7)$ hold. If (y_n) is an eventually positive solution of Equation (E), then there exist positive constant C_1 and C_2 and integer N such that*

$$C_1 \leq y_n \leq C_2 Q_{n,N} \tag{2}$$

for enough large n.

Proof. See proof of Lemma 3 in [7]. □

Let us denote

$$z_n = b_n \Delta(c_n \Delta y_n). \tag{3}$$

Lemma 2.3. *Assume that conditions (H1), (H2), (H5), (H6), and (H7) hold. Then every positive solution (y_n) of Equation (E) fulfill the inequality*

$$y_n > R_{n,N} \Delta z_n,$$

for enough large n.

Proof. Suppose first that (y_n) is a solution of Equation (E) that satisfies condition (i) from Lemma 1.1 and N is so large that (i) holds, for $n \geq N$. Then since $\Delta(a_n \Delta z_n) < 0$, $(a_{n+1}\Delta^2 z_n + \Delta a_n \Delta z_n < 0)$ and from $(H5)$ we get $\Delta^2 z_n < 0$. We see that

$$z_n \geq \sum_{i=N}^{n-1} \Delta z_i > \sum_{i=N}^{n-1} \Delta z_n = (n-N)\Delta z_n.$$

Thus

$$\Delta(c_n \Delta y_n) > \frac{n-N}{b_n} \Delta z_n,$$

and

$$\begin{aligned}
c_n \Delta y_n &> \textstyle\sum_{i=N}^{n-1} \frac{i-N}{b_i} \Delta z_i \\
\Delta y_n &> \textstyle\frac{1}{c_n} \sum_{i=N}^{n-1} \frac{i-N}{b_i} \Delta z_i \\
y_n &> \textstyle\sum_{j=N}^{n-1} \frac{1}{c_j} \sum_{i=N}^{j-1} \frac{i-N}{b_i} \Delta z_i \\
y_n &> R_{n,N} \Delta z_n.
\end{aligned}$$

Next we suppose that (y_n) is a solution of (E) and satisfies condition (ii) of Lemma 1.1. Let N be so large that condition (ii) holds for $n \geq N$. Multiplying $0 > \Delta^2 z_i$ by $R_{i+1,N}$ and summing the result equation from N to $n-1$, by $(H7)$, we obtain

$$
\begin{aligned}
0 &> \sum_{i=N}^{n-1} R_{i+1,N}\Delta^2 z_i = R_{i,N}\Delta z_i|_{i=N}^{n} - \sum_{i=N}^{n-1} \Delta R_{i,N}\Delta z_i \\
&= R_{n,N}\Delta z_n - \sum_{i=N}^{n-1}\left(\tfrac{1}{c_i}\sum_{j=N}^{i-1}\tfrac{j-N}{b_j}\right)\Delta z_i \\
&\geq R_{n,N}\Delta z_n - \tfrac{1}{K_1}\sum_{i=N}^{n-1}\left(\sum_{j=N}^{i-1}\tfrac{j-N}{b_j}\Delta z_i\right) \\
&= R_{n,N}\Delta z_n - \tfrac{1}{K_1}\left(\sum_{j=N}^{i-2}\tfrac{j-N}{b_j} z_i|_{i=N}^{n} - \sum_{i=N}^{n-1}\tfrac{i-N-1}{b_{i-1}} z_i\right) \\
&> R_{n,N}\Delta z_n - \tfrac{1}{K_1}\sum_{j=N}^{n-2}\tfrac{j-N}{b_j} z_n + \tfrac{1}{K_1}\sum_{i=N}^{n-1}\tfrac{i-N-1}{b_{i-1}} z_{i-1} \\
&> R_{n,N}\Delta z_n + \tfrac{1}{K_1}\sum_{i=N}^{n-2}(i-N)\Delta(c_i\Delta y_i) \\
&= R_{n,N}\Delta z_n + \tfrac{1}{K_1}\left[(i-N)c_i\Delta y_i|_{i=N}^{n-1} - \sum_{i=N}^{n-2} c_{(i+1)}\Delta y_{(i+1)}\right] \\
&\geq R_{n,N}\Delta z_n + \tfrac{1}{K_1}(n-N)c_n\Delta y_n - \tfrac{1}{K_1}\sum_{i=N}^{n-2} c_{(i+1)}\Delta y_{(i+1)} \\
&\geq R_{n,N}\Delta z_n - \tfrac{1}{K_1}\sum_{i=N}^{n-1} c_i\Delta y_i \geq R_{n,N}\Delta z_n - \tfrac{K_2}{K_1}\sum_{i=N}^{n-1}\Delta y_i \\
&= R_{n,N}\Delta z_n - \tfrac{K_2}{K_1}(y_n - y_N) = R_{n,N}\Delta z_n - \tfrac{K_2}{K_1}y_n + \tfrac{K_2}{K_1}y_N \\
&> R_{n,N}\Delta z_n - \tfrac{K_2}{K_1}y_n.
\end{aligned}
$$

So, we have

$$\frac{K_1}{K_2}R_{n,N}\Delta z_n < y_n, \text{ where } \frac{K_1}{K_2} \leq 1.$$

Hence

$$y_n > R_{n,N}\Delta z_n;$$

then we obtain the thesis of our lemma. □

Let

$$g_n = \max(n, \delta_n). \tag{4}$$

Theorem 2.4. *Assume that Equation (E) is strongly sublinear and conditions (H1), (H2), (H4), (H5), (H6), and (H7) hold. Suppose further that for all nonzero constants c,*

$$\sum_{n=1}^{\infty} \frac{Q_{n,N}}{a_n Q_{g_n,N}}|f(n, cQ_{g_n,N})| = \infty, \tag{5}$$

then every solution of Equation (E) is oscillatory.

Proof. Suppose to the contrary that (y_n) is an eventually positive solution of Equation (E) such that $y_n > 0$, $\Delta y_n > 0$ and $\Delta z_n > 0$, for $n > N$, where z_n is define by (3). From Equation (E) we obtain

$$\Delta(a_n \Delta z_n) = -f(n, y_{\delta_n}),$$

so,

$$\Delta z_n = \frac{1}{a_n} \sum_{i=n}^{\infty} f(i, y_{\delta_i}). \tag{6}$$

In view of Lemma 2.3, we have

$$y_n \geq R_{n,N} \Delta z_n.$$

From the above inequality and (1),

$$y_n \geq Q_{n,N} \Delta z_n,$$

and by definition of g_n, (6) and (H4) we get

$$y_{g_n} \geq y_n \geq \frac{1}{a_n} Q_{n,N} \sum_{i=n}^{\infty} f(i, y_{\delta_i}) \geq \frac{1}{a_n} Q_{n,N} \sum_{i=n}^{\infty} f(i, y_{g_i}). \tag{7}$$

In view of Lemma 1.1 and Lemma 2.2,

$$y_{g_n} \leq C_2 Q_{g_n,N},$$

for all large n.

By $(H5)$ and $(H6)$ we have

$$\frac{1}{a_n} \frac{Q_{n,N}}{Q_{g_n,N}} \leq 1.$$

We apply Lemma 2.1 to (7) with

$$u_n = y_{g_n}, w_n = Q_{g_n,N}, \mu_n = \frac{1}{a_n} \frac{Q_{n,N}}{Q_{g_n,N}} \text{ and } \nu_n = 1.$$

Then, we obtain

$$\sum_{n=1}^{\infty} \frac{Q_{n,N}}{a_n Q_{g_n,N}} |f(n, c Q_{g_n,N})| < \infty,$$

which is contrary to (5). Hence the thesis of this theorem holds. □

Let $\delta_n \leq n$, then $g_n = n$, and

$$\frac{Q_{n,N}}{Q_{g_n,N}} = 1.$$

By $(H5)$ and $(H6)$, condition (5) takes the form

$$\sum_{n=1}^{\infty} |f(n, cQ_{n,N})| = \infty.$$

Corollary 2.5. *Assume that the equation*

$$\Delta(a_n \Delta(b_n \Delta(c_n \Delta y_n))) + f(n, y_n) = 0, \quad n \in \mathcal{N} \tag{E1}$$

is strongly sublinear and conditions (H1), (H2), (H5), (H6), and (H7) hold. Suppose further that for all nonzero constants c,

$$\sum_{n=1}^{\infty} |f(n, cQ_{n,N})| = \infty. \tag{8}$$

Then every solution of Equation $(E1)$ *is oscillatory.*

Theorem 2.6. *Let conditions* $(H1)$*-*$(H7)$ *hold and Equation* $(E1)$ *be strongly sublinear. Then the necessary and sufficient condition for Equation* $(E1)$ *to be oscillatory is that condition* (8) *holds for all nonzero constants c.*

Proof. The proof follows from Corollary 1 and Theorem 3 from [7]. □

References

[1] R. P. Agarwal, *Difference Equations and Inequalities. Theory, Methods and Applications*, Marcel Dekker, Inc., New York, 1992.

[2] S.S. Cheng, On a class of fourth order linear recurrence equations, *Internat. J. Math. and Math. Sci.*, **7** (1984), 131–149.

[3] J.R. Graef, E. Thandapani, Oscillatory and asymptotic behavior of fourth order nonlinear delay difference equations, *Fasc. Math.*, **31** (2001), 23–36.

[4] J.W. Hooker, W.T. Patula, Growth and oscillation properties of solutions of a fourth order linear difference equation, *J. Aust. Math. Soc.* Ser. B **26** (1985), 310–328.

[5] W.G. Kelly, A.C. Peterson, *Difference Equations*, Academic Press, Inc., Boston-San Diego, 1991.

[6] B. Liu, J. Yan, Oscillatory and asymptotic behavior of fourth order nonlinear difference equations, *Acta Math. Sin.*, **13** (1997), 105–115.

[7] M. Migda, A. Musielak, E. Schmeidel, On a class of fourth order difference equations, *Advances in Difference Equations* (to appear).

[8] J. Popenda, E. Schmeidel, On the solution of fourth order difference equations, *Rocky Mountain J. Math.*, **25** (1995), 1485–1499.

[9] E. Schmeidel, Oscillation and nonoscillation theorems for fourth order difference equations, *Rocky Mountain J. Math.* (to appear).

[10] E. Schmeidel, Nonoscillation and oscillation properties for fourth order nonlinear difference equations, *Proceedings of the Sixth International Conference on Difference Equations*, 531–538, CRC, Boca Raton, FL, 2004.

[11] E. Schmeidel, Nonoscillation and oscillation theorems for a fourth order nonlinear difference equations, *Fields Institute Communications* (in press).

[12] E. Schmeidel, B. Szmanda, Oscillatory and asymptotic behavior of certain difference equation, *Nonlinear Anal.*, **47** (2001), 4731–4742.

[13] B. Smith, and W.E. Taylor, Oscillatory and asymptotic behavior of certain fourth order difference equations, *Rocky Mountain J. Math.*, **16** (1986), 403–406.

[14] B. Smith, and W.E. Taylor, Oscillation properties of fourth order linear difference equations, *Tamkang J. Math.*, **18** (1987), 89–95.

[15] W.E. Taylor, Oscillation properties of fourth order difference equations, *Port. Math.*, **45** (1988), 105–144.

[16] W.E. Taylor, and M. Sun, Oscillation properties of nonlinear difference equations, *Port. Math.*, **52** (1995), 15–24.

[17] E. Thandapani, and M. Arockiasamy, Oscillatory and assymptotic properties of solution of nonlinear fourth order difference equations, *Glas. Mat.*, **37** (2002), 119–131.

[18] B.G. Zhang, and S.S. Cheng, On a class of nonlinear difference equations, *J. Difference Equ. Appl.*, **1** (1995), 391–411.

Iterates of the Tangent Map – The Bifurcation Scheme

H. OLIVEIRA and J. SOUSA RAMOS
Department of Mathematics, Instituto Superior Técnico
Lisbon, Portugal

Abstract We study in this paper iterates of odd mappings in the real line with two infinite discontinuities and symmetric asymptotic values at plus and minus infinity. We study the bifurcations that occur and the bifurcation scheme when we change the parameter.

Keywords Symbolic dynamics, combinatorial dynamics, maps of the interval, tangent map, bifurcation
AMS Subject Classification 37B10, 37E05, 37E15, 37G35

1 Introduction and basic facts

In this paper we study iterates of the family $g_\beta(x) = -\beta \tan(\beta \tanh(x))$, for $\frac{\pi}{2} < \beta < \frac{3\pi}{2}$. It is a family of odd mappings with two infinite discontinuities and symmetric asymptotic values at plus and minus infinity, where $g_\beta(x) = -g_\beta(-x)$, $\lim_{x\to\pm\infty} g_\beta(x) = \pm d = \mp\beta\tan(\beta)$, $\lim_{x\to c^\pm} g_\beta(x) = \pm\infty$ and $\lim_{x\to -c^\pm} g_\beta(x) = \pm\infty$, where $c = \tanh^{-1}\left(\frac{\pi}{2\beta}\right)$. The maps of this family result from the restriction of the maps of the complex family $\lambda\tan(z)$ to the imaginary axis, with a real parameter $\lambda = i\beta$, considering the return map to the imaginary axis, or the even iterates of $\lambda\tan(z)$ when the previous restrictions are made [5]. In this paper we are concerned with the maximum number of stable orbits and the bifurcations that occur when we change the parameter.

The complex family of the tangent was studied by Devaney and Keen [1] and by Keen and Kotus [2]. In this paper we are not interested in the complex plane, but only with real one-dimensional dynamics. The family g_β has the schwarzian derivative

$$S(g_\beta(x)) = \frac{f_\beta'''(x)}{f_\beta'(x)} - \frac{3}{2}\left(\frac{f_\beta''(x)}{f_\beta'(x)}\right)^2 = 2\frac{\beta^2 - \cosh^4(x)}{\cosh^4(x)},$$

which does not have a fixed sign.

When $\beta > 1$, $Sg_\beta(x)$ is positive for $x \in \left]-\cosh^{-1}\left(\sqrt{\beta}\right), \cosh^{-1}\left(\sqrt{\beta}\right)\right[$, with maximum at the origin which is $2\left(\beta^2 - 1\right)$ and is nonpositive in the complement of the previous set in $\mathbb{R}$, with the horizontal asymptote $y = -2$. The family $\xi_\beta(x) = -\beta \tanh(\beta \tan(x))$ could be used instead of the family g_β because these families are topologically conjugated as we will see in the next section.

We define the orbit of a real point x_0 as a sequence of numbers $O(x_0) = \{x_j\}_{j=0,1,...}$ such that $x_j = g_\beta^j(x_0)$ where g_β^j is the composition of order j of g_β with itself. Any point x is periodic with period $n > 0$ if the condition $g_\beta^n(x) = x$ is fulfilled with n minimal.

Because g_β is odd (like ξ_β) the orbit of any point x is symmetric to the orbit of $-x$.

Consider the alphabet $\mathcal{A} = \{L, A, M, B, R\}$, the address $\mathtt{A}(x) : \mathcal{A} \mapsto \mathcal{A}$ of a real point x is defined such that $\mathtt{A}(x) = L$ if $x < -c$, $\mathtt{A}(x) = A$ if $x = -c$, $\mathtt{A}(x) = M$ if $-c < x < c$, $\mathtt{A}(x) = B$ if $x = c$, and $\mathtt{A}(x) = R$ if $x > c$. We can apply this function to an orbit of a given real point x_0, we associate with that orbit one infinite symbolic sequence $I(x_0) = \mathtt{A}(x_0)\,\mathtt{A}(x_1)\,\mathtt{A}(x_2)...\mathtt{A}(x_n)...$, the symbolic itinerary of x_0. Take the order naturally induced from the order in the real axis $L \prec A \prec M \prec B \prec R$. A parity function $\rho(S)$, for any finite sequence S, is defined such that is $+1$ if S has an even number of symbols, and -1 otherwise. As usual there is an ordering $\prec$ on the set $\mathcal{A}^{\mathbb{N}}$ such that: given two symbolic sequences $P = P_1P_2...$ and $Q = Q_1Q_2...$ let n be the first integer such that $P_n \neq Q_n$. Denote by $S = S_1S_2...S_{n-1}$ the common first subsequence of both P and Q. Then, $P \prec Q$ if $P_n \prec Q_n$ and $\rho(S) = +1$ or $Q_n \prec P_n$ if $\rho(S) = -1$. If no such n exists then $P = Q$. This ordering is originated by the fact that when $x < y$ then $I(x) \preceq I(y)$.

In the present situation, discontinuous maps, the orbits that are considered in the determination of the combinatorial structure of the iterates of g_β are the orbits of the images of the lateral limits of the discontinuity points, which are the orbits of $-\infty$ and $+\infty$. Note that $g_\beta(-c^-) = g_\beta(c^-) = -\infty$ and $g_\beta(-c^+) = g_\beta(-c) = g_\beta(c^+) = g_\beta(c) = +\infty$. The orbit of $O(+\infty)$ is $\{x_j^{(2)} : x_j^{(2)} = g_\beta^j(+\infty),\ j = 0, 1, \ldots\}$, with $g_\beta(+\infty) = d$.

The kneading sequences are defined as the symbolic itineraries of these orbits, the kneading pair is the ordered pair formed by these two symbolic sequences.

To state the rules of admissibility the shift operator σ will be used, defined as usual. The orbit of $+\infty$ has the symbolic itinerary $I(+\infty)$. The sequence $I(+\infty)$ is maximal (resp. $I(-\infty)$ is minimal) in the ordering defined in this section. Maximal in the sense that every shift of the sequence $I(+\infty)$ is less or equal than $I(+\infty)$. Every orbit with the initial condition x_0 is symmetric to the orbit with the initial condition $-x_0$, thus any orbit beginning with $+\infty$ is accompanied by a symmetric orbit started by $-\infty$, therefore we shall focus the admissibility rules for kneading sequences only on the itineraries with the first symbol R (corresponding to $+\infty$).

Given a particular sequence Q the operator $\tau : \mathcal{A}^{\mathbb{N}} \mapsto \mathcal{A}^{\mathbb{N}}$ is defined such that when applied to the sequence Q, this operator interchanges the symbols L and R, leaving the symbols M unchanged. For instance $\tau((RLMR)^{\infty}) =$

$(LRML)^{\infty}$.

Given any itinerary of $+\infty$ denoted by S, the corresponding itinerary of $-\infty$ is $\tau(S)$. The kneading pair is $(S, \tau(S))$. When we know the kneading sequence S, corresponding to the orbit of $+\infty$, to obtain the kneading pair is straightforward. By some abuse of notation, sometimes (mainly in the examples) we use only the kneading sequence S instead of the kneading pair, in that manner we avoid overloading the notation, but the meaning is precise.

Admissibility rules: Let S be a given sequence of symbols and $(S, \tau(S))$ be a pair of sequences. $(S, \tau(S))$ is a kneading pair and S is a kneading sequence, if S satisfies the admissibility condition: $\tau(S) \preceq \sigma^k(S) \preceq S$, for every integer k.

Given a finite sequence P with length n, the sequence $S = P^{\infty}$ is called an n-periodic sequence.

A bistable periodic orbit contains both the orbit of $+\infty$ and the orbit of $-\infty$. Any bistable orbit has an itinerary $S = P^{\infty} = (Q\tau(Q))^{\infty}$ or shortly $P = Q\tau(Q)$, as a consequence this type of orbit and associated symbolic itineraries must have an even period.

2 Singer-like theorem

We will prove in this section that g_{β} or ξ_{β} have at most two stable orbits, a result similar to the Singer's theorem [7]. In this case the schwarzian derivative of g_{β} does not have a fixed sign, but we will use the fact that ξ_{β} and g_{β} are topologically conjugated. In this situation we cannot use the techniques of Kozlovski [3] because g_{β} has two flat points in $\pm\infty$.

Proposition 2.1. *The maps of the families g_{β} and ξ_{β} are topologically conjugated, given the same parameter $\beta > 0$.*

Given the functions $q_{\beta} = -\beta \tanh(x)$ and $\widehat{q}_{\beta} = \beta \tan(x)$ we have $g_{\beta} = \widehat{q}_{\beta} \circ q_{\beta}$ and $\xi_{\beta} = q_{\beta} \circ \widehat{q}_{\beta}$, where $q_{\beta} : \overline{\mathbb{R}} \rightarrow [-\beta, \beta]$ is a homeomorphism, so $g_{\beta} = q_{\beta}^{-1} \circ \xi_{\beta} \circ q_{\beta}$. This construction works in this case because g_{β} has limits in $\pm\infty$.

Proposition 2.2. *If p is a stable periodic point of ξ_{β}, with $\frac{\pi}{2} < \beta < \frac{3\pi}{2}$, the orbit of p is the ω-limit set of $\xi_{\beta}(\frac{\pi}{2})$ or $\xi_{\beta}(-\frac{\pi}{2})$, i.e., of the images of the discontinuity points of ξ_{β}. Consequently the number of stable periodic orbits of f is finite and less or equal than 2.*

A maximal interval of continuity of ξ_{β}^{n} is called a lap of ξ_{β}^{n}. The endpoints of any lap of ξ_{β}^{n} are the pre-images of the discontinuity points $\frac{\pi}{2}$ or $-\frac{\pi}{2}$.

The schwarzian derivative of ξ_{β} is $2\left(\cos^4(x) - \beta^2\right)\sec^4(x)$, which is always negative for $\beta > 1$; the rest of the proof follows the same ideas of [7].

Corollary 2.3. *The maps in the family g_{β} have at most 2 stable periodic orbits.*

We easily see that $\lambda(\xi_{\beta}^{n}(p)) = \xi_{\beta}^{\prime n}(p)$, where p is a fixed point of ξ_{β}^{n}, is equal to $\lambda(g_{\beta}^{n}(q_{\beta}^{-1}(p)))$, where $q_{\beta}^{-1}(p)$ is a fixed point of g_{β}^{n}.

3 Types and scheme of bifurcations

3.1 Types of bifurcations

Notation: $F(x,\beta) = g_{\beta}(x) = -\beta\tan(\beta\tanh(x))$ and $F^{n}(x,\beta) = g_{\beta}^{n}(x)$, we denote $\partial_{x}F$ or F_{x} and $\partial_{\beta}F$ or F_{β} to denote the derivatives with respect to x and β. The points of discontinuity of $F(x,\beta)$ are $c_{1} = -\tan^{-1}\frac{\pi}{2\beta}$ and $c_{2} = \tan^{-1}\frac{\pi}{2\beta}$.

B1. **Saddle-node bifurcation.** In the family g_{β}, there are the usual saddle node (type B_{1}), the period doubling (B_{2}) and the pitchfork (B_{3}) bifurcations.

In the case B_{1} we have the well-known bifurcation equations: $F^{n}(x,\beta) = x$, $\partial_{x}F^{n}(x,\beta) = 1$. In this paper we use p and μ to represent the solutions of bifurcation equations. We have also the nondegeneracy conditions $\partial_{\beta}F^{n}(p,\mu) > 0$ and $\partial_{xx}F^{n}(p,\mu) < 0$.

Lemma 3.1. *The necessary condition for the occurrence of the saddle node bifurcation is* $n = 2(2l-1)$, *with* $l \geq 1$ *a positive integer. The orbits obtained in this type of bifurcation are symmetric, i.e.,* $F^{2l-1}(p,\mu) = -p$.

Proof. The value of n must be even because only the even iterates can have a positive derivative $(+1)$, so $n = 2k$. In the case where the orbit of p is symmetric we have $F^{k}(p,\mu) = -p$; in that case $-p$ is a point of the orbit of p. In the case where the orbit is nonsymmetric $F^{k}(p,\mu) \neq -p$. We assume that p corresponds to a nonsymmetric orbit, thereafter there are two distinct symmetrical (one relative to another) orbits, p belonging to the first one and $-p$ belonging to the second, because $F(x,\beta)$ is an odd function in x. In that case it is clear that $F^{2k}(p,\mu) = -p$, which is a contradiction, because F^{2k} is an even function in x.

We prove that $F^{k}(p,\mu) = -p$, which implies that k is odd, finally $k = 2l-1$. □

This type of bifurcation cannot occur for the origin because $\partial_{\beta}F^{n}(0,\beta) = 0$, as a matter of fact the origin never bifurcates because it is always an unstable fixed point.

The orbits with symbolic itinerary $RMLLMR$ or $RLMLRM$ are examples of orbits generated by saddle-node bifurcations.

We must confirm the nondegeneracy conditions. We must verify that by virtue of the symmetry of the problem the values of $\partial_{\beta}F^{n}(p,\mu)$ and $\partial_{xx}F^{n}(p,\mu)$ are not trivially zero in the case of the saddle-node bifurcation (in the pitchfork case we shall see that these quantities are automatically zero).

Let the initial condition be $x_0 = p$. The points of this type of orbit satisfy $x_j = -x_{j+k}$, where k is an odd integer and $0 \leq j \leq k$. Because the derivative $F_\beta(x,\beta)$ with respect to β is an odd function in x we have $F_\beta(x_j,\mu) = -F_\beta(x_{j+k},\mu)$, the derivative $F_x(x,\beta)$ is an even function in x: $F_x(x_j,\mu) = F_x(x_{j+k},\mu) < 0$, the derivative $F_{xx}(x,\beta)$ is an odd function in x: $F_{xx}(x_j,\mu) = -F_{xx}(x_{j+k},\mu)$. Using the chain rule we compute $F^{2n}(x,\mu)$: $\partial_x F^{2k}(x,\beta) = \prod_{j=0}^{2n-1} F_x\left(F^j(x,\beta),\beta\right)$. In the bifurcation point we have $\partial_x F^{2n}(x_0,\mu) = \prod_{j=0}^{2n-1} F_x(x_j,\mu) = 1$. By the symmetry of the problem and the fact that k is odd, we get $\prod_{j=0}^{k-1} F_x(x_k,\mu) = -1$ or $\prod_{j=l}^{k+l-1} F_x(x_j,\mu) = -1$, $0 < l < k$. We must study $\partial_\beta F^{2k}(x_0,\mu)$ and $\partial_{xx}F^{2k}(x_0,\mu)$.

We start by studying

$$\partial_\beta F^{2k}(x,\beta) = \sum_{j=0}^{2k-1} F_\beta\left(F^j(x,\beta),\beta\right) \prod_{l=j+1}^{2k-1} F_x\left(F^l(x,\beta),\beta\right).$$

Notation: $\prod_{l=b}^{a} c_l = 1$ if $b > a$.
In the bifurcation point we have

$$\partial_\beta F^{2k}(x_0,\mu) = \sum_{j=0}^{2k-1} F_\beta(x_j,\mu) \prod_{l=j+1}^{2k-1} F_x(x_l,\mu),$$

which is $-2\sum_{j=0}^{k-1} F_\beta(x_j,\mu) \prod_{l=j+1}^{k-1} F_x(x_l,\mu)$. The computation is now restricted to the iterates of the first half of the trajectory, so the symmetry of the problem does not impose any degeneracy condition trivially.

We now study $\partial_{xx}F^{2n}(x,\beta)$ in the bifurcation point:

$$F_{xx}^{2k}(x_0,\mu) = \sum_{j=0}^{2k-1} F_{xx}(x_j,\mu) \prod_{m=0}^{j-1} F_x(x_m,\mu) \prod_{\substack{l=0\\ l\neq j}}^{2k-1} F_x(x_l,\mu);$$

after some computations we get

$$F_{xx}^{2n}(x_0,\mu) = -2\sum_{j=0}^{k-1} F_{xx}(x_j,\mu) \prod_{l=0}^{j-1} F_x(x_l,\mu) \prod_{\substack{l=0\\ l\neq j}}^{k-1} F_x(x_l,\mu),$$

which is not trivially zero by the same motives as before. The symbolic itinerary of the orbits generated in this bifurcation is of type $S\tau S$.

B2. **Period doubling bifurcation.** The bifurcation conditions for the period doubling bifurcation, case B_2, are well known $F^n(x,\beta) = x$ and $\partial_x F^n(x,\beta) = -1$. The nondegeneracy conditions are: $\partial_\beta F^n(p,\mu) \neq 0$ and $\partial_{xx}F^n(p,\mu) \neq 0$.

This type of bifurcation does not repeat itself in a cascade because of the next lemma.

Lemma 3.2. *The period doubling bifurcations occur only for periods* $n = 2k+1$, *with* $k \geq 1$*integer.*

Proof. The value of n must be odd because only the odd iterates have a negative derivative, in this case -1. □

B3. **Pitchfork bifurcation.**

Lemma 3.3. *For every symmetric periodic orbit with period* $2n$, *such that* n *is even, there is a pitchfork bifurcation. This type of orbit has a symbolic itinerary* $S\tau S$ *with the number of symbols* $\#S = n$ *even. The bifurcation conditions are* $F^n(x,\beta) = x$ *and* $\partial_x F^n(x,\beta) = 1$, *the nondegeneracy conditions for the solution* (p,μ) *of the above equations are* $S_x F^{2n}(p,\mu) < 0$, *where* S_x *is the schwarzian derivative with respect to* x, *and* $\left(F^{2n}(p,\mu)\right)_{\beta x} > 0$.

Proof. Let the initial condition be $x_0 = p$. This type of orbit is symmetric, i.e., $x_j = -x_{j+n}$, for $0 \leq j \leq n$. Because $F_x(x,\beta)$ is negative and even (as a function of x) we have $F_x(x_j,\mu) = F_x(x_{j+n},\mu) < 0$. $F_\beta(x,\beta)$ is an odd function of x, so $F_\beta(x_j,\mu) = -F_\beta(x_{j+n},\mu)$. In the bifurcation point is we have $\partial_x F^{2n}(x_0,\mu) = \prod_{j=0}^{2n-1} F_x(x_0,\mu) = 1$. Because of the symmetry of the problem and the fact that n is even we get $\prod_{j=0}^{n-1} F_x(x_j,\mu) = 1$ or $\prod_{j=k}^{n+k-1} F_x(x_j,\mu) = 1$, for $0 < k < n$. We must show that $F_\beta^{2n}(x_0,\mu) = 0$, $F_{xx}^{2n}(x_0,\mu) = 0$, $F_{xxx}^{2n}(x_0,\mu) < 0$ and $F_{x\beta}^{2n}(x_0,\mu) > 0$, classical conditions of the pitchfork bifurcation. We first work on the derivative with respect to the parameter: $\partial_\beta F^{2n}(x,\beta)$, in the bifurcation point we have $\partial_\beta F^{2n}(x_0,\mu) = \sum_{j=0}^{2n-1} F_\beta(x_j,\mu) \prod_{k=j+1}^{2n-1} F_x(x_k,\mu)$, which is

$$\sum_{j=0}^{n-1} F_\beta(x_j,\mu) \prod_{k=j+1}^{n-1} F_x(x_k,\mu) - \sum_{j=0}^{n-1} F_\beta(x_j,\mu) \prod_{k=j+1}^{2n-1} F_x(x_k,\mu) = 0.$$

Now we study $\partial_{xx} F^{2n}(x,\beta)$ in the bifurcation point

$$\begin{aligned} \partial_{xx} F^{2n}(x_0,\mu) &= \sum_{j=0}^{2n-1} F_{xx}(x_j,\mu) \prod_{l=0}^{j-1} F_x(x_k,\mu) \prod_{\substack{k=0\\k\neq j}}^{2n-1} F_x(x_k,\mu) \\ &= \sum_{j=0}^{n-1} F_{xx}(x_j,\mu) \prod_{l=0}^{j-1} F_x(x_k,\mu) \prod_{\substack{k=0\\k\neq j}}^{n-1} F_x(x_k,\mu) \\ &\quad - \sum_{j=0}^{n-1} F_{xx}(x_j,\mu) \prod_{l=0}^{j-1} F_x(x_k,\mu) \prod_{\substack{k=0\\k\neq j}}^{2n-1} F_x(x_k,\mu) = 0. \end{aligned}$$

In the bifurcation point we have $\left(F^{2n}(x_0,\mu)\right)_x = 1$ and $\left(F^{2n}(x_0,\mu)\right)_{xx} = 0$, so $F_{xxx}^{2n}(x_0,\mu) = S_x F^{2n}(x_0,\mu) < 0$, one of the classical conditions of the pitchfork bifurcation. Finally we study

$$F^n_{x\beta}(x,\beta) = \sum_{j=0}^{2n-1} \left(F_x\left(F^j(x,\beta),\beta\right)\right)_\beta \prod_{\substack{l=0\\ l\neq j}}^{2n-1} F_x\left(F^l(x,\beta),\beta\right),$$

which is

$$\sum_{j=0}^{2n-1} \left(F_{x\beta}\left(F^j(x,\beta),\beta\right) + F_{xx}\left(F^j(x,\beta),\beta\right) \sum_{m=0}^{j-1} F_\beta\left(F^m(x,\beta),\beta\right)\right.$$
$$\left.\times \prod_{k=m+1}^{j-1} F_x\left(F^k(x,\beta),\beta\right)\right) \prod_{\substack{l=0\\ l\neq j}}^{2n-1} F_x\left(F^l(x,\beta),\beta\right).$$

In the bifurcation point we have

$$\sum_{j=0}^{2n-1} \left(F_{x\beta}(x_j,\mu) + F_{xx}(x_j,\mu) \sum_{m=0}^{j-1} F_\beta(x_m,\mu) \prod_{k=m+1}^{j-1} F_x(x_k,\mu)\right)$$
$$\times \prod_{\substack{l=0\\ l\neq j}}^{2n-1} F_x(x_l,\mu).$$

After some computations we obtain:

$$\sum_{j=0}^{n-1} \left(2F_{x\beta}(x_j,\mu) + F_{xx}(x_j,\mu) \sum_{m=0}^{j-1} F_\beta(x_m,\mu) \prod_{k=m+1}^{j-1} F_x(x_k,\mu)\right)$$
$$\times \prod_{\substack{l=0\\ l\neq j}}^{n-1} F_x(x_l,\mu).$$

We reduce the computation to the first half of the trajectory, by consequence $F^n_{x\beta}(x_0,\mu)$ is not trivially zero by the symmetry of the problem. □

B4. **Nonclassical bifurcation.** There is a bifurcation originated by the noncontinuous nature of g (type B_4). This type of bifurcation is obtained when a symmetric pair of points, each one a member of two distinct symmetric periodic orbits, crosses the discontinuity pair when we vary the parameter β. In this type of bifurcation two superstable (containing an infinity point) periodic orbits merge together in only one periodic orbit with the period doubled. We remember that the right discontinuity point is $c = \tan^{-1}\frac{\pi}{2\beta}$. The bifurcation equations are $F^k(+\infty,\beta) = c$ and $F^k(-\infty,\beta) = -c$ or

$F^k(+\infty, \beta) = -c$ and $F^k(-\infty, \beta) = c$ for an integer k. The nondegeneracy condition is $\partial_\beta F^k(\pm\infty, \beta) \neq 0$, with opposite sign for $+\infty$ and $-\infty$.

The two points that cross the discontinuity must be: a pre-image of $+\infty$ and a pre-image of $-\infty$. After crossing the discontinuity the image of each point interchanges with the image of the other point, the pre-image of $+\infty$ is now the pre-image of de $-\infty$ and vice versa. The consequence of this process is the merging of two orbits with the creation of a new bistable orbit with the period doubled.

Example 3.4. *When $\beta \simeq 2.94186$, this type of bifurcation occurs, symbolically $RR \to RB \to RM$, and at the same time $LL \to LA \to LM$. A new orbit results from the merging of RM with LM. The new orbit has the itinerary $RMLM$. This type of bifurcation always generates orbits of the type $S\tau S$.*

There is a bifurcation similar to B_4 but occurring in reverse order, i.e., a bistable orbit can split apart in two stable symmetric orbits. We call this type of bifurcation B_4'. This situation occurs when the middle point of the orbit becomes a pre-image of infinity. The process of bifurcation is exactly the same as in case B_4 if we reverse the order in which we change the parameter. This process occurs only for bistable orbits with a period $2(2k+1)$ with $1 \leq k$.

Example 3.5. *$RMLLMR$ splits in RMM and LMM, or $RLMLRM$ splits in RLR and LRL.*

3.2 Global scheme of bifurcations

We can see in Figure 1 the bifurcation diagram of the map g_β; this diagram is the motivation of the next theorem.

Theorem 3.6. *Let g_β be the family of maps defined in Section 1, with $\frac{\pi}{2} < \beta < \frac{3\pi}{2}$. Whenever a saddle-node bifurcation occurs there is a sequence of bifurcations of the type*

$$B_1 \to B_4' \to B_2 \to B_4 \to B_3 \to B_4 \to B_3 \to \ldots \to B_4 \to B_3 \to \ldots.$$

The only exception to this process is the first sequence of bifurcations, which starts by a classical period doubling B_2, where the pair of stable fixed points of g_β becomes unstable, having the sequence $B_2 \to B_4 \to B_3 \to B_4 \to B_3 \to \ldots \to B_4 \to B_3 \to \ldots$, generating orbits with periods $2, 4, 8, \ldots$.

Let l be a positive integer, by this process there is a bifurcation sequence giving orbits with periods $2(2l+1)$, $2l+1$, $2(2l+1)$, $4(2l+1)$, $4(2l+1)$, $\ldots$.

Proof. The proof follows from the result limiting the maximum number of stable orbits of Section 2 and the lemmas of paragraph 3.1 of this note, combined with the isomorphism theorem of [6]. □

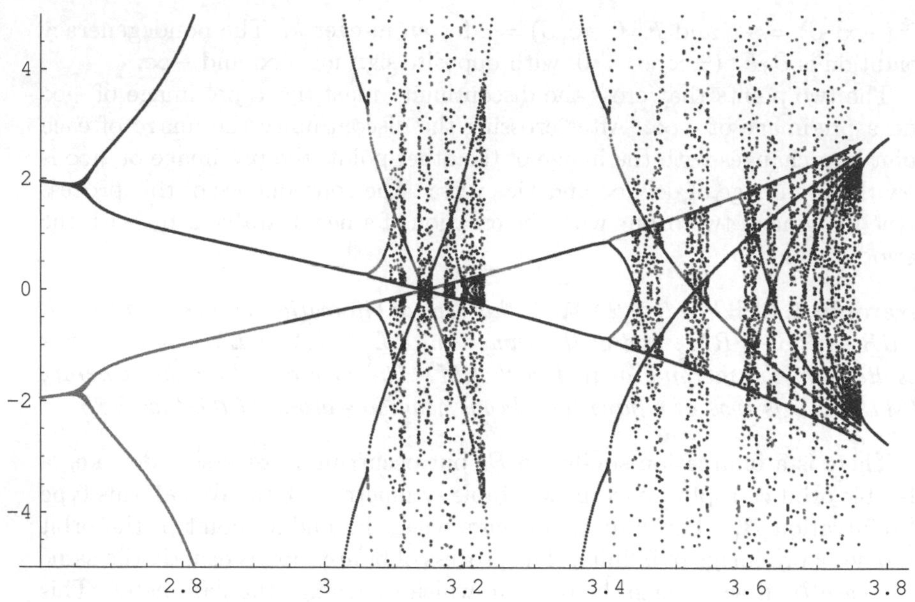

Figure 1: Bifurcation diagram of the family g_β, in order to the parameter $2.6 < \beta < 3.8$.

References

[1] Devaney, R. and Keen, L., Dynamics of Tangent, in *Dynamical Systems: Proc. Univ. of Maryland 1986–1987*, Lecture Notes in Mathematics **1342** (Springer, Berlin, New York), 105–111, 1989.

[2] Keen, L. and Kotus, J., Dynamics of the Family $\lambda \tan z$, *Conformal Geometry and Dynamics*, Vol I, (1997), 28–57.

[3] Kozlovski, O. S., Getting rid of the negative Schwarzian derivative condition. *Ann. of Math.* **152** (2000), 743–762.

[4] Milnor, J. and Thurston, W., On iterated maps of the interval, in *Dynamical Systems: Proc. Univ. of Maryland 1986–1987*, Lecture Notes in Mathematics **1342** (Springer, Berlin, New York), 465–563, 1988.

[5] Oliveira, H. and Sousa Ramos, J., Dynamics of the tangent map, *New Progress in Difference Equations: Proc. of ICDEA 2001*, Taylor & Francis, London, 199–206, 2003.

[6] Oliveira, H. and Sousa Ramos, J., Symbolic dynamics of the tangent map, *Int. J. Pure Appl. Math.* **8** (2003), 183–208.

[7] Singer, D., Stable orbits and bifurcations of maps of the interval, *SIAM J. Appl. Math.* **35** (1978), 260–267.

Difference Equations for Photon-Number Distribution in the Stationary Regime of a Random Laser

VLASTA PEŘINOVÁ AND ANTONÍN LUKŠ[1]
Laboratory of Quantum Optics
Faculty of Natural Sciences
Palacký University, Olomouc
Czech Republic

Abstract The theory of a laser has provided many semiclassical and quantum models of the laser. Differential-difference equations play a prominent role in the description of the dynamics of the laser. Investigation of the steady state leads to a simplification of these equations to the difference equations. Recent studies of random lasers have aroused interest in the description of traditional lasers.

Keywords Photon statistics, difference equations, stable equilibrium

AMS Subject Classification 39A11

1 Introduction

The equations of population dynamics can be applied in many fields [1]. In particular, they describe a traditional laser. While in classical populations, a transition from a continuous approximation to a discrete description is conceivable, in the laser such a transition is intrinsic to the quantum description [2, 3]. These equations appear in the model for the random laser [5], but completed with Langevin terms, which corresponds to stochastic dynamics. The random laser is described similarly as a multimode traditional laser, but the coupling constants between light and matter and the photon escape rates become random. This kind of stochasticity will not be treated in this paper. As usual, the joint probability distribution $p(n_1, \ldots, n_{N_\mathrm{p}}, N_1, \ldots, N_{N_\mathrm{s}}, t)$ of photon numbers n_j, $j = 1, \ldots, N_\mathrm{p}$, and of densities of excited atoms N_k, $k = 1, \ldots, N_\mathrm{s}$, obeys rate equations, i.e., a differential-difference equation and an initial condition at $t = t_0$. The steady state of the process, i.e., the joint probability distribution of the steady state, e.g., a limit of those for $t_0 \to -\infty$, obeys a difference equation.

[1]Supported by the Ministry of Education of the Czech Republic, project No. LN00A015.

2 Model of the traditional laser

In this paper we will deal with the differential-difference equation [4]

$$\frac{\partial}{\partial t}p(n,N,t) = L_{\mathrm{att}}p(n,N,t) + L_{\mathrm{amp}}p(n,N,t) + L_{\mathrm{nln}}p(n,N,t), \tag{1}$$

where

$$\begin{aligned} L_{\mathrm{att}}p(n,N,t) &= g[(n+1)p(n+1,N,t) - np(n,N,t)], & (2)\\ L_{\mathrm{amp}}p(n,N,t) &= P[p(n,N-1,t) - p(n,N,t)], & (3)\\ L_{\mathrm{nln}}p(n,N,t) &= K[n(N+1)p(n-1,N+1,t) - (n+1)Np(n,N,t)], & (4) \end{aligned}$$

with g, P positive, and K positive (nonnegative) numbers. For this equation we solve the initial-value problem

$$p(n,N,t)|_{t=t_0} = p_0(n,N,t_0). \tag{5}$$

Equation (1) has the property of rate equations, i.e., if the initial condition (5) represents a probability distribution, the solution is a probability distribution for all times. Let us recall that $p(n,N,t)$ ($p_0(n,N,t_0)$) is a probability distribution if and only if

$$p(n,N,t) \geq 0, \quad \sum_{n=0}^{\infty}\sum_{N=0}^{\infty} p(n,N,t) = 1 \tag{6}$$

and similarly for $p_0(n,N,t_0)$.

When $g = 0$, the existence (uniqueness) of solution for $t \in [t_0,\infty)$ is obvious. Let $w(t) > 0$, $W(t) > 0$ be functions defined on a subset of $[t_0,\infty)$. We will use Poissonian for

$$p(n,N,t) = \frac{1}{n!m!}\exp[-w(t) - W(t)][w(t)]^n[W(t)]^N. \tag{7}$$

Then the marginal distributions

$$p_1(n,t) = \sum_{N=0}^{\infty} p(n,N,t), \quad p_2(N,t) = \sum_{n=0}^{\infty} p(n,N,t) \tag{8}$$

are Poissonian [6] and it holds that

$$p(n,N,t) = p_1(n,t)p_2(N,t). \tag{9}$$

3 Essence of linearization

Another method of solution of Equation (1), which is useful for $g > 0$, is the method of characteristic function. The characteristic function of a function $p(n,N,t)$ is defined as follows:

$$C^{(n,N)}(is,iS,t) = \sum_{n=0}^{\infty}\sum_{N=0}^{\infty} p(n,N,t)\exp(isn + iSN), \tag{10}$$

which can be inverted as

$$p(n, N, t) = \frac{1}{(2\pi)^2} \int_{-\pi}^{\pi} \int_{-\pi}^{\pi} \exp(-ins - iNS) C^{(n,N)}(is, iS, t)\, ds\, dS. \tag{11}$$

Using the usual Fourier analysis, we can derive that the function (10) obeys the partial differential equation

$$\frac{\partial}{\partial t} C^{(n,N)}(is, iS, t) = L_{\mathrm{att}}^{(\mathrm{F})} C^{(n,N)}(is, iS, t) + L_{\mathrm{amp}}^{(\mathrm{F})} C^{(n,N)}(is, iS, t) + L_{\mathrm{nln}}^{(\mathrm{F})} C^{(n,N)}(is, iS, t) \tag{12}$$

and the initial condition

$$C^{(n,N)}(is, iS, t)\Big|_{t=t_0} = C^{(n,N)}(is, iS, t_0), \tag{13}$$

where F stands for Fourier and

$$L_{\mathrm{att}}^{(\mathrm{F})} C^{(n,N)}(is, iS, t) = g(e^{-is} - 1) \frac{\partial}{\partial(is)} C^{(n,N)}(is, iS, t), \tag{14}$$

$$L_{\mathrm{amp}}^{(\mathrm{F})} C^{(n,N)}(is, iS, t) = P(e^{iS} - 1) C^{(n,N)}(is, iS, t), \tag{15}$$

$$L_{\mathrm{nln}}^{(\mathrm{F})} C^{(n,N)}(is, iS, t) = K(e^{is-iS} - 1) \left(\frac{\partial}{\partial(is)} + 1 \right) \times \frac{\partial}{\partial(iS)} C^{(n,N)}(is, iS, t). \tag{16}$$

The essence of linearization is evident when it will be applied to equations of motion

$$\frac{d}{dt} \underline{n} = -g\underline{n} + \underline{\Gamma} + K(\underline{n} + 1)\underline{N} + \underline{\Psi}, \tag{17}$$

$$\frac{d}{dt} \underline{N} = P + \underline{\Phi} - K(\underline{n} + 1)\underline{N} - \underline{\Psi}, \tag{18}$$

with the initial conditions

$$\underline{n}(t)|_{t=t_0} = \underline{n}(t_0) \equiv \underline{n}_0, \quad \underline{N}(t)|_{t=t_0} = \underline{N}(t_0) \equiv \underline{N}_0, \tag{19}$$

where $\underline{n} \equiv \underline{n}(t)$ and $\underline{N} \equiv \underline{N}(t)$ are integer random functions of the time and the joint distribution of integer random variables $\underline{n}_0$ and $\underline{N}_0$ is $p_0(n, N, t_0)$. As the Markovian property and the rate equations (1) describe the stochastic processes $\underline{n}(t)$ and $\underline{N}(t)$ completely, we can (apart from some doubt of the theoretical value of the equations (17) and (18)) define processes $\underline{n}_{\mathrm{att}}(t)$, $\underline{n}_{\mathrm{nln}}(t)$, $\underline{N}_{\mathrm{amp}}(t)$, and $\underline{N}_{\mathrm{nln}}(t)$ in the same fashion, seeing to the (nonunique) decomposition

$$\underline{n}(t) = \underline{n}_{\mathrm{att}}(t) + \underline{n}_{\mathrm{nln}}(t), \quad \underline{N}(t) = \underline{N}_{\mathrm{amp}}(t) + \underline{N}_{\mathrm{nln}}(t). \tag{20}$$

Then

$$\begin{aligned}\underline{\Gamma} &= \frac{d}{dt}\underline{n}_{\mathrm{att}} + g\underline{n}, \;\; \underline{\Phi} = \frac{d}{dt}\underline{N}_{\mathrm{amp}} - P, \\ \underline{\Psi} &= \frac{d}{dt}\underline{n}_{\mathrm{nln}} - K(\underline{n}+1)\underline{N} = -\frac{d}{dt}\underline{N}_{\mathrm{nln}} - K(\underline{n}+1)\underline{N}. \end{aligned} \tag{21}$$

The linearization consists of a replacement of (17), (18) by the equations

$$\frac{d}{dt}\langle\underline{n}\rangle = -g\langle\underline{n}\rangle + K(\langle\underline{n}\rangle+1)\langle\underline{N}\rangle, \tag{22}$$

$$\frac{d}{dt}\langle\underline{N}\rangle = P - K(\langle\underline{n}\rangle+1)\langle\underline{N}\rangle, \tag{23}$$

$$\frac{d}{dt}\delta\underline{n} = -g\delta\underline{n} + \underline{\Gamma} + K(\langle\underline{n}\rangle+1)\delta\underline{N} + K\delta\underline{n}\langle\underline{N}\rangle + \underline{\Psi}_{\mathrm{appr}}, \tag{24}$$

$$\frac{d}{dt}\delta\underline{N} = \underline{\Phi} - K(\langle\underline{n}\rangle+1)\delta\underline{N} - K\delta\underline{n}\langle\underline{N}\rangle - \underline{\Psi}_{\mathrm{appr}}, \tag{25}$$

$$\underline{n} = \langle\underline{n}\rangle + \delta\underline{n}, \quad \underline{N} = \langle\underline{N}\rangle + \delta\underline{N}, \tag{26}$$

where $\delta\underline{n} \equiv \delta\underline{n}(t)$ and $\delta\underline{N} \equiv \delta\underline{N}(t)$ are random functions of time that assume, respectively, the values $-\langle\underline{n}\rangle$, $\pm1-\langle\underline{n}\rangle$, ... and $-\langle\underline{N}\rangle$, $\pm1-\langle\underline{N}\rangle$, ... Rate equations can be generated easily. As the initial-value problem for (22) and (23), the Markovian property, and the rate equations describe the stochastic processes $\delta\underline{n}(t)$ and $\delta\underline{N}(t)$ completely, we can define means $\langle\underline{n}_{\mathrm{att}}\rangle(t)$, $\langle\underline{n}_{\mathrm{nln\,appr}}\rangle(t)$, $\langle\underline{N}_{\mathrm{amp}}\rangle(t)$, $\langle\underline{N}_{\mathrm{nln\,appr}}\rangle(t)$ and processes $\delta\underline{n}_{\mathrm{att}}(t)$, $\delta\underline{n}_{\mathrm{nln\,appr}}(t)$, $\delta\underline{N}_{\mathrm{amp}}(t)$, $\delta\underline{N}_{\mathrm{nln\,appr}}(t)$ similarly, respecting the (nonunique) decompositions

$$\langle\underline{n}\rangle(t) = \langle\underline{n}_{\mathrm{att}}\rangle(t) + \langle\underline{n}_{\mathrm{nln\,appr}}\rangle(t), \; \langle\underline{N}\rangle(t) = \langle\underline{N}_{\mathrm{amp}}\rangle(t) + \langle\underline{N}_{\mathrm{nln\,appr}}\rangle(t), \tag{27}$$

$$\delta\underline{n}(t) = \delta\underline{n}_{\mathrm{att}}(t) + \delta\underline{n}_{\mathrm{nln\,appr}}(t), \; \delta\underline{N}(t) = \delta\underline{N}_{\mathrm{amp}}(t) + \delta\underline{N}_{\mathrm{nln\,appr}}(t). \tag{28}$$

Then

$$\underline{\Gamma} = \frac{d}{dt}\delta\underline{n}_{\mathrm{att}} + g\delta\underline{n}, \;\; \underline{\Phi} = \frac{d}{dt}\delta\underline{N}_{\mathrm{amp}}, \tag{29}$$

$$\underline{\Psi}_{\mathrm{appr}} = \frac{d}{dt}\delta\underline{n}_{\mathrm{nln\,appr}} - K(\langle\underline{n}\rangle+1)\delta\underline{N} - K\delta\underline{n}\langle\underline{N}\rangle \tag{30}$$

$$= -\frac{d}{dt}\delta\underline{N}_{\mathrm{nln\,appr}} - K(\langle\underline{n}\rangle+1)\delta\underline{N} - K\delta\underline{n}\langle\underline{N}\rangle. \tag{31}$$

The initial conditions (19) are replaced by

$$\langle\underline{n}\rangle(t)|_{t=t_0} = \langle\underline{n}\rangle(t_0) \equiv \langle\underline{n}\rangle_0, \quad \langle\underline{N}\rangle(t)|_{t=t_0} = \langle\underline{N}\rangle(t_0) \equiv \langle\underline{N}\rangle_0, \tag{32}$$

$$\delta\underline{n}(t)|_{t=t_0} = \underline{n}(t_0) - \langle\underline{n}\rangle(t_0), \quad \delta\underline{N}(t)|_{t=t_0} = \underline{N}(t_0) - \langle\underline{N}\rangle(t_0). \tag{33}$$

Let us note that

$$\langle\underline{n}\rangle = \frac{P}{g}, \; \langle\underline{N}\rangle = \frac{Pg}{K(P+g)} \tag{34}$$

are solutions of the algebraic equations

$$\begin{aligned} 0 &= -g\langle \underline{n}\rangle + K(\langle \underline{n}\rangle + 1)\langle \underline{N}\rangle, \qquad (35)\\ P &= K(\langle \underline{n}\rangle + 1)\langle \underline{N}\rangle, \;\; \langle \underline{n}\rangle \geq 0, \;\; \langle \underline{N}\rangle \geq 0. \qquad (36) \end{aligned}$$

Considering a small perturbation about the equilibrium point (34), we arrive at the stability matrix

$$\begin{pmatrix} -\frac{g^2}{P+g}, & \frac{P+g}{g}K \\ -\frac{Pg}{P+g}, & -\frac{P+g}{g}K \end{pmatrix}. \qquad (37)$$

With Equations (35), (36) we study the ordinary differential equations (22), (23). The stationary solutions (34) are locally asymptotically stable for $t \to \infty$ and they can be classified as a node/focus, when (if and only if)

$$\left(\frac{P}{g}+1\right)^2 \left(\frac{K}{g}\right)^2 - 2\left(2\frac{P}{g}+1\right)\frac{K}{g} + \frac{1}{\left(\frac{P}{g}+1\right)^2} \gtrless 0. \qquad (38)$$

Proposition. Let $g > 0$, $P > 0$, $K \geq 1$, $w_0 \geq 0$, $W_0 \geq 0$. Then the solution of the initial-value problem for equations for $x(t)$, $y(t)$,

$$\begin{aligned} \dot{x} &= -gx + K(x+1)y, \qquad (39)\\ \dot{y} &= P - K(x+1)y, \qquad (40) \end{aligned}$$

with initial conditions

$$x(t)|_{t=t_0} = x(t_0) \equiv x_0, \quad y(t)|_{t=t_0} = y(t_0) \equiv y_0, \qquad (41)$$

is defined for all $t \geq t_0$ and

$$\lim_{t\to\infty} x(t) = \frac{P}{g}, \quad \lim_{t\to\infty} y(t) = \frac{Pg}{K(P+g)}. \qquad (42)$$

For the proof see Section 4.

We pay attention to the moments

$$\langle \underline{n}^k \underline{N}^l\rangle(t) = \sum_{n=0}^{\infty}\sum_{N=0}^{\infty} n^k N^l p(n,N,t). \qquad (43)$$

Particularly, $\langle \underline{n}\rangle(t)$, $\langle \underline{N}\rangle(t)$, and these "means" must be nonnegative. Moreover, we take into account $\langle(\Delta\underline{n})^2\rangle(t)$, $\langle(\Delta\underline{N})^2\rangle(t)$, $\langle\Delta\underline{n}\Delta\underline{N}\rangle(t)$. Appropriately arranged, these "variances" and "covariance" must form a positive semidefinite matrix. Especially, the variances $\langle(\Delta\underline{n})^2\rangle(t)$, $\langle(\Delta\underline{N})^2\rangle(t)$ are nonnegative.

Whereas the ordinary differential equations (22), (23) are equations of population dynamics, Equations (24), (25), (26) suggest a linearization of (12), i.e., the equation

$$\frac{\partial}{\partial t} C^{(\delta n, \delta N)}(is, iS, t)$$

$$= \left[g(e^{-is} - 1)\frac{\partial}{\partial(is)} + K(e^{is-iS} - 1)\langle \underline{N} \rangle \frac{\partial}{\partial(is)} \right] C^{(\delta n, \delta N)}(is, iS, t)$$
$$+ K(e^{is-iS} - 1)(\langle \underline{n} \rangle + 1)\frac{\partial}{\partial(iS)} C^{(\delta n, \delta N)}(is, iS, t)$$
$$+ \left[g(e^{-is} - 1 + is)\langle \underline{n} \rangle + P(e^{iS} - 1 - iS) \right.$$
$$\left. + K(e^{is-iS} - 1 - is + iS)(\langle \underline{n} \rangle + 1)\langle \underline{N} \rangle \right] C^{(\delta n, \delta N)}(is, iS, t) \qquad (44)$$

for the function

$$\begin{aligned} C^{(\delta n, \delta N)}(is, iS, t) &= \sum_{\delta n = -\langle \underline{n} \rangle}^{\infty} \sum_{\delta N = -\langle \underline{N} \rangle}^{\infty} p(\langle \underline{n} \rangle + \delta n, \langle \underline{N} \rangle + \delta N, t) \exp(is\delta n + iS\delta N) \\ &= \exp(-is\langle \underline{n} \rangle - iS\langle \underline{N} \rangle) C^{(n,N)}(is, iS, t) \qquad (45) \end{aligned}$$

and the appropriately transformed initial condition (13) is taken into account.

The linearization is the favored method in the physical community. Equation (44) with an initial condition

$$C^{(\delta n, \delta N)}(is, iS, t)\Big|_{t=t_0} = C^{(\delta n, \delta N)}(is, iS, t_0) \qquad (46)$$

can be solved by the method of characteristics and even by that in its pure form.

4 Global asymptotic stability

In the following we assume that $g = 1$. In studying the global stability of the equilibrium solutions to Equations (22), (23), we try to construct a generalized Lyapunov function. A first generalization would be the Lyapunov function $V_{\mathrm{L}}(\mathbf{q})$ for vector field $\mathbf{F}(\mathbf{q})$ and equilibrium point $\mathbf{q}_{\mathrm{e}}$. The definition for $\mathbf{q}_{\mathrm{e}} = \mathbf{0}$ is known [1]. A Lyapunov function according to this definition can be denoted by $V_{\mathrm{L0}}(\mathbf{q})$. The function $V_{\mathrm{L}}(\mathbf{q})$ is called the Lyapunov function for vector field $\mathbf{F}(\mathbf{q})$ and point $\mathbf{q}_{\mathrm{e}}$, if it has the form

$$V_{\mathrm{L}}(\mathbf{q}) = U_{\mathrm{L0}}(\mathbf{q} - \mathbf{q}_{\mathrm{e}}), \qquad (47)$$

where $U_{\mathrm{L0}}(\mathbf{q})$ is the Lyapunov function for vector field $\mathbf{G}(\mathbf{q}) = \mathbf{F}(\mathbf{q} + \mathbf{q}_{\mathrm{e}})$ and point $\mathbf{0}$.

We recourse to a phase portrait for the equations

$$\dot{x} = K(x - P)\left(y - \frac{P}{K(P+1)}\right), \qquad (48)$$

$$\dot{y} = -K(x - P)\left(y - \frac{P}{K(P+1)}\right). \qquad (49)$$

The sets $\left\{(x,y) : x > P, y = \frac{P}{K(P+1)}\right\}$ and $\left\{(x,y) : x = P, y < \frac{P}{K(P+1)}\right\}$ are attractors and $\left\{(x,y) : x < P, y = \frac{P}{K(P+1)}\right\}$ and $\left\{(x,y) : x = P, y < \frac{P}{K(P+1)}\right\}$ are repellers. On passing to Equations (39), (40), the character of the field is partly changed and partly conserved.

Let $K \geq 1$. Then a trapezium $ABCD$ can be constructed. We choose A, B, C, D in the form

$$A = \left(\xi, \frac{1}{K}\frac{\xi}{\xi+1}\right), B = \left(P\frac{\xi+1}{\xi} - 1, \frac{1}{K}\frac{\xi}{\xi+1}\right),$$

$$D = \left(\xi, \frac{P}{\xi} - 1 + \frac{1}{K}\frac{\xi}{\xi+1}\right), C = \left(P, \frac{P}{\xi} - 1 + \frac{1}{K}\frac{\xi}{\xi+1}\right), \tag{50}$$

where the parameter ξ obeys a relation $0 < \xi < P$. Point A lies on the hyperbola a, $y = \frac{1}{K}\frac{x}{x+1}$, point B lies on the hyperbola b, $y = \frac{1}{K}\frac{P}{x+1}$, point C lies on the straight line c, $x = P$, point D lies on the cubic curve d, $y = \frac{P}{x} - 1 + \frac{1}{K}\frac{x}{x+1}$. On side AB it holds that $\dot{y} \geq 0$, the equality occurring at point B. On side CD, $\dot{y} < 0$. On side AD, $\dot{x} \geq 0$, the equality occurring at point A. On side BC, $(x+y)^{\cdot} \leq 0$, the equality occurring at point C. We summarize that on the perimeter of the trapezium, the field is directed inside this figure except for points B, A, and C. At point $B \in b$, $\ddot{y} > 0$, at point $A \in a$, $\ddot{x} > 0$, and at point $C \in c$, $(x+y)^{\cdot\cdot} < 0$. The half-axis $x \geq 0$ can be used as any side AB without B and the half-axis $y \geq 0$ as any side AD without D. Therefore, the quadrant $x \geq 0$, $y \geq 0$ can be added to the family of trapezia under consideration and assigned to $\xi = 0$. Provided that a solution of Equations (39), (40) starts at point $(x(t_0) \geq 0, x(t_0) > 0) \neq \left(P, \frac{1}{K}\frac{P}{P+1}\right)$, to each of its points $(x(t), y(t))$, $t \geq t_0$, a trapezium or the whole quadrant $x \geq 0$, $y \geq 0$ can be assigned. We conjecture that the stability theorem and the asymptotic stability theorem can be reformulated with weaker assumptions. As to the construction of the generalized Lyapunov function, let $v(\eta)$ be an increasing once-differentiable function of η, $\eta \geq 0$, such that $v(0) = 0$. Then $V_{\mathrm{L}}(x,y) = v(P - \xi)$ is a generalized Lyapunov function for vector field $(-x + K(x+1)y, P - K(x+1)y)$ and equilibrium point $\left(P, \frac{1}{K}\frac{P}{P+1}\right)$. It is sufficient to choose $v(\eta) \equiv \eta$. Because the parameter ξ is determined by the "parameters" of the trapezium sides in terms of strictly increasing functions, we can investigate it as the increasing function of time for $t \geq t_0$. Since the function $\xi(t)$ is increasing and obeys the inequality $\xi(t) < P$, there exists $\lim_{t\to\infty} \xi(t) \leq P$. Let us suppose for a moment that $\lim_{t\to\infty} \xi(t) = \xi(\infty) < P$. We define $\frac{d\xi}{dt} = a(\xi)$ at all the points of the solution that are not vertices of a trapezium. Let G denote the union of the open intervals included in the set $t \geq t_0$, when the solution is "near" the vertices B, A, C, i.e., let the Lebesgue measure $\lambda(G)$ of this union is finite and its image H after the function $\xi(t)$ is a union of open intervals included in $\{\xi : 0 \leq \xi \leq \xi(\infty)\}$. Then the difference $\langle 0, \xi(\infty)\rangle \setminus H$ is a compact set on which $\min a(\xi) > 0$ can be considered. Let m denote it. We have

$$\xi(\infty) = \int_0^\infty a[\xi(t)]dt > \int_{\langle 0,\infty)\setminus G} m dt = \infty, \tag{51}$$

which is a contradiction.

5 Conclusion

We have analyzed a differential-difference equation for probability distribution of the number of photons and the number of excited atoms. A second-order equation can be obtained by the method of characteristic function; it simplifies to a first-order equation, and the initial-value problem can be solved upon linearization of the underlying physical process.

References

[1] Haken, H. *Synergetics. An Introduction*, Springer Verlag, New York, 1977, pp. 121, 124.

[2] Mandel, L. and Wolf, E., *Optical Coherence and Quantum Optics*, Cambridge University Press, Cambridge, 1995.

[3] Scully, M.O. and Zubairy, M.S. *Quantum Optics*, Cambridge University Press, Cambridge, 1997.

[4] Rice, P.R. and Carmichael, H.J., Photon statistics of a cavity-QED laser: A comment on the laser–phase-transition analogy, *Physical Review* A, **50** (1994), 4318–4329.

[5] Patra, M., Theory for photon statistics of random lasers, *Physical Review* A, **65** (2002), 043809, 1–9.

[6] Van Kampen, N. G., *Stochastic Processes in Physics and Chemistry*, North-Holland, Amsterdam, 1981.

Delay Equations on Measure Chains: Basics and Linearized Stability

CHRISTIAN PÖTZSCHE[1]
Department of Mathematics, University of Augsburg
Augsburg, Germany

Abstract We introduce the notion of a dynamic delay equation, which includes differential and difference equations with possibly time-dependent backward delays. After proving a basic global existence and uniqueness theorem for appropriate initial value problems, we derive a criterion for the asymptotic stability of such equations in case of bounded delays.

Keywords Dynamic delay equation, stability, time scale, measure chain

AMS Subject Classification 34K05, 34D20, 39A12

1 Introduction and preliminaries

In this paper we briefly introduce dynamic equations on measure chains (or time scales), where time-dependent backward delays are present. Our approach provides a framework sufficiently flexible to include ordinary differential and difference equations without delays ($\dot{x}(t) = F(t, x(t))$ for $t \in \mathbb{R}$ and $\Delta x(t) = F(t, x(t))$ for $t \in \mathbb{Z}$, resp.), equations with constant delays ($\dot{x}(t) = F(t, x(t), x(t-r))$ for $t \in \mathbb{R}$, $r > 0$, and $\Delta x(t) = F(t, x(t), x(t-r))$ for $t, r \in \mathbb{Z}$, $r > 0$, resp.), as well as equations with proportional delays, like, e.g., the *pantograph equation* $\dot{x}(t) = A(t)x(t) + B(t)x(qt)$, $q \in (0, 1)$.

We prove an existence and uniqueness theorem for initial value problems of such equations under global Lipschitz conditions, which basically extends [Hil90, Section 5], who considers equations without delays. Section 3 contains sufficient conditions for the exponential decay of solutions for semi-linear equations and bounded delays. On this occasion, the delay term is interpreted as a perturbation of a linear delay-free dynamic equation, since we avoid the use of a general variation of constants formula for linear delay equations.

From now on, $\mathbb{Z}$ stands for the integers, $\mathbb{R}$ for the reals, and $\mathbb{R}_+$ for the nonnegative real numbers. Throughout this paper, Banach spaces $\mathcal{X}$ are all real or complex and their norm is denoted by $\|\cdot\|$. The closed ball in $\mathcal{X}$ with center 0 and radius $r > 0$ is given by $\bar{B}_r := \{x \in \mathcal{X} : \|x\| \leq r\}$. If I

[1]Research supported by the "Graduiertenkolleg: Nichtlineare Probleme in Analysis, Geometrie und Physik" (GRK 283) financed by the DFG and the State of Bavaria.

is a topological space, then $\mathcal{C}(I, \mathcal{X})$ are the continuous functions between I and $\mathcal{X}$. Finally, we write $D_{(2,3)}f$ for the partial Fréchet derivative of a mapping $f : \mathbb{T} \times \mathcal{X} \times \mathcal{X} \to \mathcal{X}$ w.r.t. the variables in $\mathcal{X} \times \mathcal{X}$, provided it exists.

We also sketch the basic terminology from the calculus on measure chains (cf. [Hil90, BP01]). In all the subsequent considerations we deal with a *measure chain* $(\mathbb{T}, \preceq, \mu)$, i.e., a conditionally complete totally ordered set $(\mathbb{T}, \preceq)$ (see [Hil90, Axiom 2]) with growth calibration $\mu : \mathbb{T} \times \mathbb{T} \to \mathbb{R}$ (see [Hil90, Axiom 3]). The most intuitive and relevant examples of measure chains are *time scales*, where $\mathbb{T}$ is a canonically ordered closed subset of $\mathbb{R}$ and μ is given by $\mu(t,s) = t - s$. Continuing, $\sigma : \mathbb{T} \to \mathbb{T}$, $\sigma(t) := \inf\{s \in \mathbb{T} : t \prec s\}$ defines the *forward jump operator* and the *graininess* $\mu^* : \mathbb{T} \to \mathbb{R}$ is defined by $\mu^*(t) := \mu(\sigma(t), t)$. If $\mathbb{T}$ has a left-scattered maximum m, we set $\mathbb{T}^\kappa := \mathbb{T}\backslash\{m\}$ and $\mathbb{T}^\kappa := \mathbb{T}$ otherwise. For $\tau, t \in \mathbb{T}$ we abbreviate $\mathbb{T}_\tau^+ := \{s \in \mathbb{T} : \tau \preceq s\}$, $\mathbb{T}_\tau^- := \{s \in \mathbb{T},\ s \preceq \tau\}$ and $[\tau, t]_\mathbb{T} := \{s \in \mathbb{T} : \tau \preceq s \preceq t\}$. Any other notation concerning measure chains is taken from [Hil90].

2 Dynamic delay equations

Let $\theta : \mathbb{T}^\kappa \to \mathbb{T}$ be a nondecreasing function satisfying $\theta(t) \preceq t$ for all $t \in \mathbb{T}^\kappa$. Then we denote θ as a *delay function* and say an equation of the form

$$\boxed{x^\Delta(t) = F\big(t, x(t), x(\theta(t))\big)} \tag{1$_F$}$$

is a *dynamic delay equation* with a *right-hand side* $F : \mathbb{T}^\kappa \times \mathcal{X} \times \mathcal{X} \to \mathcal{X}$. With given $\tau \in \mathbb{T}$, we abbreviate $\mathcal{C}_\tau(\theta) := \mathcal{C}([\theta(\tau), \tau]_\mathbb{T}, \mathcal{X})$. For $\phi_\tau \in \mathcal{C}_\tau(\theta)$, a continuous function $\nu : I \to \mathcal{X}$ is said to solve the *initial value problem* (IVP)

$$x^\Delta(t) = F\big(t, x(t), x(\theta(t))\big), \qquad (\tau, \phi_\tau), \tag{2}$$

if I is a $\mathbb{T}$-interval with $[\theta(\tau), \tau]_\mathbb{T} \subseteq I$, $\nu(t) = \phi_\tau(t)$ for all $t \in [\theta(\tau), \tau]_\mathbb{T}$ and $\nu^\Delta(t) = F\big(t, \nu(t), \nu(\theta(t))\big)$ for $t \in I$, $\tau \preceq t$ holds, where $\nu^\Delta(\tau) \in \mathcal{X}$ is understood as a right-sided derivative of ν in case of a right-dense $\tau \in \mathbb{T}$. Any solution satisfying the IVP (2) will be denoted by $\varphi(\cdot; \tau, \phi_\tau)$.

A tool solely important for the proof of Theorem 2.4 is given by means of the mapping $F^{\tau]} : \mathbb{T}_\tau^- \times \mathcal{X} \times \mathcal{X} \to \mathcal{X}$, which is defined for a fixed $\tau \in \mathbb{T}^\kappa$ by

$$F^{\tau]}(t,x,y) := \begin{cases} F(t,x,y) & \text{for } t \prec \tau,\ (x,y) \in \mathcal{X} \times \mathcal{X} \\ \lim\limits_{\substack{(s,\xi,\eta) \to (\tau,x,y) \\ s \prec \tau}} F(s,\xi,\eta) & \text{for } t = \tau,\ (x,y) \in \mathcal{X} \times \mathcal{X}. \end{cases}$$

Lemma 2.1. *Suppose $\theta : \mathbb{T}^\kappa \to \mathbb{T}$ is a continuous delay function, let I be a $\mathbb{T}$-interval, $\tau, r \in I$ with $\tau \preceq r$ and define $I_\tau := [\theta(\tau), \tau]_\mathbb{T} \cup I$. Then a function $\nu : I_\tau \to \mathcal{X}$ is a (unique) solution of the IVP* (2)*, if and only if*

(i) $\nu_1 := \nu|_{\mathbb{T}_\tau^- \cap I_\tau}$ *is a (unique) solution of the IVP* $(1)_{F^{\tau]}}$, (τ, ϕ_τ),

(ii) $\nu|_{\mathbb{T}_\tau^+ \cap I_\tau}$ *is a (unique) solution of the IVP* $(1)_F$, $(r, \nu_1|_{[\theta(r),r]_{\mathbb{T}}})$.

Proof. The proof is similar to [Hil90, Theorem 5.3] and is omitted here. $\square$

Lemma 2.2. *Suppose* $\theta : \mathbb{T}^\kappa \to \mathbb{T}$ *is a continuous delay function and define* $I := [a, b]_{\mathbb{T}}$ *for* $a, b \in \mathbb{T}$, $a \prec b$. *Moreover, let* $\ell : I \to \mathbb{R}_+$ *be rd-continuous,*

$$\int_a^b \ell(s)\,\Delta s < 1 \tag{3}$$

and assume that the rd-continuous mapping $F : \mathbb{T}^\kappa \times \mathcal{X} \times \mathcal{X} \to \mathcal{X}$ *satisfies*

$$\|F(t,x,y) - F(t,\bar{x},\bar{y})\| \leq \ell(t) \left\| \begin{pmatrix} x - \bar{x} \\ y - \bar{y} \end{pmatrix} \right\| \quad \text{for all } x, y, \bar{x}, \bar{y} \in \mathcal{X} \tag{4}$$

and $t \in I$. *Then, for any* $\tau \in I$ *and any* $\phi_\tau \in \mathcal{C}_\tau(\theta)$, *the IVP* (2) *possesses exactly one solution* $\nu : [\theta(\tau), \tau]_{\mathbb{T}} \cup I \to \mathcal{X}$.

Proof. Let $\tau \in I$ and $\phi_\tau \in \mathcal{C}_\tau(\theta)$. We define the $\mathbb{T}$-interval $I_\tau := I \cup [\theta(\tau), \tau]_{\mathbb{T}}$ and $\mathcal{C}(I_\tau, \mathcal{X})$ is complete w.r.t. the norm $\|\nu\|_{\mathcal{C}(I_\tau,\mathcal{X})} := \max_{t \in I_\tau} \|\nu(t)\|$. Now consider the operator $\mathcal{T}_\tau : \mathcal{C}(I_\tau, \mathcal{X}) \to \mathcal{C}(I_\tau, \mathcal{X})$,

$$\mathcal{T}_\tau(\nu)(t) := \begin{cases} \phi_\tau(t) & \text{for } \theta(\tau) \preceq t \prec \tau \\ \phi_\tau(\tau) + \int_\tau^t F\big(s, \nu(s), \nu(\theta(s))\big)\,\Delta s & \text{for } \tau \preceq t \end{cases}, \tag{5}$$

which is well defined due to [Hil90, Theorem 4.4]. Then $\nu \in \mathcal{C}(I_\tau, \mathcal{X})$ is a fixed point of $\mathcal{T}_\tau$, if and only if ν solves the IVP (2).

Because of [Hil90, Theorem 4.3(iii)], and for $\nu, \bar{\nu} \in \mathcal{C}(I_\tau, \mathcal{X})$, one obtains

$$\begin{aligned} \|\mathcal{T}_\tau(\nu)(t) - \mathcal{T}_\tau(\bar{\nu})(t)\| &\overset{(5)}{\leq} \int_\tau^t \big\|F\big(s, \nu(s), \nu(\theta(s))\big) - F\big(s, \bar{\nu}(s), \bar{\nu}(\theta(s))\big)\big\|\,\Delta s \\ &\overset{(4)}{\leq} \int_\tau^t \ell(s) \left\| \begin{pmatrix} \nu(s) - \bar{\nu}(s) \\ \nu(\theta(s)) - \bar{\nu}(\theta(s)) \end{pmatrix} \right\| \Delta s \\ &\leq \int_a^b \ell(s)\,\Delta s\, \|\nu - \bar{\nu}\|_{\mathcal{C}(I_\tau,\mathcal{X})} \quad \text{for all } t \in I_\tau,\ \tau \preceq t \end{aligned}$$

and by passing over to the least upper bound for $t \in I_\tau$, we get

$$\|\mathcal{T}_\tau(\nu) - \mathcal{T}_\tau(\bar{\nu})\|_{\mathcal{C}(I_\tau,\mathcal{X})} \leq \int_a^b \ell(s)\,\Delta s\, \|\nu - \bar{\nu}\|_{\mathcal{C}(I_\tau,\mathcal{X})}.$$

Using (3), we know that $\mathcal{T}_\tau$ is a contraction on $\mathcal{C}(I_\tau, \mathcal{X})$ and the contraction mapping principle yields that $\mathcal{T}_\tau$ possesses exactly one fixed point ν. $\square$

To show, e.g., the continuous dependence of solutions on the initial functions, we need a generalized version of Gronwall's inequality.

Lemma 2.3. *Let $\tau \in \mathbb{T}$, suppose $\theta : \mathbb{T}^\kappa \to \mathbb{T}$ is a continuous delay function, $C \geq 0$ and $b_1, b_2 : \mathbb{T}_\tau^+ \to \mathbb{R}_+$, $y : \mathbb{T}_{\theta(\tau)}^+ \to \mathbb{R}_+$ are rd-continuous. Then*

$$y(t) \leq C + \int_\tau^t b_1(s)y(s)\,\Delta s + \int_\tau^t b_2(s)y(\theta(s))\,\Delta s \quad \text{for all } t \in \mathbb{T}_\tau^+ \tag{6}$$

implies that $y(t) \leq Ce_{b_1+b_2}(t,\tau)$ for all $t \in \mathbb{T}_\tau^+$ with $\tau \preceq \theta(t)$.

Proof. The function $z : \mathbb{T}_\tau^+ \to \mathbb{R}$, $z(t) := \int_\tau^t b_1(s)y(s)\Delta s + \int_\tau^t b_2(s)y(\theta(s))\Delta s$ satisfies $z(\tau) = 0$ and is nondecreasing. Furthermore, we have

$$\begin{aligned} z^\Delta(t) &\leq b_1(t)y(t) + b_2(t)y(\theta(t)) \\ &\overset{(6)}{\leq} C(b_1(t) + b_2(t)) + b_1(t)z(t) + b_2(t)z(\theta(t)) \\ &\leq C(b_1(t) + b_2(t)) + (b_1(t) + b_2(t))z(t) \quad \text{for all } t \in \mathbb{T}_\tau^+,\ \tau \preceq \theta(t) \end{aligned}$$

and [BP01, p. 255, Theorem 6.1] yields

$$z(t) \leq C \int_\tau^t e_{b_1+b_2}(t,\sigma(s))(b_1(s) + b_2(s))\,\Delta s = C\left[e_{b_1+b_2}(t,\tau) - 1\right]$$

for all $t \in \mathbb{T}_\tau^+$, $\tau \preceq \theta(t)$. Hence the claim follows because of $y(t) \leq C + z(t)$. □

Theorem 2.4 (global existence and uniqueness). *Suppose $\theta : \mathbb{T}^\kappa \to \mathbb{T}$ is a continuous delay function, $L_1, L_2 : \mathbb{T}^\kappa \to \mathbb{R}_+$ are rd-continuous, and that the rd-continuous mapping $F : \mathbb{T}^\kappa \times \mathcal{X} \times \mathcal{X} \to \mathcal{X}$ satisfies the condition:*

For each $t \in \mathbb{T}^\kappa$ there exists a compact $\mathbb{T}$-neighborhood U_t of t such that

$$\begin{aligned} \left\|F^{t]}(s,x,y) - F^{t]}(s,\bar{x},y)\right\| &\leq L_1(t)\left\|x - \bar{x}\right\|, \\ \left\|F^{t]}(s,x,y) - F^{t]}(s,x,\bar{y})\right\| &\leq L_2(t)\left\|y - \bar{y}\right\| \end{aligned} \tag{7}$$

for all $s \in U_t^\kappa$, $x, \bar{x}, y, \bar{y} \in \mathcal{X}$ hold.

Then, for any $\tau \in \mathbb{T}^\kappa$ and $\phi_\tau \in \mathcal{C}_\tau(\theta)$, the IVP (2) admits exactly one solution $\varphi(\cdot;\tau,\phi_\tau) : \mathbb{T}_{\theta(\tau)}^+ \to \mathcal{X}$. Moreover, for $\phi_\tau, \bar{\phi}_\tau \in \mathcal{C}_\tau(\theta)$ and $t \in \mathbb{T}_\tau^+$ we have

$$\begin{aligned} &\left\|\varphi(t;\tau,\phi_\tau) - \varphi(t;\tau,\bar{\phi}_\tau)\right\| \\ &\leq \begin{cases} e_{L_1+L_2}(t,\tau)\left\|\phi_\tau(\tau) - \bar{\phi}_\tau(\tau)\right\| & \text{for } \tau \preceq \theta(t) \\ e_{L_1}(t,\tau)\left(1 + \int_\tau^t L_2(s)\,\Delta s\right)\sup_{s \in [\theta(\tau),\tau]_\mathbb{T}}\left\|\phi_\tau(s) - \bar{\phi}_\tau(s)\right\| & \text{for } \theta(t) \preceq \tau \end{cases} \end{aligned} \tag{8}$$

Proof. Let $\tau \in \mathbb{T}^\kappa$ and $\phi_\tau \in \mathcal{C}_\tau(\theta)$ be given arbitrarily.

(I) To show the existence and uniqueness of solutions, we apply the induction principle (cf. [Hil90, Theorem 1.4(c)] for $r \in (\mathbb{T}_\tau^+)^\kappa$ to the statement:

$$\mathcal{A}(r) : \begin{cases} \text{The IVP} \\ \qquad x^\Delta(t) = F^{r]}(t, x(t), x(\theta(t))), \qquad (\tau, \phi_\tau) \\ \text{possesses exactly one solution } \nu_r : [\theta(\tau), r]_\mathbb{T} \to \mathcal{X}. \end{cases} \tag{9}$$

(i): Obviously there exists a unique continuous mapping $\nu_\tau : [\theta(\tau), \tau]_\mathbb{T} \to \mathcal{X}$ satisfying $\nu_\tau(t) = \phi_\tau(t)$ for $t \in [\theta(\tau), \tau]_\mathbb{T}$ and $\nu_\tau^\Delta(t) = F^{\tau]}(t, \nu_\tau(t), \nu_\tau(\theta(t)))$ for all $t \in \{\tau\}^\kappa = \emptyset$.
(ii): Let r be a right-scattered point. Using the induction hypothesis $\mathcal{A}(r)$, the IVP in (9) possesses exactly one solution $\nu_r : [\theta(\tau), r]_\mathbb{T} \to \mathcal{X}$. We define its continuous extension $\nu_{\sigma(r)} : [\theta(\tau), \sigma(r)]_\mathbb{T} \to \mathcal{X}$ as

$$\nu_{\sigma(r)}(t) := \begin{cases} \nu_r(t) & \text{for } t \in [\theta(\tau), r]_\mathbb{T} \\ \nu_r(r) + \mu^*(r) F(r, \nu_r(r), \nu_r(\theta(r))) & \text{for } t = \sigma(r) \end{cases},$$

which, by Lemma 2.1, is the unique solution of the above IVP, since the restriction on $[\theta(\tau), r]_\mathbb{T}$ is the unique solution of (9) and the restriction on $[\theta(r), \sigma(r)]_\mathbb{T}$ is the unique solution of $(1)_F$, $(r, \nu_r|_{[\theta(r), r]_\mathbb{T}})$ on $[\theta(r), \sigma(r)]_\mathbb{T}$.
(iii): Let r be right-dense. Due to the induction hypothesis $\mathcal{A}(r)$ we have a unique solution ν_r of (9). Let $[a_r, b_r]_\mathbb{T} \subseteq U_r$ be a compact $\mathbb{T}$-neighborhood of r, such that the function $\ell : \mathbb{T}^\kappa \to \mathbb{R}_+$, $\ell(t) := \max\{L_1(r), L_2(r)\}$ for all $t \in [a_r, b_r]_\mathbb{T}$ from Lemma 2.2 satisfies $\int_{a_r}^{b_r} \ell(s)\,\Delta s = \ell(r)\mu(b_r, a_r) < 1$. Now Lemma 2.2 guarantees that the IVP $(1)_{F^{s]}}$, $(r, \nu_r|_{[\theta(r), r]_\mathbb{T}})$ has exactly one solution $\nu : [\theta(r), s]_\mathbb{T} \to \mathcal{X}$ for any $s \in [a_r, b_r]_\mathbb{T}$. Because of Lemma 2.1, the function $\nu_s : [\theta(\tau), s]_\mathbb{T} \to \mathcal{X}$, defined by

$$\nu_s(t) := \begin{cases} \nu_r(t) & \text{for } t \in [\theta(\tau), r]_\mathbb{T} \\ \nu(t) & \text{for } t \in [r, s]_\mathbb{T} \end{cases},$$

is the unique solution of (9) for $r = s$. Hence, the statement $\mathcal{A}(s)$ holds for all $s \in [a_r, b_r]_\mathbb{T} \cap \mathbb{T}_r^+$.
(iv): Let r be left-dense and we choose a $\mathbb{T}$-interval $[a_r, b_r]_\mathbb{T}$ as in (iii). Then there exists an $s \in [a_r, b_r]_\mathbb{T}$, $s \prec r$. Using the induction hypothesis $\mathcal{A}(s)$, as well as Lemma 2.2, one shows the existence and uniqueness of the solution $\nu_r : [\theta(\tau), r]_\mathbb{T} \to \mathcal{X}$ of (9) exactly as in step (iii). Since on every interval $[\theta(\tau), r]_\mathbb{T}$, $\tau \preceq r$, there exists exactly one solution ν_r, there is one on $\mathbb{T}_{\theta(\tau)}^+$.

(II) It remains to prove estimate (8). Thereto, let $\phi_\tau, \bar{\phi}_\tau \in \mathcal{C}_\tau(\theta)$. The solution $\varphi(\cdot; \tau, \phi_\tau)$ of $(1)_F$ satisfies the integral equation

$$\varphi(t; \tau, \phi_\tau) = \phi_\tau(\tau) + \int_\tau^t F(s, \varphi(s; \tau, \phi_\tau), \varphi(\theta(s); \tau, \phi_\tau))\,\Delta s \quad \text{for all } t \in \mathbb{T}_\tau^+,$$

yielding the estimate

$$\begin{aligned} \|\varphi(t; \tau, \phi_\tau) - \varphi(t; \tau, \bar{\phi}_\tau)\| &\overset{(7)}{\leq} \|\phi_\tau(\tau) - \bar{\phi}_\tau(\tau)\| \\ &\quad + \int_\tau^t L_1(s) \|\varphi(s; \tau, \phi_\tau) - \varphi(s; \tau, \bar{\phi}_\tau)\|\,\Delta s \\ &\quad + \int_\tau^t L_2(s) \|\varphi(\theta(s); \tau, \phi_\tau) - \varphi(\theta(s); \tau, \bar{\phi}_\tau)\|\,\Delta s \end{aligned}$$

for all $t \in \mathbb{T}_\tau^+$, and with Lemma 2.3 we obtain

$$\|\varphi(t;\tau,\phi_\tau) - \varphi(t;\tau,\bar{\phi}_\tau)\| \leq e_{L_1+L_2}(t,\tau)\,\|\phi_\tau(\tau) - \bar{\phi}_\tau(\tau)\|$$

for all $t \in \mathbb{T}_\tau^+$, $\tau \preceq \theta(t)$. On the other hand, in case of $\theta(t) \preceq \tau$, one has

$$\begin{aligned}
&\|\varphi(t;\tau,\phi_\tau) - \varphi(t;\tau,\bar{\phi}_\tau)\| \\
&\leq \|\phi_\tau(\tau) - \bar{\phi}_\tau(\tau)\| + \int_\tau^t L_2(s)\,\|\phi_\tau(\theta(s)) - \bar{\phi}_\tau(\theta(s))\|\,\Delta s \\
&\quad + \int_\tau^t L_1(s)\,\|\varphi(s;\tau,\phi_\tau) - \varphi(s;\tau,\bar{\phi}_\tau)\|\,\Delta s \\
&\leq \|\phi_\tau(\tau) - \bar{\phi}_\tau(\tau)\| + \int_\tau^t L_2(s)\,\Delta s \sup_{s\in[\theta(\tau),\tau]_{\mathbb{T}}} \|\phi_\tau(s) - \bar{\phi}_\tau(s)\| \\
&\quad + \int_\tau^t L_1(s)\,\|\varphi(s;\tau,\phi_\tau) - \varphi(s;\tau,\bar{\phi}_\tau)\|\,\Delta s
\end{aligned}$$

and Gronwall's Lemma (cf. [BP01, p. 256, Theorem 6.4]) implies the second inequality in (8). This concludes the present proof. □

3 Linearized asymptotic stability

Throughout this section, let $\mathbb{T}$ be unbounded above. Moreover, $\mathcal{C}_{rd}^+\mathcal{R}(\mathbb{T},\mathbb{R})$ is the set of rd-continuous functions $a : \mathbb{T} \to \mathbb{R}$ with $1 + \mu^*(t)a(t) > 0$ for $t \in \mathbb{T}$.

Lemma 3.1. *Let $\tau \in \mathbb{T}$, $K \geq 1$, $a \in \mathcal{C}_{rd}^+\mathcal{R}(\mathbb{T},\mathbb{R})$, suppose $\theta : \mathbb{T} \to \mathbb{T}$ is a continuous delay function, $A : \mathbb{T} \to \mathcal{L}(\mathcal{X})$ and $f : \mathbb{T} \times \mathcal{X} \times \mathcal{X} \to \mathcal{X}$ are rd-continuous. Consider the dynamic delay equation*

$$\boxed{x^\Delta(t) = A(t)x(t) + f\big(t, x(t), x(\theta(t))\big)} \tag{10$_f$}$$

under the following assumptions:

(i) The transition operator of $x^\Delta(t) = A(t)x(t)$ satisfies

$$\|\Phi_A(t,s)\| \leq K e_a(t,s) \quad \text{for all } \tau \preceq s \preceq t, \tag{11}$$

(ii) $f(t,0,0) \equiv 0$ on $\mathbb{T}$, and there exist reals $L_1, L_2 \geq 0$ such that we have

$$\begin{aligned}
\|f(t,x,y) - f(t,\bar{x},y)\| &\leq L_1\,\|x - \bar{x}\|, \\
\|f(t,x,y) - f(t,x,\bar{y})\| &\leq L_2\,\|y - \bar{y}\|
\end{aligned} \tag{12}$$

for all $t \in \mathbb{T}$, $x, \bar{x}, y, \bar{y} \in \mathcal{X}$.

Then the solution $\varphi(\cdot;\tau,\phi_\tau)$ of $(10)_f$ satisfies

$$\|\varphi(t;\tau,\phi_\tau)\| \leq K e_{\bar{a}}(t,\tau)\,\|\phi_\tau(\tau)\| \quad \text{for all } t \in \mathbb{T}_\tau^+,\ \tau \preceq \theta(t), \tag{13}$$

initial functions $\phi_\tau \in \mathcal{C}_\tau(\theta)$, and $\bar{a}(t) := a(t) + K\left(L_1 + L_2 e_a(\theta(t),t)\right)$.

Proof. Let $\tau \in \mathbb{T}$. Due to our present assumptions, one can apply Theorem 2.4 to the dynamical delay equation $(10)_f$ and consequently all solutions $\varphi(\cdot;\tau,\phi_\tau)$ with $\phi_\tau \in \mathfrak{C}_\tau(\theta)$ exist on $\mathbb{T}^+_{\theta(\tau)}$. Furthermore, the variation of constants formula (cf. [Pöt02, p. 56, Satz 1.3.11]) implies the identity

$$\varphi(t;\tau,\phi_\tau) = \Phi_A(t,\tau)\phi_\tau(\tau) + \int_\tau^t \Phi_A(t,\sigma(s))f(s,\varphi(s;\tau,\phi_\tau),\varphi(\theta(s);\tau,\phi_\tau))\,\Delta s$$

for all $t \in \mathbb{T}^+_\tau$, and from $f(t,0,0) \equiv 0$ we obtain

$$\begin{aligned}\|\varphi(t;\tau,\phi_\tau)\| &\overset{(11)}{\leq} Ke_a(t,\tau)\,\|\phi_\tau(\tau)\| \\ &\quad + K\int_\tau^t e_a(t,\sigma(s))\,\|f(s,\varphi(s;\tau,\phi_\tau),\varphi(\theta(s);\tau,\phi_\tau))\|\,\Delta s \\ &\overset{(12)}{\leq} Ke_a(t,\tau)\,\|\phi_\tau(\tau)\| + KL_1\int_\tau^t e_a(t,\sigma(s))\,\|\varphi(s;\tau,\phi_\tau)\|\,\Delta s \\ &\quad + KL_2\int_\tau^t e_a(t,\sigma(s))\,\|\varphi(\theta(s);\tau,\phi_\tau)\|\,\Delta s \quad \text{for all } t \in \mathbb{T}^+_\tau,\end{aligned}$$

which, in turn, yields (cf. [Hil90, Theorem 6.2])

$$\begin{aligned}\|\varphi(t;\tau,\phi_\tau)\|\,e_a(\tau,t) \leq K\,\|\phi_\tau(\tau)\| &+ \int_\tau^t \frac{K_1L}{1+\mu^*(s)a(s)}e_a(\tau,s)\,\|\varphi(s;\tau,\phi_\tau)\|\,\Delta s \\ &+ KL_2\int_\tau^t e_a(\theta(s),\sigma(s))e_a(\tau,\theta(s))\,\|\varphi(\theta(s);\tau,\phi_\tau)\|\,\Delta s\end{aligned}$$

for all $t \in \mathbb{T}^+_\tau$. Then Lemma 2.3 gives us the desired estimate (13). □

Theorem 3.2. *Let $\tau \in \mathbb{T}$, suppose $\theta : \mathbb{T} \to \mathbb{T}$ is a continuous delay function, $A : \mathbb{T} \to \mathcal{L}(\mathcal{X})$ is rd-continuous, $F : \mathbb{T} \times \mathcal{X} \times \mathcal{X} \to \mathcal{X}$ is rd-continuous and continuously differentiable w.r.t. the variables in $\mathcal{X}\times\mathcal{X}$. Consider the dynamic delay equation $(10)_f$ under the following assumptions:*

(i) The transition operator of $x^\Delta(t) = A(t)x(t)$ satisfies the estimate (11) with $\sup_{s\in\mathbb{T}^+_\tau} a(s) < 0$ and $\sup_{s\in\mathbb{T}^+_\tau} e_a(\theta(s),s) < \infty$,

(ii) $f(t,0,0) \equiv 0$ on $\mathbb{T}$, and we have

$$\lim_{(x,y)\to(0,0)} D_{(2,3)}f(t,x,y) = 0 \quad \text{uniformly in } t \in \mathbb{T}. \tag{14}$$

Then there exists a $\rho > 0$ such that all solutions $\varphi(\cdot,\tau,\phi_\tau)$ of $(10)_f$ with initial functions $\phi_\tau \in \mathfrak{C}_\tau(\theta)$, $\sup_{t\in[\theta(\tau),\tau]_\mathbb{T}} \|\phi_\tau(t)\| \leq \rho$ exist uniquely on $\mathbb{T}^+_{\theta(\tau)}$ and decay to 0 exponentially.

Proof. Let $\tau \in \mathbb{T}$. Due to hypothesis (i) there exists a $L > 0$ such that

$$KL\Big(1 + \sup_{s \in \mathbb{T}_\tau^+} e_a(\theta(s), s)\Big) < \inf_{s \in \mathbb{T}_\tau^+} (-a(s)) \tag{15}$$

holds, and the limit relation (14) guarantees that there is a $\rho_1 > 0$ with $\left\|D_{(2,3)} f(t,x,y)\right\| \leq \frac{1}{2} L$ for all $t \in \mathbb{T}$, $x, y \in \bar{B}_{\rho_1}$. Now the mean value inequality implies that $\|f(t,x,y) - f(t,\bar{x},\bar{y})\| \leq \frac{1}{2} L \left\|\binom{x-\bar{x}}{y-\bar{y}}\right\|$ for $t \in \mathbb{T}$, $x, \bar{x}, y, \bar{y} \in B_{\rho_1}$. Using the radial retraction $R_\rho : \mathcal{X} \to \bar{B}_\rho$, defined by $R_\rho(x) := x$ for $\|x\| \leq \rho$ and $R_\rho(x) := \frac{\rho}{\|x\|} x$ for $\|x\| \geq \rho$, it is well known that the modified mapping $\tilde{f} : \mathbb{T} \times \mathcal{X} \times \mathcal{X} \to \mathcal{X}$, $\tilde{f}(t,x,y) := f(t, R_{\rho_1}(x), R_{\rho_1}(y))$ coincides with f on the set $\mathbb{T} \times \bar{B}_{\rho_1} \times \bar{B}_{\rho_1}$ and satisfies $\left\|\tilde{f}(t,x,y) - \tilde{f}(t,\bar{x},\bar{y})\right\| \leq L \left\|\binom{x-\bar{x}}{y-\bar{y}}\right\|$ for all $t \in \mathbb{T}$, $x, \bar{x}, y, \bar{y} \in \mathcal{X}$. Therefore, from Theorem 2.4 we get that all solutions $\tilde{\varphi}(\cdot; \tau, \phi_\tau)$, $\phi_\tau \in \mathcal{C}_\tau(\theta)$, of $(10)_{\tilde{f}}$ exist and are unique on $\mathbb{T}^+_{\theta(\tau)}$. Furthermore, from Lemma 3.1 we have the inequality

$$\|\tilde{\varphi}(t; \tau, \phi_\tau)\| \overset{(13)}{\leq} K e_{\bar{a}}(t, \tau) \|\phi_\tau(\tau)\| \quad \text{for all } t \in \mathbb{T}_\tau^+,\ \tau \preceq \theta(t) \tag{16}$$

with $\bar{a}(t) := a(t) + KL(1 + e_a(\theta(t), t))$ and (15) yields $\sup_{s \in \mathbb{T}_\tau^+} \bar{a}(s) < 0$. This implies $\|\tilde{\varphi}(t; \tau, \phi_\tau)\| \leq K \|\phi_\tau(\tau)\| \leq \rho_1$ for all $t \in \mathbb{T}_\tau^+$, $\tau \preceq \theta(t)$, $\phi_\tau \in \bar{B}_{\frac{K}{\rho_1}}$, and from Theorem 2.4 we additionally get

$$\|\tilde{\varphi}(t; \tau, \phi_\tau)\| \overset{(8)}{\leq} e_L(t, \tau) \left(1 + \int_\tau^t L(s)\, \Delta s\right) \sup_{s \in [\theta(\tau), \tau]_{\mathbb{T}}} \|\phi_\tau(s)\|$$

for all $t \in \mathbb{T}_\tau^+$, $\theta(t) \preceq \tau$, which yields the existence of a $\rho_2 > 0$ such that $\|\tilde{\varphi}(t; \tau, \phi_\tau)\| \leq \rho_1$ for all $t \in \mathbb{T}_\tau^+$, $\phi_\tau \in \bar{B}_{\rho_2}$. If we choose $\rho := \min\left\{\frac{\rho_1}{K}, \rho_2\right\}$, then any solution $\tilde{\varphi}(\cdot; \tau, \phi_\tau)$ of $(10)_{\tilde{f}}$ with $\phi_\tau \in \bar{B}_\rho$ is also a solution of $(10)_f$ and together with (16) our assertion follows. □

References

[BP01] M. Bohner and A. Peterson, *Dynamic Equations on Time Scales — An Introduction with Applications*, Birkhäuser, Boston, 2001.

[Hil90] S. Hilger, Analysis on measure chains — a unified approach to continuous and discrete calculus, *Results in Mathematics* **18** (1990), 18–56.

[Pöt02] C. Pötzsche, *Langsame Faserbündel dynamischer Gleichungen auf Maßketten* (in German), Logos-Verlag, Berlin, 2002.

The System of Two Difference Equations

$$x_{n+1} = \frac{p + y_{n-k}}{y_n}, \quad y_{n+1} = \frac{q + x_{n-k}}{x_n}$$

C.J. SCHINAS and G. PAPASCHINOPOULOS

Department of Electrical and Computer Engineering
Democritus University of Thrace
Xanthi, Greece

Abstract In this paper we study the boundedness, the periodicity and the asymptotic behavior of the positive solutions of the system of difference equations

$$x_{n+1} = \frac{p + y_{n-k}}{y_n}, \quad y_{n+1} = \frac{q + x_{n-k}}{x_n},$$

where p, q are positive constants, $k \in \{2, 3, ...\}$ and the initial conditions $x_{-k}, y_{-k}, ..., x_0, y_0$ are positive numbers.

Keywords System of two difference equations, boundedness, periodicity, convergence, asymptotic behavior

AMS Subject Classification 39A10

1 Introduction

Consider, initially, the third-order difference equation of the form

$$x_{n+1} = \frac{\alpha + \beta x_n + \gamma x_{n-1} + \delta x_{n-2}}{A + Bx_n + Dx_{n-2}}, \quad n = 0, 1, ... \tag{1}$$

where the parameters $\alpha, \beta, \gamma, \delta, A, B, D$ and the initial conditions are non-negative real numbers such that $A + Bx_n + Dx_{n-2} > 0$ for all n. In case $\delta = D = 0$, Eq. (1) reduces to

$$x_{n+1} = \frac{\alpha + \beta x_n + \gamma x_{n-1}}{A + Bx_n}, \quad n = 0, 1, ..., \tag{2}$$

which was investigated in [3] by Gibbons, Kulenovic, Ladas and it was shown that it possesses period-2 trichotomy. More detailed, they proved that:

(a) Every positive solution of Eq. (2) has a finite limit if and only if $\gamma < \beta + A$.
(b) Every positive solution of Eq. (2) converges to a period-2 solution if and only if $\gamma = \beta + A$.
(c) Eq. (2) has positive unbounded solutions if and only if $\gamma > \beta + A$.

Eq. (1) for $\beta = B = 0$, $D = 1$ and $\gamma + \delta + A > 0$ becomes

$$x_{n+1} = \frac{\alpha + \gamma x_{n-1} + \delta x_{n-2}}{A + x_{n-2}}, \quad n = 0, 1, ... \tag{3}$$

This equation was studied in [2] where Camouzis, Ladas, and Voulov proved that Eq. (3) possesses period-2 trichotomy. More precisely, they proved that:
(a) Every positive solution of Eq. (3) has a finite limit if and only if $\gamma < \delta + A$.
(b) Every positive solution of Eq. (3) converges to a period-2 solution if and only if $\gamma = \delta + A$.
(c) Eq. (3) has positive unbounded solutions if and only if $\gamma > \delta + A$.

In paper [1] the authors investigated the difference equation

$$x_{n+1} = \frac{p + x_{n-2}}{x_n}, \quad n = 0, 1, ..., \tag{4}$$

for which they proved that:
(a) Every positive solution of Eq. (4) has a finite limit if and only if $p \geq 2$.
(b) Every positive solution of Eq. (4) converges to a period-2 solution if and only if $p = 1$.
(c) Eq. (4) has positive unbounded solutions if and only if $0 < p < 1$.
In addition, it was shown that there exist unbounded solutions of Eq. (4) when k is odd.

2 General Properties

Consider the system of difference equations

$$x_{n+1} = \frac{p + y_{n-k}}{y_n}, \quad y_{n+1} = \frac{q + x_{n-k}}{x_n}, \quad n = 0, 1, ..., \tag{5}$$

where p, q are positive constants, $k \in \{2, 3, ...\}$ and the initial conditions $x_{-k}, y_{-k}, ..., x_0, y_0$ are positive real numbers. For any values of the parameters p and q, system (5) possesses a unique positive equilibrium $(\overline{x}, \overline{y})$ where

$$\overline{x} = \frac{p - q + 1 + \sqrt{(q - p - 1)^2 + 4q}}{2},$$

$$\overline{y} = \frac{q - p + 1 + \sqrt{(p - q - 1)^2 + 4p}}{2},$$

and each one of them satisfies the equations

$$\overline{x}^2 + (q - p - 1)\overline{x} - q = 0,$$

$$\overline{y}^2 + (p - q - 1)\overline{y} - p = 0.$$

Obviously, when $p = q$, then

$$\overline{x} = \overline{y} = \frac{1 + \sqrt{1 + 4p}}{2}.$$

Theorem 2.1. *System* (5) *has no nontrivial periodic solutions of period k.*

Proof. Suppose that

$$(a_1, b_1),\ (a_2, b_2),\ ...,\ (a_k, b_k),\ (a_1, b_1), ... \tag{6}$$

is a periodic solution of system (5) of period k. Then

$$a_1 = \frac{p + b_k}{b_k},\ a_2 = \frac{p + b_1}{b_1},\ ...,\ a_k = \frac{p + b_{k-1}}{b_{k-1}},$$

$$b_1 = \frac{q + a_k}{a_k},\ b_2 = \frac{q + a_1}{a_1},\ ...,\ b_k = \frac{q + a_{k-1}}{a_{k-1}},$$

Then (6) is also a period k solution of the following system

$$x_{n+1} = \frac{p + y_n}{y_n}, \quad y_{n+1} = \frac{q + x_n}{x_n}. \tag{7}$$

But, it is shown (see [5]) that every solution of system (7) converges to the unique positive eqilibrium $(\overline{x}, \overline{y})$ and so system (7) has no nontrivial periodic solutions. Thus the periodic solution (6) must be the equilibrium solution $(\overline{x}, \overline{y})$. The proof is complete. □

Theorem 2.2. *Let* $\{(x_n, y_n)\}_{n=-k}^{\infty}$ *be a solution of system* (5). *If either*

$$x_n \geq \overline{x} \ \text{ and } \ y_n \geq \overline{y} \ \text{ for all } \ n \geq -k\ . \tag{8}$$

or

$$x_n < \overline{x} \ \text{ and } \ y_n < \overline{y} \ \text{ for all } \ n \geq -k\ . \tag{9}$$

holds, then $\{(x_n, y_n)\}_{n=-k}^{\infty}$ *converges to the unique equilibrium* $(\overline{x}, \overline{y})$.

Proof. Consider that $\{(x_n, y_n)\}_{n=-k}^{\infty}$ is a solution and assume that (8) holds. We first claim that

$$x_{n-k} \geq x_n, \quad \text{for all } \ n = 0, 1, ... \tag{10}$$

For the sake of contradiction suppose that there exists $N \geq 0$ such that $x_{N-k} < x_N$. By (5) we obtain

$$y_{N+1} = \frac{q + x_{N-k}}{x_N} < \frac{q + x_N}{x_N} \leq \frac{q + \overline{x}}{\overline{x}} = \overline{y},$$

which is a contradiction. Thus

$$\overline{x} \leq x_n \leq x_{n-k}, \quad \text{for all } \; n = 0, 1, ... \tag{11}$$

Similarly

$$\overline{y} \leq y_n \leq y_{n-k}, \quad \text{for all } \; n = 0, 1, ... \tag{12}$$

Using (11) and (12), we get that there exist (a_i, b_i) for $i = 0, 1, ..., k-1$ such that

$$\lim_{n\to\infty} x_{nk+i} = a_i, \quad \lim_{n\to\infty} y_{nk+i} = b_i.$$

But then (a_0, b_0), (a_1, b_1), ..., (a_{k-1}, b_{k-1}) is a periodic solution of period k. According to Theorem 2.1, system (5) has no nontrivial periodic solutions of period k.

Similarly, we can show the theorem when (9) holds. So the proof is complete. □

3 The case k is odd and $p = q$

Consider the system of difference equations

$$x_{n+1} = \frac{p + y_{n-k}}{y_n}, \quad y_{n+1} = \frac{p + x_{n-k}}{x_n}, \quad n = 0, 1, ..., \tag{13}$$

where p is positive constant, $k \in \{2, 3, ...\}$ and the initial conditions x_{-k}, y_{-k}, ..., x_0, y_0 are positive real numbers.

Theorem 3.1. *Let $\{(x_n, y_n)\}_{n=-k}^{\infty}$ be a positive solution of system* (13). *Suppose that k is odd and that there exists $N \geq 0$ such that*

$$x_{N-k}, x_{N-k+2}, ..., x_{N-3}, x_{N-1} < \overline{x} < x_N, x_{N-2}, ..., x_{N-K+3}, x_{N-K+1},$$

$$y_{N-k}, y_{N-k+2}, ..., y_{N-3}, y_{N-1} < \overline{x} < y_N, y_{N-2}, ..., y_{N-K+3}, y_{N-K+1}.$$

Then

$$\lim_{i\to\infty} x_{N+2i+1} = 0, \quad \lim_{i\to\infty} x_{N+2i} = \infty,$$

$$\lim_{i\to\infty} y_{N+2i+1} = 0, \quad \lim_{i\to\infty} y_{N+2i} = \infty.$$

Proof. We have

$$x_{N+1} = \frac{p + y_{N-k}}{y_N} < \frac{p + \overline{x}}{\overline{x}} = \overline{x},$$

$$x_{N+2} = \frac{p + y_{N-k+1}}{y_{N+1}} = \frac{p + y_{N-k+1}}{\frac{p + x_{N-k}}{x_N}} = \frac{p + y_{N-k+1}}{p + x_{N-k}} x_N > x_N > \overline{x}.$$

Similarly it can be proved

$$x_{N+3} < x_{N+1} < \overline{x} < x_{N+2} < x_{N+4}.$$

Working inductively we can show that

$$.... < x_{N+5} < x_{N+3} < x_{N+1} < \overline{x} < x_{N+2} < x_{N+4} < x_{N+6} < .... \tag{14}$$

Similarly we can prove

$$.... < y_{N+5} < y_{N+3} < y_{N+1} < \overline{x} < y_{N+2} < y_{N+4} < y_{N+6} < .... \tag{15}$$

Let

$$\lim_{i \to \infty} x_{N+2i+1} = l \neq 0, \quad \lim_{i \to \infty} x_{N+2i} = L \neq \infty,$$

$$\lim_{i \to \infty} y_{N+2i+1} = m \neq 0, \quad \lim_{i \to \infty} y_{N+2i} = M \neq \infty.$$

Then

$$(l, m), \ (L, M), \ (l, m), \ (L, M), ...$$

is a periodic solution of system (13) of period 2. In view of (13) we obtain

$$l = M$$

and

$$m = L,$$

which contradicts (14) and (15) and so the proof is complete. □

4 The case $k = 2$ and $p = q = 1$

If $k = 2$ and $p = q = 1$ then system (5) becomes

$$x_{n+1} = \frac{1 + y_{n-2}}{y_n}, \quad y_{n+1} = \frac{1 + x_{n-2}}{x_n}, \quad n = 0, 1, ... \tag{16}$$

We shall prove that every positive solution of system (16) converges to a period-10 solution of this system. For this aim we show that there exist infinitely many period-10 solutions of system (16). So, we state the following theorem.

Theorem 4.1. *A solution* $\{(x_n, y_n)\}_{n=-2}^{\infty}$ *of system* (16) *is periodic with period 10 if*

$$x_{-2} = \alpha, \quad y_{-2} = \beta,$$
$$x_{-1} = \gamma, \quad y_{-1} = \delta,$$
$$x_0 = \tfrac{1+\alpha}{\alpha\delta - 1}, \quad y_0 = \tfrac{1+\beta}{\beta\gamma - 1},$$

with $\alpha, \beta, \gamma, \delta \in (0, \infty)$, $\alpha\delta > 1$, $\beta\gamma > 1$.

Proof. From (16) we get

$$x_1 = \frac{1 + y_{-2}}{y_0} = \frac{1+\beta}{\frac{1+\beta}{\beta\gamma-1}} = \beta\gamma - 1,$$

$$y_1 = \frac{1 + x_{-2}}{x_0} = \frac{1+\alpha}{\frac{1+\alpha}{\alpha\delta-1}} = \alpha\delta - 1,$$

$$x_2 = \frac{1 + y_{-1}}{y_1} = \frac{1+\delta}{\alpha\delta - 1},$$

$$y_2 = \frac{1 + x_{-1}}{x_1} = \frac{1+\gamma}{\beta\gamma - 1},$$

$$x_3 = \frac{1 + y_0}{y_2} = \beta = y_{-2},$$

$$y_3 = \frac{1 + x_0}{x_2} = \alpha = x_{-2}.$$

Working inductively we obtain

$$x_{n+3} = y_{n-2}, \quad y_{n+3} = x_{n-2}, \quad \text{for all } n = 0, 1, ...$$

and so

$$x_{n+10} = x_n, \quad y_{n+10} = y_n, \quad \text{for all } n = 0, 1, ...$$

The proof is complete. □

Lemma 4.2. *Every positive solution* $\{(x_n, y_n)\}_{n=-2}^{\infty}$ *of system* (16) *satisfies the following identities:*

$$x_{n+4} - y_{n-1} = \frac{1}{y_{n+3}}(y_{n+3} - x_{n-2}).$$

Similarly

$$y_{n+4} - x_{n-1} = \frac{1}{x_{n+3}}(x_{n+3} - y_{n-2}).$$

Proof. Using (16) we have

$$y_{n+2} = \frac{1 + x_{n-1}}{x_{n+1}},$$

$$y_{n+4} = \frac{1 + x_{n+1}}{x_{n+3}}$$

and so

$$\begin{aligned} x_{n-1} &= y_{n+2}x_{n+1} - 1 \\ &= y_{n+2}(y_{n+4}x_{n+3} - 1) - 1 \\ &= x_{n+3}y_{n+2}y_{n+4} - y_{n+2} - 1. \end{aligned}$$

Similarly

$$y_{n-1} = x_{n+2}x_{n+4}y_{n+3} - x_{n+2} - 1.$$

Thus

$$\begin{aligned} x_{n-2} &= x_n y_{n+1} - 1 \\ &= (y_{n+3}x_{n+2} - 1)(x_{n+4}y_{n+3} - 1) - 1 \\ &= x_{n+2}x_{n+4}y_{n+3}^2 - x_{n+2}y_{n+3} - x_{n+4}y_{n+3} \\ &= y_{n+3}(x_{n+2}x_{n+4}y_{n+3} - x_{n+2} - 1 - x_{n+4}) + y_{n+3} \\ &= y_{n+3}(y_{n-1} - x_{n+4}) + y_{n+3} \end{aligned}$$

and so

$$x_{n-2} - y_{n+3} = y_{n+3}(y_{n-1} - x_{n+4}).$$

Therefore

$$x_{n+4} - y_{n-1} = \frac{1}{y_{n+3}}(y_{n+3} - x_{n-2}).$$

Similarly

$$y_{n+4} - x_{n-1} = \frac{1}{x_{n+3}}(x_{n+3} - y_{n-2}).$$

The proof is complete. □

Theorem 4.3. *Every positive solution* $\{(x_n, y_n)\}_{n=-2}^{\infty}$ *of system* (16) *converges to a period 10 solution of System* (16).

Proof. We prove the theorem for each one of the following five cases.
Case 1. Suppose $x_{-2} \geq y_3 > 0$ and $y_{-2} \geq x_3 > 0$. Then applying Lemma 4.2, we obtain

$$0 < x_{n+8} \leq y_{n+3} \leq x_{n-2}, \quad n = 0, 1, \dots ,$$

and

$$0 < y_{n+8} \leq x_{n+3} \leq y_{n-2}, \quad n = 0, 1, \dots .$$

Hence the solution $\{(x_n, y_n)\}_{n=-2}^{\infty}$ has ten decreasing positive subsequences. If a subsequence of $\{x_n\}_{n=-2}^{\infty}$ or a subsequence of $\{y_n\}_{n=-2}^{\infty}$, converges to 0, then the next subsequence must diverges to infinity. Thus all ten subsequences are decreasing and bounded from below and so $\{(x_n, y_n)\}_{n=-2}^{\infty}$ converges to a period 10 solution.

Case 2. Suppose $x_{-2} = y_3 > 0$ and $y_{-2} = x_3 > 0$. Then using Lemma 4.2 and arguing as in Case 1 we obtain

$$0 < x_{n+8} = y_{n+3} = x_{n-2}, \quad n = 0, 1, ...,$$

and

$$0 < y_{n+8} = x_{n+3} = y_{n-2}, \quad n = 0, 1, ...$$

Therefore $\{(x_n, y_n)\}_{n=-2}^{\infty}$ is periodic of period 10.

Case 3. If $0 < x_{-2} \leq y_3$ and $0 < y_{-2} \leq x_3$ then the solution $\{(x_n, y_n)\}_{n=-2}^{\infty}$ consists of ten increasing subsequences

$$\{(x_{10n-2}, y_{10n-2})\}_{n=-2}^{\infty},$$

$$\{(x_{10n-1}, y_{10n-1})\}_{n=-2}^{\infty},$$

$$....,$$

$$\{(x_{10n+7}, y_{10n+7})\}_{n=-2}^{\infty}.$$

It suffices to show that $\{(x_n, y_n)\}_{n=-2}^{\infty}$ is bounded. In view of Lemma 4.2 and relation (16) we have

$$x_{n+13} - y_{n+8} = (x_{n+3} - y_{n-2}) \left(\prod_{j=0}^{4} x_{n+2j+3}^{-1} \right) \left(\prod_{j=0}^{4} y_{n+2j+4}^{-1} \right).$$

Similarly,

$$y_{n+13} - x_{n+8} = (y_{n+3} - x_{n-2}) \left(\prod_{j=0}^{4} x_{n+2j+3}^{-1} \right) \left(\prod_{j=0}^{4} y_{n+2j+4}^{-1} \right).$$

For the sake of contradiction, suppose that one of the ten subsequences diverges to infinity. Without loss of generality assume that $\{x_{10N+3}\}_{n=0}^{\infty}$ diverges to infinity. Then

$$\left\{ \left(\prod_{j=0}^{4} x_{10n+2j+3} \right) \left(\prod_{j=0}^{4} y_{10n+2j+4} \right) \right\}_{n=0}^{\infty}$$

will also diverges to infinity. Therefore there exists $N \geq -1$ and $k \in (0, 1)$ such that

$$\frac{1}{\left(\prod_{j=0}^{4} x_{10n+2j+3}\right) \left(\prod_{j=0}^{4} y_{10n+2j+4}\right)} < k \text{ for any } n \geq N.$$

For $m \geq 0$ we have

$$x_{10N+23+10m} - y_{10N+18+10m} < k^{m+1}(x_{10N+13} - y_{10N+8}),$$

and so, from Lemma 4.2, we obtain

$$x_{10N+23+10m} - y_{10N+18} < \sum_{j=1}^{m+1} k^j (x_{10N+13} - y_{10N+8}).$$

Hence

$$x_{10N+23+10m} - y_{10N+18} < \frac{k}{k-1}(x_{10N+13} - y_{10N+8}).$$

So $\{x_{10N+23+10m}\}_{m=0}^{\infty}$ is a bounded subsequence of $\{x_{10n+3}\}_{n=-1}^{\infty}$ which is a contradiction.

Case 4. Assume that $x_{-2} \geq y_3 > 0$ and $0 < y_{-2} \leq x_3$. Then working in a same fashion as in Case 3, we can show the theorem.

Case 5. Suppose that $y_{-2} \geq x_3 > 0$ and $0 < x_{-2} \leq y_3$. Then working in a same fashion as in Case 3, we can show the theorem.

The proof is complete. □

References

[1] E. Camouzis, R. DeVault, and W. Kosmala, On the dynamics of $x_{n+1} = \frac{p+x_{n-2}}{x_n}$ (in press).

[2] E. Camouzis, G. Ladas, and H.D. Voulov, On the dynamics of $x_{n+1} = \frac{\alpha+\gamma x_{n-1}+\delta x_{n-2}}{A+x_{n-2}}$, *J. Difference Equ. Appl.* **9** (2003), 731–738.

[3] C.H. Gibbons, M.R.S. Kulenovic, and G. Ladas, On the recursive equence $x_{n+1} = \frac{\alpha+\beta x_{n-1}}{\gamma x_n}$, *Math. Sci. Res. Hot-Line* **24** (2000), 1–11.

[4] V.L. Kocic and G. Ladas, *Global Behavior of Nonlinear Difference Equations of Higher Order With Applications*, Kluwer Academic Publishers, Dordrecht, 1993.

[5] G. Papaschinopoulos and B.K. Papadopoulos, On the fuzzy difference equation $x_{n+1} = A + \frac{B}{x_n}$, *Soft Computing* **6** (2002), 456–461.

Asymptotic Behavior of Solutions of Certain Second-Order Nonlinear Difference Equations

EWA SCHMEIDEL

Institute of Mathematics
Faculty of Electrical Engineering
Poznań University of Technology
Poznań, Poland

Abstract We consider a second-order nonlinear difference equation

$$\Delta(r_n \Delta y_n) = a_n y_{n+1} + f(n, y_n). \tag{E}$$

The necessary conditions under which there exists a solution of Equation (E), which can be written in the following form:

$$y_n = \alpha_n \frac{u_n}{r_n} + \beta_n \frac{v_n}{r_n},$$

are given. Here (u_n) and (v_n) are two linearly independent solutions of the linear equation

$$\Delta(r_n \Delta y_n) = a_n y_{n+1}, \quad (\lim_{n\to\infty} \alpha_n = \alpha \quad \text{and} \quad \lim_{n\to\infty} \beta_n = \beta).$$

Also, two special cases of equation (E) are considered, and necessary conditions for some asymptotic behavior are obtained.

Keywords Difference equation, second-order, asymptotic behavior, nonlinear

AMS Subject Classification 39A10

1 Introduction

We consider the difference equation

$$\Delta(r_n \Delta y_n) = a_n y_{n+1} + f(n, y_n), \tag{E}$$

where N denotes the set of positive integers, R the set of real numbers, R_+ the set of nonnegative numbers, Δ the forward difference operator defined by $\Delta y_n = y_{n+1} - y_n$. Througout this paper we assume that

$$r : N \to R_+ \text{ is bounded away from zero.} \tag{1}$$

Denote

$$\sum_{i=1}^{n-1} \frac{1}{r_i} = R_n. \tag{2}$$

In the last few years there has been an increasing interest in the study of asymptotic behavior of solutions of difference equations, in particular second-order difference equations (see, for example, [1]–[10]).

2 Main result

We define $F : R \to R$ as continuous, nondecreasing function, such that $F(x) \neq 0$ for $x \neq 0$ and

$$\int_0^\epsilon \frac{ds}{F(s)} = \infty, \tag{3}$$

where ϵ is a positive constant.
Moreover, let the function $B : N \times R_+ \to R_+$ be continuous in the second component for each $n \in N$, and such that

$$B(n, z_1) \leq B(n, z_2) \text{ for } 0 \leq z_1 \leq z_2, \tag{4}$$

and

$$B(n, a_n z) \leq F(a_n) B(n, z) \text{ for any } a : N \to R_+. \tag{5}$$

We start with the following theorem.

Theorem 2.1. *Let (u_n) and (v_n) are linearly independent solutions of the linear equation*

$$\Delta(r_n \Delta z_n) = a_n z_{n+1}, \quad n \in N. \tag{6}$$

Assume that $f : N \times R \to R$, conditions (3), (4), and (5) holds, and

$$|f(n,x)| \leq B(n, |x|), \quad \textit{for all} \quad x \in R \quad \textit{and} \quad \textit{fixed} \quad n \in N. \tag{7}$$

Denote

$$U_j = \max\{|u_j|, |v_j|, |u_{j+1}|, |v_{j+1}|\}. \tag{8}$$

If

$$\sum_{j=1}^{\infty} U_j B(j, U_j) < \infty, \tag{9}$$

then there exists a solution (y_n) *of Equation* (E)*, which can be written in the form*

$$y_n = \alpha_n \frac{u_n}{r_n} + \beta_n \frac{v_n}{r_n},$$

where

$$\lim_{n\to\infty} \alpha_n = \alpha \quad \text{and} \quad \lim_{n\to\infty} \beta_n = \beta, \quad (\alpha, \beta - \textit{constants}).$$

Proof. See the proof of Theorem 1 from [7]. □

In the following theorem we assume that $a_n \equiv 0$ in Equation (E). Then Equation (E) takes the following form:

$$\Delta(r_n \Delta y_n) = f(n, y_n), \quad n \in N. \tag{10}$$

Theorem 2.2. *Assume that* (3), (4), (5), (7), *and* (8) *hold. If*

$$\sum_{j=1}^{\infty} j^2 B(j,j) < \infty, \tag{11}$$

then there exists a solution (y_n) *of Equation* (10)*, which can be written in the form*

$$y_n = \frac{C_1}{r_n} + \frac{C_2}{r_n} R_n + \varphi(n),$$

where C_1 *and* C_2 *are constants,* (R_n) *is defined by* (2)*, and*

$$\lim_{n\to\infty} \varphi(n) = 0.$$

Proof. The equation

$$\Delta(r_n \Delta y_n) = 0 \tag{12}$$

has two linearly independent solutions $u_n = 1$ and $v_n = R_n$.
From (8), we have

$$U_j = \max\{1, \sum_{i=1}^{j} \frac{1}{r_i}\}. \tag{13}$$

From (1), there exists a positive constant c such that $r_i \geq c > 0$. Hence $\frac{1}{r_i} \leq \frac{1}{c}$. Therefore

$$U_j \leq \max\{1, \frac{1}{c} \sum_{i=1}^{j} 1\} = \max\{1, \frac{1}{c} j\}.$$

Without losing the generality we can assume that $\max\{1, \frac{1}{c}j\} = \frac{1}{c}j$, so

$$U_j \leq \frac{1}{c_j}. \tag{14}$$

Because

$$U_jB(j,U_j) \le \frac{1}{c}jB(j,\frac{1}{c}j) \le \frac{1}{c}jF(\frac{1}{c})B(j,j) \le \frac{1}{c}F(\frac{1}{c})j^2B(j,j),$$

we see that (11) implies (9).
In a similar way as the proof of Theorem 1 from [7], we get

$$y_n = \frac{1}{r_n}(c_1A_n + c_3B_n) + \frac{1}{r_n}(c_2A_n + c_4B_n)R_n, \tag{15}$$

where

$$A_n = A_1 + \sum_{j=1}^{n-1} v_{j+1}f(j, y_j),$$

and

$$B_n = B_1 - \sum_{j=1}^{n-1} u_{j+1}f(j, y_j).$$

Here (u_n) and (v_n) denote linearly independent solutions of Equation (12) such that

$$\begin{vmatrix} v_n & u_n \\ \Delta v_n & \Delta u_n \end{vmatrix} = 1, \quad \text{for all} \quad n \in N.$$

Hence

$$\begin{aligned} |y_n| &\le \tfrac{1}{r_n}|c_1A_n + c_3B_n| + \tfrac{1}{r_n}|c_2A_n + c_4B_n||R_n| \\ &\le \tfrac{\bar{c}}{r_n}(|A_n| + |B_n|) + \tfrac{\bar{c}}{r_n}(|A_n| + |B_n|)|R_n|, \end{aligned}$$

where

$$\bar{c} = \max\{c_1, c_2, c_3, c_4\}.$$

In a similar way as in the proof of Theorem 1 from [7], we show that finite limits of (A_n) and (B_n) exist. Then there exists the constant $M \in R$ such that $|A_n| + |B_n| < M$. Hence

$$|y_n| \le \frac{\bar{c}M}{r_n}(1 + |R_n|) \le 2\frac{\bar{c}M}{r_n}U_n \le 2\frac{\bar{c}M}{c}U_n.$$

We have (see (8) in [7])

$$A_n = A_1 + \sum_{j=1}^{n-1} v_{j+1}f(j, y_j) \quad \text{and} \quad \lim_{n\to\infty} A_n = a.$$

Therefore

$$|A_n - a| = |\sum_{j=n}^{\infty} v_{j+1}f(j, y_j)|.$$

From (4), (5), (7), (13), and (14), we obtain

$$\begin{aligned}|A_n - a| &\leq \sum_{j=n}^{\infty} U_j B(j, |y_j|) \\ &\leq \sum_{j=n}^{\infty} U_j B(j, \tfrac{2\bar{c}M}{c} U_j) \\ &\leq F(\tfrac{2\bar{c}M}{c}) \sum_{j=n}^{\infty} U_j B(j, U_j) \\ &\leq \tfrac{1}{c} F(\tfrac{2\bar{c}M}{c}) F(\tfrac{1}{c}) \sum_{j=n}^{\infty} jB(j, j).\end{aligned}$$

From (1) and (13), we have $R_n \leq U_n$. From above and (14), we obtain

$$R_n|A_n - a| \leq \frac{1}{c^2} F(\frac{2\bar{c}M}{c}) F(\frac{1}{c}) \sum_{j=n}^{\infty} j^2 B(j, j).$$

From (11), we get

$$\lim_{n\to\infty} \sum_{j=n}^{\infty} j^2 B(j, j) = 0.$$

Then $\lim_{n\to\infty} R_n|A_n - a| = 0$. Similary we prove that $\lim_{n\to\infty} R_n|B_n - b| = 0$. From (15), we obtain

$$\begin{aligned}r_n y_n &= c_1(A_n - a) + c_3(B_n - b) + c_2(A_n - a)R_n + c_4(B_n - b)R_n \\ &\quad + c_1 a + c_2 a R_n + c_3 b + c_4 b R_n \\ &= (c_1 a + c_3 b) + (c_2 a + c_4 b)R_n + r_n \varphi(n),\end{aligned}$$

where

$$r_n \varphi(n) = c_1(A_n - a) + c_2(A_n - a)R_n + c_3(B_n - b) + c_4(B_n - b)R_n.$$

It is clear that

$$\lim_{n\to\infty} c_1(A_n - a) + c_2(A_n - a)R_n + c_3(B_n - b) + c_4(B_n - b)R_n = 0.$$

Therefore

$$\lim_{n\to\infty} r_n \varphi(n) = 0.$$

From (1), we get

$$\lim_{n\to\infty} \varphi(n) = 0.$$

Denote

$$c_1 a + c_3 b = C_1 \text{ and } c_2 a + c_4 b = C_2.$$

Then

$$y_n = \frac{C_1}{r_n} + \frac{C_2}{r_n} R_n + \varphi(n),$$

and we get the thesis of Theorem 2. □

Remark 2.3. *Assume that* (3), (4), (5), (7), *and* (8) *hold. If*

$$\lim_{n\to\infty} R_n < \infty,$$

and

$$\sum_{j=1}^{\infty} jB(j,j) < \infty,$$

then the conclusion of Theorem 2 holds.

Proof. Under assumption $\lim_{n\to\infty} R_n < \infty$, condition $\lim_{n\to\infty} R_n|A_n - a| = 0$ implies directly from $\lim_{n\to\infty} |A_n - a| = 0$. The other part of this proof is similar to the proof of Theorem 2 and will be omitted. □

We get theorems proved by Schmeidel in [6] as special cases of these results. To obtain theorems contained therein we take $r_n \equiv 1$ in Equation (E).

In the following theorem we assume that $a_n \equiv -(r_{n+1} + r_n)$ in Equation (E), so Equation (E) take the following form

$$\Delta(r_n \Delta y_n) = -(r_{n+1} + r_n)y_{n+1} + f(n, y_n). \tag{16}$$

Theorem 2.4. *Assume that* (3), (4), (5), (7) *and* (8) *hold. Let*

$$(r_n) \text{ is nondecreasing sequence.} \tag{17}$$

If

$$\sum_{j=1}^{\infty} B(j,1) < \infty, \tag{18}$$

then there exists a solution of Equation (16), *which can be written in the form*

$$y_n = \frac{C_3}{r_n} u_n + \frac{C_4}{r_n} v_n + \xi(n),$$

where C_1 and C_2 are constants, and

$$u_n = \begin{cases} 0, & \text{for } n \text{ even} \\ (-1)^{\frac{n-1}{2}} \prod_{i=1}^{\frac{n-1}{2}} \frac{r_{2i-1}}{r_{2i}}, & \text{for } n \text{ odd,} \end{cases} \tag{19}$$

$$v_n = \begin{cases} (-1)^{\frac{n}{2}-1} \prod_{i=1}^{\frac{n}{2}-1} \frac{r_{2i}}{r_{2i+1}}, & \text{for } n \text{ even} \\ 0, & \text{for } n \text{ odd,} \end{cases} \tag{20}$$

and

$$\lim_{n\to\infty} \xi(n) = 0.$$

Proof. Let us consider the equation

$$\Delta(r_n \Delta y_n) = -(r_{n+1} + r_n) y_{n+1},$$

which can be written in the form $r_{n+1}y_{n+1} + r_n y_n = 0$. It is easy to verify that sequences (u_n) and (v_n), defined by (19) and (20), are two linearly independent solutions of the above equation.
From (17), we have

$$\frac{r_{2i-1}}{r_{2i}} \leq 1 \text{ and } \frac{r_{2i}}{r_{2i+1}} \leq 1.$$

Hence

$$|(-1)^{\frac{n-1}{2}} \prod_{i=1}^{\frac{n-1}{2}} \frac{r_{2i-1}}{r_{2i}}| \leq 1$$

and

$$|(-1)^{\frac{n}{2}-1} \prod_{i=1}^{\frac{n}{2}-1} \frac{r_{2i}}{r_{2i+1}}| \leq 1.$$

We have

$$|u_n| \leq 1 \text{ and } |v_n| \leq 1, \text{ for all } n \in N.$$

From above and (8), we get $U_j \leq 1$.
From

$$U_j B(j, U_j) \leq B(j, 1),$$

by (18), we see that condition (9) holds.
Hence, by Theorem 1, we get the conclusion of Theorem 3. □

Remark 2.5. *Assume that* (3), (4), (5), (7), *and* (8)*hold.*
If

$$\lim_{n\to\infty} r_n = r < \infty,$$

and

$$\sum_{j=1}^{\infty} K^j B(j, K^j) < \infty,$$

for some constant K, then the conclusion of Theorem 3 holds.

Proof. From $\lim_{n\to\infty} r_n = r$ we have $\lim_{n\to\infty} \frac{r_{2i}}{r_{2i+1}} = 1 = \lim_{n\to\infty} \frac{r_{2i-1}}{r_{2i}}$. Then there exists a constant K^2 such that $\frac{r_{2i}}{r_{2i+1}} \leq K^2$ and $\frac{r_{2i-1}}{r_{2i}} \leq K^2$. Hence $U_j \leq (K^2)^{\frac{j}{2}} = K^j$. The other part of this proof is similar to the proof of Theorem 3 and will be omitted.

□

References

[1] S.S. Cheng, H.J. Li, W.T. Patula, Bounded and zero convergent solutions of second order difference equations, *J. Math. Anal. Appl.* **141** (1989), 463–483.

[2] A. Drozdowicz, J. Popenda, Asymptotic behavior of solutions of difference equations of second order, *J. Comput. Appl. Math.* **47** (1993), 141–149.

[3] R. Medina, M. Pinto, Asymptotic behavior of solutions of second order nonlinear difference equations, *Nonlinear Anal.* **19** (1992), 187–195.

[4] J. Migda, M. Migda, Asymptotic properties of the solutions of second order difference equation, *Arch. Math.* **34** (1998), 467–476.

[5] M. Migda, Asymptotic behavior of solutions of nonlinear delay difference equations, *Fasc. Math.* **31** (2001), 57–62.

[6] E. Schmeidel, Asymptotic behaviour of solutions of the second order difference equations, *Demonstratio Math.* **25** (1993), 811–819.

[7] E. Schmeidel, On the asymptotic behavior of solutions of second order nonlinear difference equations, *Folia Facultatis Scientiarium Naturalium Universitatis Masarykianae Brunensis, Mathematica* **13** (2003), 287–293.

[8] E. Schmeidel, M. Migda, M. Zbaszyniak, Some asymptotic properties of solutions of second order nonlinear difference equations, *Functional Differential Equations* (in press).

[9] E. Thandapani, M.M.S. Manuel, J.R. Graef, P.W. Spikes, Monotone properties of certain classes of solutions of second order difference e-quations, Advances in Difference Equations, II., *Comput.Math.Appl.* **36** (1998), 291–297.

[10] E. Thandapani, L. Rampuppillai, Oscillatory and asymptotic behaviour of perturbed quasilinear second order difference equations, *Arch.Math.* **34** (1998), 455–466.

Asymptotic Behavior of the Solutions of the Fuzzy Difference Equation

$$x_{n+1} = \frac{A + \sum_{i=0}^{k} a_i x_{n-i}}{B + \sum_{i=0}^{k} b_i x_{n-i}}$$

G. STEFANIDOU and G. PAPASCHINOPOULOS

Department of Electrical and Computer Engineering

Democritus University of Thrace Xanthi, Greece

Abstract In this paper we study the existence, the uniqueness, the boundedness, and the asymptotic dehavior of the positive solutions of the following fuzzy difference equation

$$x_{n+1} = \frac{A + \sum_{i=0}^{k} a_i x_{n-i}}{B + \sum_{i=0}^{k} b_i x_{n-i}}, n = 0, 1, \ldots,$$

where $k \in \{1, 2, \ldots\}$, A, B, a_i, b_i, $i \in \{0, 1, \ldots, k\}$ are positive fuzzy numbers and x_i, $i \in \{-k, -k+1, \ldots, 0\}$ are positive fuzzy numbers.

Keywords Fuzzy difference equations, boundedness, persistence, asymptotic dehavior, fuzzy number, a-cuts

AMS Subject Classification 39A10

1 Introduction

The various applications of difference equations in applied sciences and espesially in computer sciences have, as an anticipated effect, the progress of the theory of difference equations (for the detailed study of difference equations see [1]).

In recent years the fuzzy difference equations have been attracting increasing attention (see [6] and the references cited therein).

As we can see in paper [6] in order to study the behavior of a parametric fuzzy difference equation we investigate the behavior of a related family of systems of parametric difference equations.

Some previous papers are related to the system of ordinary difference equations studied in this paper, follow:

In [7] and [8] Pielou has studied the equation

$$H_{t+1} = \frac{aH_t}{1 + bH_{t-k}}$$

$a \in (1, \infty)$, $b \in (0, \infty)$ and $k \in N$, as a discrete analogous to the delay logistic equation of the form

$$H'(t) = rH(t)\Big(1 - \frac{H(t-\tau)}{p}\Big),$$

where r,τ and p are positive constants.

In [3] Kuruklis and Ladas studied the oscillation and asymptotic stability of all positive solutions of the equation

$$x_{n+1} = \frac{ax_n}{1 + bx_{n-k}}, \quad n = 0, 1, \ldots,$$

$a \in (0, \infty)$, $b \in (0, \infty)$ and $k \in N$, about its positive equilibrium.

In addition, in [2] Kocic, Ladas, and Rodrigues have studied the properties of the positive solutions of the recursive sequences

$$x_{n+1} = \frac{a + \sum_{i=0}^{k-1} b_i x_{n-i}}{x_{n-k}}, \quad x_{n+1} = \frac{a + bx_n}{A + x_{n-k}}, \quad n = 0, 1, \ldots,$$

where A, a, b, b_i, $i \in \{0, 1 \ldots, k-1\}$ are nonnegative real numbers and k is a positive integer.

Finally, in [5] Papaschinopoulos and Schinas studied the system of two rational recursive sequences

$$x_{n+1} = \frac{a + \sum_{i=0}^{k} a_i y_{n-i}}{b + \sum_{i=0}^{k} b_i y_{n-i}}, \quad y_{n+1} = \frac{c + \sum_{i=0}^{k} c_i x_{n-i}}{d + \sum_{i=0}^{k} d_i x_{n-i}}, \quad n = 0, 1, \ldots,$$

where k is a positive integer, a, b, c, d are positive real constants and a_i, b_i, c_i, d_i, $i \in \{0, 1 \ldots, k\}$ are nonnegative real constants such that

$$\sum_{i=0}^{k} a_i = \sum_{i=0}^{k} c_i = 1, \quad B = \sum_{i=0}^{k} b_i > 0, \quad D = \sum_{i=0}^{k} d_i > 0.$$

In this paper we study the existence, the boundedness, and the asymptotic behavior of the fuzzy difference equation

$$x_{n+1} = \frac{A + \sum_{i=0}^{k} a_i x_{n-i}}{B + \sum_{i=0}^{k} b_i x_{n-i}}, \tag{1}$$

where $k \in \{1, 2, \ldots\}$, A, B, a_i, b_i, $i \in \{0, 1, \ldots, k\}$ are positive fuzzy numbers.

2 Preliminaries

We need the following definitions:

For a set B we denote by $\overline{B}$ the closure of B. We say that a function A from $\mathbb{R}^+ = (0, \infty)$ into the interval $[0, 1]$ is a fuzzy number if the following conditions hold:

(i) A is normal (see [4]),

(ii) A is a convex fuzzy set (see [4]),

(iii) A is upper semicontinuous,

(iv) The support of A, supp$A = \overline{\bigcup_{a \in (0,1]} [A]_a} = \overline{\{x : A(x) > 0\}}$ is compact.

Then from Theorems 3.1.5 and 3.1.8 of [4] the a-cuts of the fuzzy number A, $[A]_a = \{x \in \mathbb{R}^+ : A(x) \geq a\}$ are closed intervals.

We say that a fuzzy number A is positive if supp$A \subset (0, \infty)$.

It is obvious that if A is a positive real number then A is a positive fuzzy number and $[A]_a = [A, A]$, $a \in (0, 1]$. Then we say that A is a trivial fuzzy number.

The fuzzy analog of the boundedness and persistence is given as follows: We say that a sequence of positive fuzzy numbers x_n persists (resp. is bounded) if there exists a positive number M (resp. N) such that

$$\text{supp}x_n \subset [M, \infty), \quad (\text{resp. supp}x_n \subset (0, N]), \quad n = 1, 2, \ldots,.$$

We say x_n is a positive solution of (1) if x_n is a sequence of positive fuzzy numbers that satisfies (1).

Let A, B be fuzzy numbers with

$$[A]_a = [A_{l,a}, A_{r,a}], \quad [B]_a = [B_{l,a}, B_{r,a}], \quad a \in (0,1].$$

We define the following metric:

$$D(A,B) = \sup \max \{|A_{l,a} - B_{l,a}|, |A_{r,a} - B_{r,a}|\},$$

where sup is taken for all $a \in (0,1]$.

Let x_n be a sequence of positive fuzzy numbers and let x be a positive fuzzy number. Suppose that

$$[x_n]_a = [L_{n,a}, R_{n,a}], \quad a \in (0,1], \quad n = n_0, n_0+1, \ldots$$

and

$$[x]_a = [L_a, R_a], \quad a \in (0,1]$$

is satisfied. We say that x_n nearly converges to x with respect to D as $n \to \infty$ if for every $\delta > 0$ there exists a measurable set T, $T \subset (0,1]$ of measure less than δ such that

$$\lim D_T(x_n, x) = 0, \quad \text{as} \quad n \to \infty,$$

where

$$D_T(x_n, x) = \sup_{a \in (0,1]-T} \left\{ \max \{|L_{n,a} - L_a|, |R_{n,a} - R_a|\} \right\}.$$

If $T = \emptyset$, we say that x_n converges to x with respect to D as $n \to \infty$.

Let E be the set of positive fuzzy numbers. From Theorem 2.1 of [9] we have that $A_{l,a}$, $B_{l,a}$ (resp. $A_{r,a}$, $B_{r,a}$) are increasing (resp. decreasing) functions on $(0,1]$. Therefore, using the condition (iv) of the definition of the fuzzy numbers there exist the Lebesque integrals

$$\int_J |A_{l,a} - B_{l,a}| da, \quad \int_J |A_{r,a} - B_{r,a}| da$$

where $J = (0,1]$. We define the function $D_1 : E \times E \to R^+$ such that

$$D_1(A,B) = \max\left\{ \int_J |A_{l,a} - B_{l,a}| da, \quad \int_J |A_{r,a} - B_{r,a}| da \right\}.$$

If $D_1(A,B) = 0$ we have that there exists a measurable set T of measure zero such that

$$A_{l,a} = B_{l,a} \quad A_{r,a} = B_{r,a} \quad \text{for all} \quad a \in (0,1] - T. \tag{2}$$

We consider, however, two fuzzy numbers A, B to be equivalent if there exists a measurable set T of measure zero such that (2) hold and if we do not distinguish between equivalent fuzzy numbers, then E becomes a metric space with metric D_1.

We say that a sequence of positive fuzzy numbers x_n converges to a positive fuzzy number x with respect to D_1 as $n \to \infty$ if

$$\lim D_1(x_n, x) = 0, \quad \text{as} \quad n \to \infty.$$

3 Main results

First, we study the existence and the uniqueness of the positive solutions of (1).

We give the following lemma (see [6]).

Lemma 3.1. *Let f be a continuous function from $\mathbb{R}^+ \times \mathbb{R}^+ \times \cdots \times \mathbb{R}^+$ into $\mathbb{R}^+$ and let $B_0, B_1, \ldots, B_k$ be fuzzy numbers. Then*

$$[f(B_0, B_1, \ldots, B_k)]_a = f([B_0]_a, [B_1]_a, \ldots, [B_k]_a), \quad a \in (0, 1].$$

Proposition 3.2. *Consider Equation (1), where $k \in \{1, 2, dots\}$, A, B, a_i, b_i, $i \in \{0, 1, \ldots, k\}$ are positive fuzzy numbers. Then for any positive fuzzy numbers $x_{-k}, x_{-k+1}, \ldots, x_0$ there exists a unique positive solution x_n of (1) with initial values x_{-k}, $x_{-k+1}, \ldots,$ x_0.*

Proof. Let x_{-k}, $x_{-k+1}, dots$, x_0 are positive fuzzy numbers such that for all $i = -k, -k+1, \ldots, 0$

$$[x_i]_a = [L_{i,a}, R_{i,a}], \quad a \in (0, 1]. \tag{3}$$

Moreover, suppose that

$$[A]_a = [A_{l,a}, A_{r,a}], \quad [B]_a = [B_{l,a}, B_{r,a}],$$

$$[a_i]_a = [a_{i,l,a}, a_{i,r,a}], \quad [b_i]_a = [b_{i,l,a}, b_{i,r,a}], \quad i = 0, 1, \ldots, k, \quad a \in (0, 1]. \tag{4}$$

Since A, B, a_i, b_i, $i = 0, 1, \ldots, k$, x_i, $i = -k, -k+1, \ldots, 0$ are positive fuzzy numbers then from Theorem 2.1 of [9] and for any $a_1, a_2 \in (0, 1]$, $a_1 \leq a_2$

$$\begin{aligned} &0 < A_{l,a_1} \leq A_{l,a_2} \leq A_{r,a_2} \leq A_{r,a_1}, \\ &0 < B_{l,a_1} \leq B_{l,a_2} \leq B_{r,a_2} \leq B_{r,a_1}, \\ &0 < a_{i,l,a_1} \leq a_{i,l,a_2} \leq a_{i,r,a_2} \leq a_{i,r,a_1}, \quad i = 0, 1, \ldots, k, \\ &0 < b_{i,l,a_1} \leq b_{i,l,a_2} \leq b_{i,r,a_2} \leq b_{i,r,a_1}, \quad i = 0, 1, \ldots, k, \end{aligned} \tag{5}$$

and

$$0 < L_{i,a_1} \leq L_{i,a_2} \leq R_{i,a_2} \leq R_{i,a_1}, \quad for \quad i = -k, -k+1, \ldots, 0. \tag{6}$$

Moreover, from Theorem 2.1 of [9], the functions $L_{i,a}, R_{i,a}, \quad i = -k, -k+1, \ldots, 0, \quad A_{l,a}, A_{r,a}, B_{l,a}, B_{r,a}, a_{i,l,a}, a_{i,r,a}, b_{i,l,a}, b_{i,r,a}, i = 0, 1, \ldots, k, \quad a \in (0,1]$ are left continuous.

In addition, since x_i, $i = -k, -k+1, \ldots, 0$, are positive fuzzy numbers there exist positive constants V_i, W_i, $i = -k, \ldots, 0$, such that for

$$[L_{i,a}, R_{i,a}] \subset \overline{\bigcup_{a \in (0,1]} [L_{i,a}, R_{i,a}]} \subset [V_i, W_i], \; i = -k, -k+1, \ldots, 0. \tag{7}$$

Also, since A, B, a_i, b_i, $i = 0, 1, \ldots, k$ are positive fuzzy numbers there exist positive constants K, L, M, N, P_i, Q_i, S_i, T_i, $i = 0, 1, \ldots, k$ such that

$$\begin{aligned}
&[A_{l,a}, A_{r,a}] \subset \overline{\bigcup_{a \in (0,1]} [A_{l,a}, A_{r,a}]} \subset [K, L], \\
&[B_{l,a}, B_{r,a}] \subset \overline{\bigcup_{a \in (0,1]} [B_{l,a}, B_{r,a}]} \subset [M, N], \\
&[a_{i,l,a}, a_{i,r,a}] \subset \overline{\bigcup_{a \in (0,1]} [a_{i,l,a}, a_{i,r,a}]} \subset [P_i, Q_i], \\
&[b_{i,l,a}, b_{i,r,a}] \subset \overline{\bigcup_{a \in (0,1]} [b_{i,l,a}, b_{i,r,a}]} \subset [S_i, T_i], \; i = 0, 1, \ldots, k, \; a \in (0,1].
\end{aligned} \tag{8}$$

We consider the family of systems of parametric ordinary difference equations

$$L_{n+1,a} = \frac{A_{l,a} + \sum_{i=0}^{k} a_{i,l,a} L_{n-i,a}}{B_{r,a} + \sum_{i=0}^{k} b_{i,r,a} R_{n-i,a}}, R_{n+1,a} = \frac{A_{r,a} + \sum_{i=0}^{k} a_{i,r,a} R_{n-i,a}}{B_{l,a} + \sum_{i=0}^{k} b_{i,l,a} L_{n-i,a}}, \; n = 0, 1, \ldots \tag{9}$$

Let $(L_{n,a}, R_{n,a})$, $n = 0, 1, \ldots$ be a positive solution of (9) with initial values $(L_{i,a}, R_{i,a})$, $i = -k, -k+1, \ldots, 0$. Then using Lemma 3.1, relations (3)–(9) and arguing as in Proposition 3.2 of [6] we can easily prove that $L_{n,a}, R_{n,a}$, $n = 1, 2, \ldots$ are left-continuous functions and satisfy analogous relations (3), (6), and (7) for $i = 1, 2, \ldots$. Then from Theorem 2.1 of [9], $[L_{n,a}, R_{n,a}]$ determines a sequence of positive fuzzy numbers x_n.

Finally, as in [6] we can easily prove that x_n is the unique positive solution of (1) with initial values $x_{-k}, x_{-k+1}, \ldots, x_0$ such that

$$[x_n]_a = [L_{n,a}, R_{n,a}], \quad a \in (0,1], \quad n = -k, -k+1, \ldots. \tag{10}$$

This completes the proof of the proposition. □

In what follows, we need the following definition and lemma. Consider the system of difference equations

$$y_{n+1} = \frac{V + \sum_{i=0}^{k} v_i y_{n-i}}{U + \sum_{i=0}^{k} u_i z_{n-i}}, \quad z_{n+1} = \frac{C + \sum_{i=0}^{k} c_i z_{n-i}}{D + \sum_{i=0}^{k} d_i y_{n-i}}, \quad n = 0, 1, 2, \dots, \tag{11}$$

where $k \in \{1, 2, \dots\}$, V, U, C, D, v_i, u_i, c_i, d_i, $i \in \{0, 1, \dots, k\}$ are positive real constants and the initial values y_i, z_i, $i = -k, -k+1, \dots, 0$ are positive real numbers. We say that a positive solution (y_n, z_n) of (11) persists (resp. is bounded) if there exists a positive constant M (resp. N) such that

$$M \le y_n, z_n \quad (\text{resp. } y_n, z_n \le N), \quad n = 1, 2, \dots.$$

In addition, we say that (y_n, z_n) is bounded and persists if there exist positive constants M, N such that

$$M \le y_n, z_n \le N, \quad n = 1, 2, \dots$$

In the following lemma we study system (11).

Lemma 3.3. *Consider system (11), where $k \in \{1, 2, \dots\}$, $V, U, C, D, v_i, u_i, c_i, d_i$, $i \in \{0, 1, \dots, k\}$ are positive real constants and the initial values y_i, z_i, $i = -k, -k+1, \dots, 0$ are positive real numbers. If*

$$\frac{\sum_{i=0}^{k} v_i}{U} < 1 \quad \textit{and} \quad \frac{\sum_{i=0}^{k} c_i}{D} < 1, \tag{12}$$

then the following statements are true.

(a) Every positive solution of (11) is bounded and persists.

(b) System (11) has a unique positive equilibrium.

(c) Every positive solution of (11) tends to the unique positive equilibrium of (11).

Proof. (a) In view of (11) we have that

$$y_{n+1} \le \frac{V + \sum_{i=0}^{k} v_i y_{n-i}}{U}. \tag{13}$$

We consider the linear difference equation

$$w_{n+1} = \frac{V + \sum_{i=0}^{k} v_i w_{n-i}}{U}, \quad n \ge 0. \tag{14}$$

Let w_n be the solution of (14) such that

$$w_s = y_s, \quad s = -k, -k+1, \ldots, 0. \tag{15}$$

From (13)–(15) it obvious that

$$y_1 - w_1 \leq \frac{\sum_{i=0}^{k} u_i(y_{-i} - w_{-i})}{U} = 0. \tag{16}$$

Since (15) and (16) hold then by induction we can prove that

$$y_n \leq w_n, \quad n \geq 1. \tag{17}$$

From (12), (14) and Remark 1.3.1 of [1] there exists a sequence g_n such that $g_n \to 0$ as $n \to \infty$ and

$$w_n = g_n + c, \quad c = \frac{V}{U - \sum_{i=0}^{k} v_i}. \tag{18}$$

So from (17) and (18), it is obvious that there exists a positive number r_1 such that $y_n \leq r_1, \quad n = 1, 2, \ldots$.

Similarly, using (11) and the second relation of (12) we can prove that there exists a positive number r_2 such that $z_n \leq r_2, \quad n = 1, 2, \ldots$.

Finally, from (11) we have that $y_n > \dfrac{V}{U + r\sum_{i=0}^{k} u_i}$ and $z_n > \dfrac{C}{D + r\sum_{i=0}^{k} d_i}$, where $r = \max\{r_1, r_2\}$.

So (y_n, z_n) is bounded and persists and the proof of statement (a) is completed.

(b) We consider the system

$$y = \frac{V + y\sum_{i=0}^{k} v_i}{U + z\sum_{i=0}^{k} u_i}, \quad z = \frac{C + z\sum_{i=0}^{k} c_i}{D + y\sum_{i=0}^{k} d_i}. \tag{19}$$

In view of (19) we have

$$\begin{aligned} &y^2 d(U - a) + y(S + T) + V(c - D) = 0, \\ &z^2 b(D - c) + z(S - T) + C(a - U) = 0, \end{aligned} \tag{20}$$

where

$$a = \sum_{i=0}^{k} v_i, \quad b = \sum_{i=0}^{k} u_i, \quad c = \sum_{i=0}^{k} c_i, \quad d = \sum_{i=0}^{k} d_i, \tag{21}$$
$$S = (c - D)(a - U), \quad T = Cb - Vd.$$

Since (12), (20), and (21) hold we can easily prove that

$$y = \frac{-S - T + \sqrt{(S+T)^2 + 4dVS}}{2(U-a)d}$$
$$z = \frac{T - S + \sqrt{(T-S)^2 + 4bCS}}{2(D-c)b} \tag{22}$$

is the unique positive solution of system (19), which is the unique positive equilibrium of system (11).

(c) Let (x_n, y_n) be a positive solution of system (11). From the statement (a) we have

$$m_1 = \liminf_{n\to\infty} x_n > 0, \quad m_2 = \liminf_{n\to\infty} y_n > 0,$$
$$L_1 = \limsup_{n\to\infty} x_n \neq \infty, \quad L_2 = \limsup_{n\to\infty} y_n \neq \infty. \tag{23}$$

From (11), (21), and (23) we obtain that

$$bC + b(c-D)m_2 \leq Vd + d(a-U)L_1,$$
$$Vd + d(a-U)m_1 \leq Cb + b(c-D)L_2$$

and so from (12),

$$Vd+d(a-U)m_1 \leq Cb+b(c-D)L_2 \leq bC+b(c-D)m_2 \leq Vd+d(a-U)L_1. \tag{24}$$

Relations (12), (21), and (24) imply that $m_1 = L_1$ and $m_2 = L_2$. This completes the proof of the lemma.

□

In the last proposition we study the boundedness of the positive solutions of Equation (1), the existence of positive equilibrium of (1), and the asymptotic behavior of the positive solutions of (1).

Proposition 3.4. *Consider Equation (1) where A, B, a_i, b_i, $i = 0, \ldots, k$ are positive fuzzy numbers. If (8) and*

$$\frac{\sum_{i=0}^{k} Q_i}{M} < 1 \tag{25}$$

hold, then the following statements are true.

(a) Every positive solution of (1) is bounded and persists.

(b) Equation (1) has a unique positive equilibrium.

(c) Every positive solution of (1) nearly converges to the unique positive equilibrium x with respect to D as $n \to \infty$ and converges to x with respect to D_1 as $n \to \infty$.

Proof. (a) Let x_n be a positive solution of (1) such that (10) holds.

We consider the system

$$H_{n+1} = \frac{K + \sum_{i=0}^{k} P_i H_{n-i}}{N + \sum_{i=0}^{k} T_i E_{n-i}}, \quad E_{n+1} = \frac{L + \sum_{i=0}^{k} Q_i E_{n-i}}{M + \sum_{i=0}^{k} S_i H_{n-i}}, \tag{26}$$

where K, N, L, M, P_i, T_i, Q_i, S_i are defined in (8).

Let (H_n, E_n) be a solution of (26) with initial values

$$H_i = V_i, \quad E_i = W_i, \quad i = -k, -k+1, \ldots, 0, \tag{27}$$

where V_i, W_i, $i = -k, -k+1, \ldots, 0$ are defined in (7).

From (7), (8), (9), (26), and (27) we have

$$H_1 = \frac{K + \sum_{i=0}^{k} P_i V_{-i}}{N + \sum_{i=0}^{k} T_i W_{-i}} \leq L_{1,a}, \quad R_{1,a} \leq \frac{L + \sum_{i=0}^{k} Q_i W_{-i}}{M + \sum_{i=0}^{k} S_i V_{-i}} = E_1. \tag{28}$$

Using (9), (26), (27), and (28), inductively we can prove that

$$H_n \leq L_{n,a}, R_{n,a} \leq E_n, n = 1, 2, \ldots. \tag{29}$$

In addition, in view of (9), (25), (26), and the statement (a) of Lemma 3.3 we have that the solution (H_n, E_n) of (26) is bounded and persists and so from (10) and (29) we also have that the solution x_n of (1) is bounded and persists. The proof of statement (a) is completed.

(b) Let x_n be a positive solution of (1) such that (10) holds. Since (9) and (25) are satisfied from statements (b) and (c) of Lemma 3.3 we have that for any $a \in (0, 1]$ there exist the $\lim_{n\to\infty} L_{n,a}$, $\lim_{n\to\infty} R_{n,a}$, and

$$\lim_{n\to\infty} L_{n,a} = L_a, \quad \lim_{n\to\infty} R_{n,a} = R_a, \quad a \in (0, 1], \tag{30}$$

where (L_a, R_a), $a \in (0,1]$ is the unique solution of the analogous system (19), which is

$$L_a = \frac{A_{l,a} + \sum_{i=0}^{k} a_{i,l,a} L_a}{B_{r,a} + \sum_{i=0}^{k} b_{i,r,a} R_a}, \quad R_a = \frac{A_{r,a} + \sum_{i=0}^{k} a_{i,r,a} R_a}{B_{l,a} + \sum_{i=0}^{k} b_{i,l,a} L_a}. \tag{31}$$

Since the fuctions $L_{i,a}, R_{i,a}$, $i = -k, -k+1, \dots, 0$, $A_{l,a}, A_{r,a}, B_{l,a}, B_{r,a}, a_{i,l,a}$, $a_{i,r,a}$, $b_{i,l,a}$, $b_{i,r,a}$, $i = 0, 1, \dots, k$, $a \in (0,1]$ are left-continuous, from the analogous relations (22) and (25) we can easily prove that L_a, R_a, $a \in (0,1]$ are left-continuous.

Since for any $a_1, a_2 \in (0,1]$ such that $a_1 \le a_2$, the analogous relation of (6) holds for $i = 1, 2, \dots$, then from (30) we have that

$$0 < L_{a_1} \le L_{a_2} \le R_{a_2} \le R_{a_1}.$$

In addition, from (8) and (31) we have that

$$R_a \le \frac{A_{r,a} + \sum_{i=0}^{k} a_{i,r,a} R_a}{B_{l,a}} \le \frac{L}{M} + \theta R_a, \tag{32}$$

where $\theta = \frac{\sum_{i=0}^{k} Q_i}{M}$. From (25) and (32) we take

$$L_a \le R_a \le \frac{L}{M(1-\theta)} = \mu, \quad a \in (0,1]. \tag{33}$$

Furthermore, from (8), (31), and (33) we have

$$L_a \ge \frac{K + \sum_{i=0}^{k} P_i L_a}{N + \sum_{i=0}^{k} T_i \mu}. \tag{34}$$

From (25) and (34) we have that

$$R_a \ge L_a \ge \frac{K}{N - \sum_{i=0}^{k} P_i + \mu \sum_{i=0}^{k} T_i} = l, \quad a \in (0,1]. \tag{35}$$

From (33) and (35) it is obvious that $\overline{\bigcup_{a\in(0,1]}[L_a,R_a]}$ is compact. Therefore, L_a, R_a determine a fuzzy number x such that $[x]_a=[L_a,R_a]$.

Finally, using (31) we see that x is the unique positive equilibrium of (1).

(c) Let x_n be a positive solution of (1) such that (10) holds. We can easily prove statement (c) using (6), (25), (30) and arguing as in part (iii) of Proposition 2 of [6]. □

References

[1] V.L.Kocic and G. Ladas, *Global Behavior of Nonlinear Difference Equations of Higher Order with Applications*, Kluwer Academic Publishers, Dordrecht, 1993.

[2] V.L. Kocic, G. Ladas, and I.W. Rodrigues, On rational recursive sequences, *J. Math. Anal. Appl.* **173** (1993), 127–157.

[3] S.A. Kuruklis and G. Ladas, Oscillations and global attractivity in a discrete delay logistic model, *Quart. Appl. Math.* **50** (1992), 227–233.

[4] H.T. Nguyen and E.A. Walker, *A First Course in Fuzzy Logic*, CRC Press, New York, London, Tokyo, 1997.

[5] G. Papaschinopoulos and J. Schinas, On Asympotic Stability of a System of Two Rational Recursive Sequences. *Pan. Math. J.* **3** (1993), 1–7.

[6] G. Papaschinopoulos and G. Stefanidou, Boundedness and asymptotic behavior of the solutions of a fuzzy difference equation, *Fuzzy Sets and Systems*, to appear.

[7] E.C. Pielou, *An Introduction to Mathematical Ecology*, Wiley Interscience, 1969.

[8] E.C. Pielou, *Population and Community Ecology*, Gordon and Breach, New York, 1974.

[9] C. Wu and B. Zhang, Embedding problem of noncompact fuzzy number space $E^{\sim}$, *Fuzzy Sets and Systems*, **105** (1999) 165–169.

The Gauss Hypergeometric Series with Roots Outside the Unit Disk

KATSUO TAKANO

Department of Mathematics, Faculty of Science
Ibaraki University
Mito, Japan

and

HIROMITSU OKAZAKI

Department of Mathematics, Faculty of Education
Kumamoto University
Kumamoto, Japan

Abstract It is shown that the Gauss hypergeometric series $F(-n, 2m; 2m + n + 1; z)$ has roots outside the unit disk.

Keywords Gauss hypergeometric series, gamma function, root, differential equation, probability distribution

AMS Subject Classification 30C15, 33C05, 60E07

1 Introduction

It is known in [10] that the normed conjugate product of gamma functions such as

$$\frac{2}{\pi}\Gamma(1-ix)\Gamma(1+ix) = \frac{2}{\pi}\frac{1}{\prod_{n=1}^{\infty}(1+x^2/n^2)} \tag{1}$$

is an infinitely divisible density. In the process showing the infinite divisibility of the probability distribution with density (1), a family of polynomials with roots outside the unit disk appeared. From the infinite divisibility of the above probability distribution and from numerical analysis of roots of the terminating hypergeometric series, we conjectured that the following density function consisting of the normed conjugate product of gamma functions is an infinitely divisible density:

$$c\left|\frac{\Gamma(m+ix)}{\Gamma(m)}\right|^2 = \frac{c}{\prod_{k=0}^{\infty}(1+x^2/(m+k)^2)} \qquad (m \in N) \tag{2}$$

(cf. [1. 6.1.25]) In this case the Gauss hypergeometric series $F(-n, 2m; 2m+n+1; z)$ appears in general form and it is much more complicated than the case $m = 1$. In this paper we show that the Gauss hypergeometric series $F(-n, 2m; 2m+n+1; z)$ has roots outside the unit disk. This result will play an important part in the proof of the infinite divisibility of the probability distribution with density function (2).

2 The hypergeometric series does not have roots on the unit circle

The hypergeometric series $F(-n, b; c; z)$ is defined by

$$F(-n, b; c; z) = \Sigma_{k=0}^{n} \frac{(-n)_k (b)_k z^k}{(c)_k k!}.$$

Here $(a)_k = a(a+1)(a+2)\cdots(a+k-1)$ denotes the Pochhammer symbol. It is often convenient for us to treat the polynomial $z^m F(-n, 2m; 2m+n+1; z)$ instead of $F(-n, 2m; 2m+n+1; z)$. Consider the unit circle $C: \ z = e^{it} \ (0 \le t \le 2\pi)$. Let

$$u(m, n; t) = \Re\, e^{imt} F(-n, 2m; 2m+n+1; e^{it}),$$

$$v(m, n; t) = \Im\, e^{imt} F(-n, 2m; 2m+n+1; e^{it}).$$

We have

$$u(m, n; t) = \Sigma_{k=0}^{n} \frac{(-n)_k (2m)_k}{(2m+n+1)_k} \frac{\cos(m+k)t}{k!} \tag{3}$$

and

$$v(m, n; t) = \Sigma_{k=0}^{n} \frac{(-n)_k (2m)_k}{(2m+n+1)_k} \frac{\sin(m+k)t}{k!}. \tag{4}$$

We note that the curve of $F(-n, 2m; 2m+n+1; e^{it})$ in the complex plane do not always make a Jordan curve when t runs through the interval $[0, 2\pi]$. It is known in [1] that the Gauss hypergeometric series $F(-n, 2m; 2m+n+1; z)$ is a solution of the hypergeometric equation, namely,

$$\begin{aligned} & z(1-z)\frac{d^2}{dz^2}F(-n, 2m; 2m+n+1; z) \\ + \quad & (2m+n+1-(2m-n+1)z)\frac{d}{dz}F(-n, 2m; 2m+n+1; z) \\ + \quad & 2mn\, F(-n, 2m; 2m+n+1; z) = 0. \end{aligned} \tag{5}$$

Lemma 2.1. *When $1 \le m$ and $1 \le n$ the functions $u(m, n; t)$ and $v(m, n; t)$ are solutions of the following differential equation:*

$$\sin\frac{t}{2}\, x''(t) - n\cos\frac{t}{2}\, x'(t) + m(m+n)\sin\frac{t}{2}\, x(t) = 0. \tag{6}$$

Proof. Let $h(z) = z^m F(-n, 2m; 2m+n+1; z)$. Then we obtain the following differential equation:

$$z^2(1-z)h''(z) + (n+1+(n-1)z)zh'(z) - m(m+n)(1-z)h(z) = 0. \quad (7)$$

If $z = e^{it}$ we obtain the following equation:

$$(1-e^{it})\frac{d^2h(e^{it})}{dt^2} + n(1+e^{it})i\frac{dh(e^{it})}{dt} + m(m+n)(1-e^{it})h(e^{it}) = 0 \quad (8)$$

and from the real part and imaginary part of the above equation we have two equations:

$$\begin{aligned} & (1-\cos t)u''(m,n;t) - n\sin t \cdot u'(m,n;t) \\ & +m(m+n)(1-\cos t)u(m,n;t) \\ = & -\sin t \cdot v''(m,n;t) + n(1+\cos t)v'(m,n;t) \\ & -m(m+n)\sin t \cdot v(m,n;t) \end{aligned} \quad (9)$$

and

$$\begin{aligned} & \sin t \cdot u''(m,n;t) - n(1+\cos t)u'(m,n;t) \\ & +m(m+n)\sin t \cdot u(m,n;t) \\ = & (1-\cos t)v''(m,n;t) - n\sin t \cdot v'(m,n;t) \\ & +m(m+n)(1-\cos t)v(m,n;t). \end{aligned} \quad (10)$$

Then we have

$$\begin{aligned} & \sin\frac{t}{2}\{\sin\frac{t}{2} \cdot u''(m,n;t) - n\cos\frac{t}{2} \cdot u'(m,n;t) \\ & +m(m+n)\sin\frac{t}{2} \cdot u(m,n;t)\} \\ = & -\cos\frac{t}{2}\{\sin\frac{t}{2} \cdot v''(m,n;t) - n\cos\frac{t}{2} \cdot v'(m,n;t) \\ & +m(m+n)\sin\frac{t}{2} \cdot v(m,n;t)\} \end{aligned} \quad (11)$$

and

$$\begin{aligned} & \cos\frac{t}{2}\{\sin\frac{t}{2} \cdot u''(m,n;t) - n\cos\frac{t}{2} \cdot u'(m,n;t) \\ & +m(m+n)\sin\frac{t}{2} \cdot u(m,n;t)\} \\ = & \sin\frac{t}{2}\{\sin\frac{t}{2} \cdot v''(m,n;t) - n\cos\frac{t}{2} \cdot v'(m,n;t) \\ & +m(m+n)\sin\frac{t}{2} \cdot v(m,n;t)\}. \end{aligned} \quad (12)$$

From the above equations we see that the functions $u(m,n;t)$ and $v(m,n;t)$ are solutions of the differential equation (6). □

By using the above Lemma 2.1 we can obtain the following result:

Theorem 2.2. *If* $1 \leq m$ *and* $1 \leq n$ *the Gauss hypergeometric series*

$$F(-n, 2m; 2m+n+1; z)$$

does not have roots on the unit circle.

Proof. In order to show that the Gauss hypergeometric series $F(-n, 2m; 2m+n+1; z)$ does not have roots on the unit circle we will show that the following relation

$$W(t) = u(m,n;t)v'(m,n;t) - u'(m,n;t)v(m,n;t) = c(1-\cos t)^n \qquad (13)$$

holds true, where c is a positive constant not depending on the variable t. If $t_0 = 0$ or 2π then $W(t_0) = 0$, while we have

$$u(m,n;t_0) = \frac{(n+1)_n}{(2m+n+1)_n}$$

by Vandermonde's formula. Let

$$\alpha(t) = 2^{-2n} \sin^{-2n}(\frac{t}{2})$$

and

$$\beta(t) = 2^{-2n} m(m+n) \sin^{-2n}(\frac{t}{2}).$$

Then the differential equation (6) can be written in the following form:

$$\{\alpha(t)x'(t)\}' + \beta(t)x(t) = 0$$

and the wronskian $W(t)$ can be expressed as the following form:

$$\alpha(t)W(t) = c_1 \ (const).$$

Therefore we obtain

$$W(t) = c_1 \, 2^{2n} \sin^{2n}(\frac{t}{2}).$$

To determine c_1 we take $t = \pi$. Then it implies that

$$W(\pi) = u(m,n;\pi)v'(m,n;\pi) - u'(m,n;\pi)v(m,n;\pi) = c_1 \, 2^{2n}.$$

We see that if m is an even positive integer then

$$\begin{aligned} & W(\pi) \\ = \ & \{\Sigma_{k=0}^{n} \frac{(-n)_k(2m)_k(-1)^k}{(2m+n+1)_k k!}\}\{\Sigma_{k=0}^{n} \frac{(-n)_k(2m)_k(m+k)(-1)^k}{(2m+n+1)_k k!}\} \\ = \ & \frac{(2m)_{n+1} 2^{2n-1}}{(2m+n+1)_n}. \end{aligned} \qquad (14)$$

Therefore we obtain $c = (2m)_{n+1} 2^{n-1}/(2m+n+1)_n$. In the same way we see that the constant c is given by the same formula as the above if m is an odd integer. □

3 The hypergeometric series has roots outside the unit disk

If $m = 1$ it is known in [9] that the roots of $F(-n, 2; n+3; z)$ appear outside the unit disk. If $n = 1$ the root of $F(-1, 2m; 2m+2; z)$ is $z_1 = (m+1)/m$ and if $n = 2$ the roots of $F(-2, 2m; 2m+3; z)$ are

$$z_1 = \frac{2m+4}{2m+1} + i\frac{1}{2m+1}\sqrt{\frac{3(m+2)}{m}}, \quad z_2 = \frac{2m+4}{2m+1} - i\frac{1}{2m+1}\sqrt{\frac{3(m+2)}{m}}$$

for all $m \in N$. These roots are outside the unit disk. Concerning the roots of the Gauss hypergeometric series $F(-n, 2m; 2m+n+1; z)$ for n larger than 2 we obtain the following result:

Theorem 3.1. *If $1 \leq m$ and $1 \leq n$ the Gauss hypergeometric series $F(-n, 2m; 2m+n+1; z)$ has roots outside the unit disk.*

Proof. Take a positive number σ in the interval (0,1] and fix it. The curve of the function $F(-n, 2m; 2m+n+1; \sigma e^{it})$ of the variable t is symmetric with respect to the real axis in the complex plane. In order to show that all the roots of the Gauss hypergeometric function $F(-n, 2m; 2m+n+1; z)$ are outside the unit disk, it suffices to show that $|F(-n, 2m; 2m+n+1; \sigma e^{it})|^2$ is positive for all t in the interval $[0, \pi]$. We make use of Watson's formula:

$$\begin{aligned} & F(-n, b; c; z)F(-n, b; c; Z) \\ = & \frac{(c-b)_n}{(c)_n} F_4[-n, b; c, 1-n+b-c; zZ, (1-z)(1-Z)] \end{aligned}$$

in Slater's book (cf. [7, formula (8.4.2)]). Let $b = 2m,\ c = 2m+n+1$. Then

$$\begin{aligned} & F_4[-n, 2m; 2m+n+1, -2n; zZ, (1-z)(1-Z)] \\ = & \Sigma_{r=0}^{n}\Sigma_{s=0}^{n}\frac{(-n)_{r+s}(2m)_{r+s}(zZ)^r[(1-z)(1-Z)]^s}{(2m+n+1)_r(-2n)_s r!s!}. \end{aligned}$$

Let $z = \sigma e^{it}$, $Z = \sigma e^{-it}$. Then $zZ = \sigma^2$ and $(1-z)(1-Z) = 1+\sigma^2-2\sigma\cos t$. Let $x = \sigma^2$ and $y = 1+\sigma^2-2\sigma\cos t$. From the above F_4 we obtain

$$\begin{aligned} & |F(-n, 2m; 2m+n+1; \sigma e^{it})|^2 \\ = & \frac{(n+1)_n}{(2m+n+1)_n}\Sigma_{r=0}^{n}\Sigma_{s=0}^{n}\frac{(-n)_{r+s}(2m)_{r+s}x^r y^s}{(2m+n+1)_r(-2n)_s r!s!} \end{aligned} \tag{15}$$

and it holds that

$$\begin{aligned} & |F(-n, 2m; 2m+n+1; \sigma e^{it})|^2 \\ = & \frac{(n+1)_n}{(2m+n+1)_n}\Sigma_{s=0}^{n}\frac{(-n)_s(2m)_s y^s}{(-2n)_s s!}\left\{\Sigma_{r=0}^{n-s}\frac{(-n+s)_r(2m+s)_r x^r}{(2m+n+1)_r r!}\right\}. \end{aligned}$$

We see that the right-hand side of the above equality is positive for all t in the interval $[0, \pi]$ since it holds that

$$
\begin{aligned}
&\Sigma_{r=0}^{n-s} \frac{(-n+s)_r (2m+s)_r x^r}{(2m+n+1)_r r!} \\
= \; & \frac{\Gamma(2m+n+1)}{\Gamma(2m+s)\Gamma(n-s+1)} \int_0^1 \tau^{2m+s-1}(1-\tau)^{n-s}(1-\tau x)^{n-s} d\tau.
\end{aligned}
$$

When $\sigma = 1$ by Vandermonde's formula we see that

$$
\begin{aligned}
&|F(-n, 2m; 2m+n+1; e^{it})|^2 \\
= \; & \frac{(n+1)_n}{(2m+n+1)_n} \Sigma_{s=0}^{n} \frac{(-n)_s (n+1-s)_{n-s} (2m)_s y^s}{(-2n)_s (2m+n+1)_{n-s} s!} \\
= \; & \Sigma_{s=0}^{n} \frac{(2m)_s}{(2m+n+1)_n (2m+n+1)_{n-s}} \binom{n}{s} \binom{2n-s}{n} (2(n-s))! y^s,
\end{aligned}
$$

where $y = 2(1 - \cos t)$. □

At last we mention the computational result of the above Theorem 3.1. The following curves are maps of the unit circle by the Gauss hypergeometric functions $F(-n, 2m; 2m+n+1; z)$ and the right side graphs are modified by a transformation of logarithm. These graphs show us that the curves do not enclose the origin. By the invariance of the region under holomorphic mapping we can see that these graphs suggest that all the roots of $F(-n, 2m; 2m+n+1; z)$ are outside the unit disk.

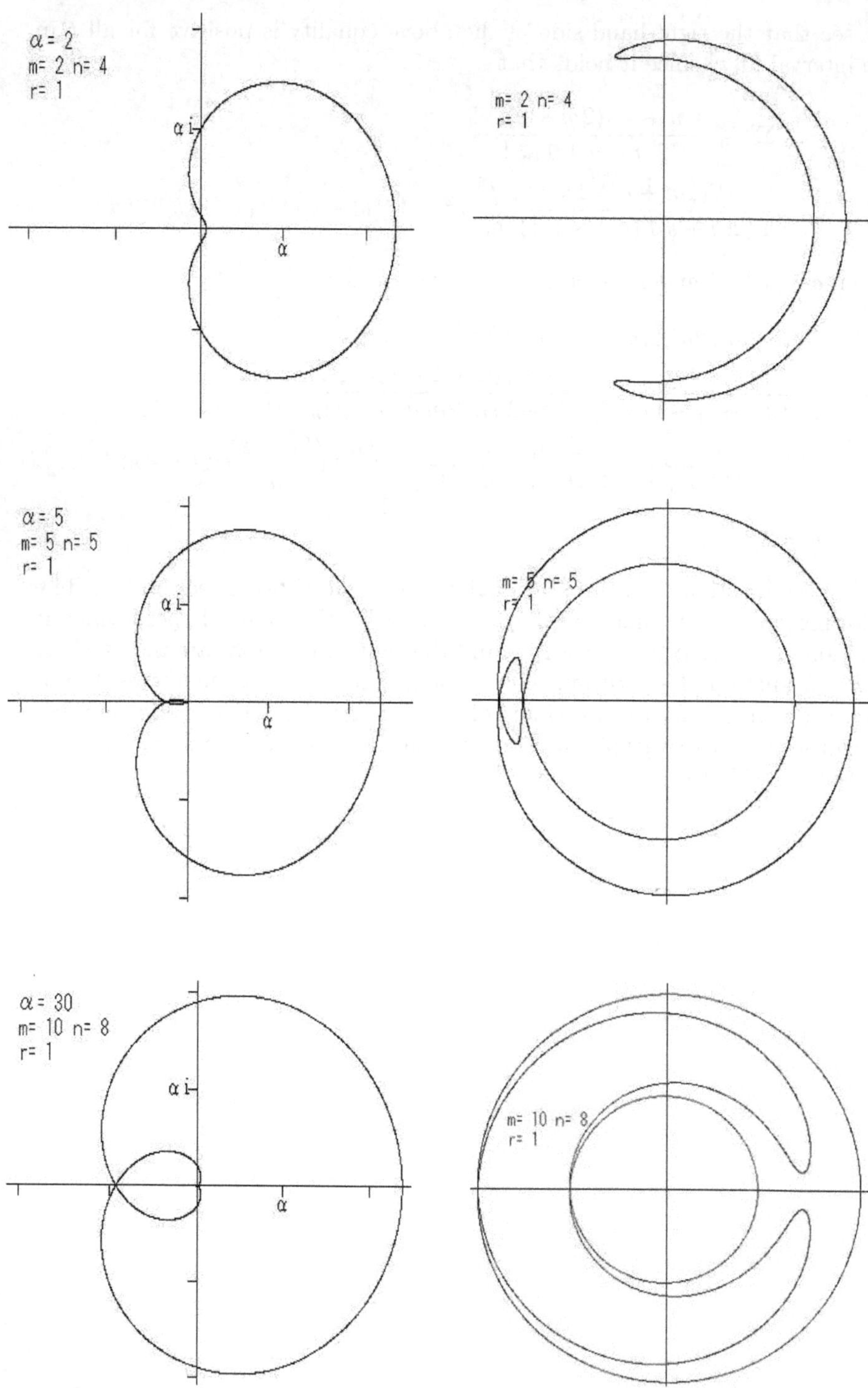
α = 2
m= 2 n= 4
r= 1
α i
α
m= 2 n= 4
r= 1
α = 5
m= 5 n= 5
r= 1
α i
α
m= 5 n= 5
r= 1
α = 30
m= 10 n= 8
r= 1
α i
α
m= 10 n= 8
r= 1

References

[1] M. Abramowitz and I.A. Stegun, *Handbook of Mathematical Functions*, New York, Dover, 1970.

[2] L. Bondesson, On the infinite divisibility of the half-Cauchy and other decreasing densities and probability functions on the nonnegative line, *Scand. Acturial J.* (1985), 225–247.

[3] E. Koelink and W. Van Assche, Orthogonal polynomials and special functions, *Lect. Notes in Math.* **1817**, Leuven 2002, Springer, 2003.

[4] W. Koepf, Hypergeometric summations, *Advanced Lecture in Mathematics*, Vieweg, Braunshweig/Wiesbaden, 1998.

[5] M.J. Goovaerts, L. D'Hooge, and N. De Pril, On the infinite divisibility of the product of two Γ-distributed stochastical variables, *Applied Mathematics and Computation* **3** (1977), 127–135.

[6] K. Sato, Class L of multivariate distributions and its subclasses, *J. Multivar. Anal.* **10** (1980), 207–232.

[7] L.J. Slater, *Generalized Hypergeometric Functions*, Cambridge University Press, 1966.

[8] F.W. Steutel, Preservation of infinite divisibility under mixing and related topics, *Math. Centre Tracts, Math. Centre, Amsterdam* **33** (1970).

[9] K. Takano, On a family of polynomials with zeros outside the unit disk, *International J. Comput. Numer. Anal. Appl.* **1** (2002), 369–382.

[10] K. Takano, On infinite divisibility of normed product of Cauchy densities, *J. Comput. Applied Math.* **150** (2003), 253–263.

Symbolic Dynamics Generated by an Idealized Time-Delayed Chua's Circuit

SANDRA VINAGRE[1]
Departamento de Matemática
Universidade de Évora
Évora, Portugal
and
J. SOUSA RAMOS
Departamento de Matemática
Instituto Superior Técnico
Lisbon, Portugal

Abstract In recent years, some attempts have been made to distinguish classes of boundary value problems (BVPs) for partial differential equations (PDEs) whose solutions are essentially determined by the iteration of a map (see, for example, [5], [8], [9], [10], [11], and [13]). The advantages are clear, since even the notion of chaos can be taken from discrete dynamical systems: we say that such a PDE system is chaotic if the map that determines its solution exhibits chaos as a discrete dynamical system. In this paper we consider the time-delayed Chua's circuit introduced in [7] the behavior of which is determined by properties of a one-dimensional map, [3], [7], [11] and [12]. We study this map in terms of symbolic dynamics that makes it possible to characterize the associated time evolution of the time-delayed Chua's circuit.

Keywords Symbolic dynamics, dynamical systems, one-dimensional maps, difference equations, boundary value problems, Chua's circuit

AMS Subject Classification 37B10, 39B12, 26A18, 35F15, 37E05, 39A11

1 Introduction

In recent years, some attempts have been made to distinguish classes of boundary value problems for partial differential equations whose solutions are essentially determined by the iteration of a map (see, for example, [5], [8], [9], [10], [11], and [13]). These classes consist mainly of problems for which the representation of the general solution is known. The reduction to a difference equation with continuous argument followed by the employment of the

[1]Supported by FCT (Portugal) through program POCTI.

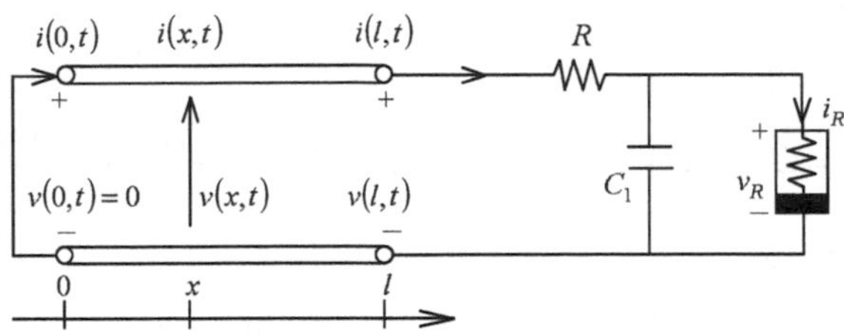

Figure 1: The time-delayed Chua's circuit.

properties of the one-dimensional map associated with the difference equation allows an insight into the properties of such chaotic solutions.

The advantages are clear, since even the notion of chaos can be taken from discrete dynamical systems: we say that such a PDE system is chaotic if the map that determines its solution exhibits chaos as a discrete dynamical system.

In this paper we consider the time-delayed Chua's circuit introduced in [7], which is an infinite-dimensional generalization of Chua's circuit, obtained by replacing the LC resonant circuit by a lossless transmission line of length l, terminated on its left, $x = 0$, by a short circuit, as shown in Figure 1.

This paper is organized as follows: in Section 2 we present the problem relating to [7]. In Section 3, we give a brief description of the kneading theory and we apply symbolic dynamical techniques to study the one-dimensional map associated with the difference equation and the solutions of difference equations, which are obtained from the class of boundary value problems by the reduction method.

With this approach we codify the levels of amplitude that occur in the solutions of PDE using the symbolic dynamics; see Theorem 3.1, Example 3.2, Figure 5, and the results obtained in [6].

2 Reduction of BVP to DE

Following [7], let the transmission line in Figure 1 be defined by the linear partial differential equations:

$$\frac{\partial v(x,t)}{\partial x} = -L\frac{\partial i(x,t)}{\partial t}, \tag{1}$$

$$\frac{\partial i(x,t)}{\partial x} = -C\frac{\partial v(x,t)}{\partial t}, \tag{2}$$

where L and C denote the inductance and the capacitance per unit length of the transmission line. The boundary conditions are given, respectively, at

$x = 0$ and $x = l$ by

$$v(0,t) = 0, \tag{3}$$

$$i(l,t) = G(v(l,t) - E - Ri(l,t)) + C_1 \frac{\partial (v(l,t) - Ri(l,t))}{\partial t}, \tag{4}$$

where G is defined by

$$G(u) = \begin{cases} m_0 u & \text{if} \quad |u| \le 1, \\ m_1 u - (m_1 - m_0)\ \operatorname{sgn} u & \text{if} \quad |u| \ge 1. \end{cases} \tag{5}$$

The general solutions of Equations (1) and (2) are of the form

$$v(x,t) = \alpha\left(t - \frac{x}{\nu}\right) - \alpha\left(t + \frac{x}{\nu}\right), \tag{6}$$

$$i(x,t) = \frac{1}{Z}\left[\alpha\left(t - \frac{x}{\nu}\right) + \alpha\left(t + \frac{x}{\nu}\right)\right], \tag{7}$$

where $\nu = \sqrt{1/LC}$ is the velocity of the incident and reflected waves, $Z = \sqrt{L/C}$ is the characteristic impedance of the transmission line and α is an arbitrary C^1-smooth function. This boundary value problem (1)–(4) is a system of two linear partial differential equations with a nonlinear boundary condition at $x = l$. Substituting (6) and (7) into the boundary condition (4) with $C_1 = 0$ and introducing the new variables $\tau = t\nu/(2l) - 1/2$ and $\beta(\tau) = \alpha(2l\tau/\nu)$, one obtains the difference equation with the continuous argument,

$$\beta(\tau + 1) = f(\beta(\tau)). \tag{8}$$

The function f is a piecewise-linear single-valued or multivalued function defined by

$$f(\beta) = A_k \beta - B_k, \quad \text{where} \quad \beta \in I_k, \quad k = \pm 1, 0 \tag{9}$$

and

$$\begin{aligned} A_k &= -1 + q_k, \\ B_k &= \frac{q_k}{2}\left[E + k\left(1 - \frac{m_0}{m_k}\right)\right], \quad q_k = \frac{2Z}{\dfrac{1}{m_k} + R + Z}, \\ I_0 &= \left\{\beta : \left|\beta - \frac{E}{2}\right| \le |\delta|\right\}, \quad I_{\pm 1} = \left\{\beta : \pm\left(\beta - \frac{E}{2}\right) > \delta\right\}, \end{aligned} \tag{10}$$

with $m_{-1} = m_{+1}$ and $\delta = m_0 Z / q_0$.

The initial values of voltage $v(x,0) = v_0(x)$ and current $i(x,0) = i_0(x)$ implies, for the difference equation (8), the following initial conditions:

$$\varphi(\tau) = \begin{cases} \dfrac{v_0(-y) + Z i_0(-y)}{2}, & y = l(1 + 2\tau) \quad \text{if} \quad -1 \le \tau < -\dfrac{1}{2}, \\ \dfrac{-v_0(y) + Z i_0(y)}{2}, & y = l(1 + 2\tau) \quad \text{if} \quad -\dfrac{1}{2} \le \tau < 0. \end{cases} \tag{11}$$

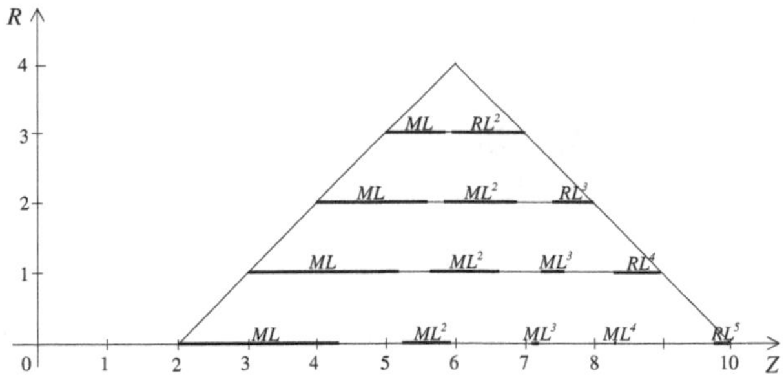

Figure 2: Some examples of the asymptotic behavior in terms of blocks of symbols of the $1D$ map for $E < 1 + Rm_0$, where $m_0 = -1/10$ and $m_1 = 1/2$.

Thus the time evolution of the time-delayed Chua's circuit with $C_1 = 0$ is governed by a scalar nonlinear difference equation with continuous argument (8) (see [7]). The qualitative behavior of this equation is determined by the properties of the one-dimensional (1D) map

$$\beta \longmapsto f(\beta), \tag{12}$$

where f is defined in (9)-(10).

From now on we will write $f_{m_0,m_1,R,Z}$ instead of f for the map in (9) and φ_c instead of φ for the initial function in (11).

3 Symbolic coding for levels of solutions of the BVP

Using the symbolic itineraries of the two turning points c_1 and c_2 of the bimodal map $f_{m_0,m_1,R,Z}$ in the $I = [a,b]$, we assign the symbols L (left), M (middle) and R (right) to each point x of the subintervals of monotonicity $[a, c_1)$, (c_1, c_2) and $(c_2, b]$, respectively, and the symbols A and B to the turning points c_1 and c_2. It is called the address of x and it is denoted $ad(x)$. By doing this, we get a correspondence between orbits of points and symbolic sequences S of the alphabet $\mathcal{A} = \{L, A, M, B, R\}$, the itinerary by the map $f_{m_0,m_1,R,Z}$,

$$it_{f_{m_0,m_1,R,Z}}(x) := \mathrm{ad}(x)\,\mathrm{ad}(f_{m_0,m_1,R,Z}(x))\,\mathrm{ad}\left(f^2_{m_0,m_1,R,Z}(x)\right)\ldots.$$

If we denote by n_M the frequency of the symbol M in a finite subsequence of S we can define the M-parity of this subsequence according to whether n_M is even or odd. In what follows we define an order relation in $\mathcal{A}^{\mathbb{N}} = \{L, A, M, B, R\}^{\mathbb{N}}$ that depends on the M-parity.

Let V be a vector space of three dimensions defined over the rationals having as a basis the formal symbols $\{L, M, R\}$, then to each sequence of

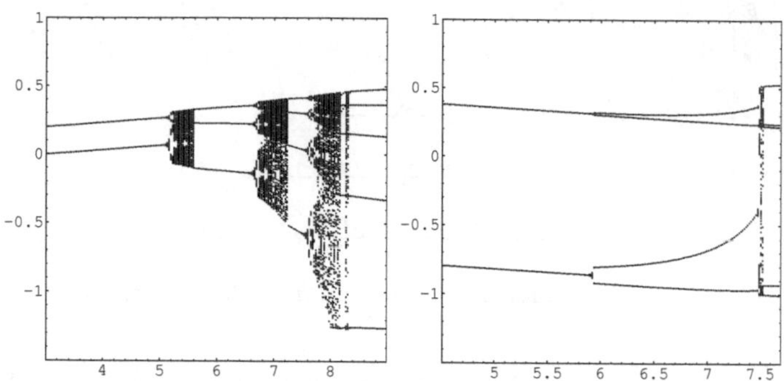

Figure 3: The bifurcation diagram of the map $f_{m_0,m_1,R,Z}$, with a) $m_0 = -1/10$, $m_1 = 1/2$, $R = 1$, $E = 0.8$, $Z \in [3,9]$, and b) $m_0 = -1/10$, $m_1 = 0.439\ldots$, $R = 2.295\ldots$, $E = 0.47\ldots$, $Z \in [4.5, 7.7]$.

symbols $S = S_1 S_2 \ldots S_j \ldots$, we can associate a sequence $\theta = \theta_0 \ldots \theta_j \ldots$ of vectors from V, setting $\theta_j = \prod_{i=0}^{j-1} \epsilon(S_i) S_j$ with $j > 0, \theta_0 = S_0$ when $i = 0$ and $\epsilon(L) = -\epsilon(M) = \epsilon(R) = 1$, where we associate the vector $(L+M)/2$ and $(M+R)/2$ to the symbols corresponding to the turning points c_1 and c_2. Thus $\epsilon(A) = \epsilon(B) = 0$. Choosing then a linear order in the vector space V in such a way that the base vectors satisfy $L \prec M \prec R$, we are able to order the sequence θ lexicographically, that is, $\theta \prec \bar{\theta}$ iff $\theta_0 = \bar{\theta}_0, \ldots, \theta_{j-1} = \bar{\theta}_{j-1}$ and $\theta_j \prec \bar{\theta}_j$ for some integer $j \geq 0$. Finally, introducing t as an undetermined variable and taking θ_j as the coefficients of a formal power series θ (invariant coordinate) we obtain $\theta = \theta_0 + \theta_1 t + \ldots = \sum_{j=0}^{\infty} \theta_j t^j$.

The sequences of symbols corresponding to periodic orbits of the turning points c_1 and c_2 are $P = AP_1P_2 \ldots P_{p-1}A \ldots$ and $Q = BQ_1Q_2 \ldots Q_{q-1}B \ldots$. The realizable itineraries of the turning points c_1 and c_2 for the maps previously defined are called by kneading sequences. The kneading invariant of the map $f_{m_0,m_1,R,Z}$ is the pair of kneading sequences, itineraries of the image of each turning point (see [4]),

$$\mathcal{K}(f_{m_0,m_1,R,Z}) := \left(it_{f_{m_0,m_1,R,Z}}(f_{m_0,m_1,R,Z}(c_1)), it_{f_{m_0,m_1,R,Z}}(f_{m_0,m_1,R,Z}(c_2))\right).$$

The significance of this symbolic topological invariant was made clear when Guckenheimer [2] presented a classification theorem of modal maps in the interval based on its kneading invariants, showing how close it is from its topological classification.

In Figure 2, a subset of elements of the period-adding phenomenon (see [12]) is illustrated and it corresponds to the sequence of blocks $ML^i, i = 1, 2, \ldots$. Between regions determined by attracting cycles of low periods, k, (which follow a rule like a period-adding phenomenon and which are given in terms of symbolic dynamics by the block of symbols ML^{k-1}, asymptotically

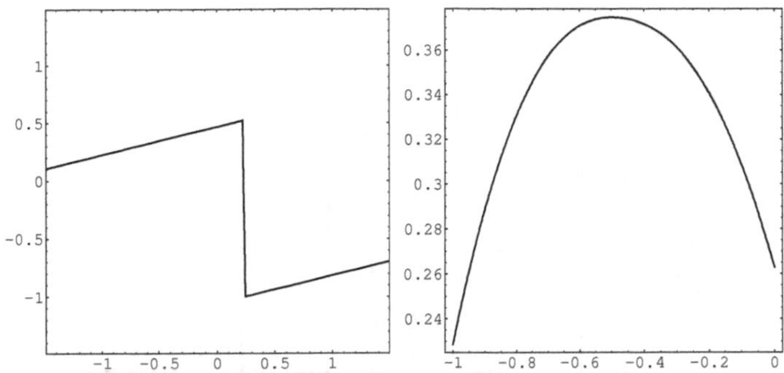

Figure 4: *a*) The graph of the map $f_{m_0,m_1,R,Z}$, with $m_0 = -1/10$, $m_1 = 0.439\ldots$, $R = 2.295\ldots$, $E = 0.47\ldots$, $Z = 7.5$. *b*) The graph of the initial function φ_c, with $v_0(x) = cx^3 + 3cx^2/2$, $i_0(x) = cx^3 + 3cx^2/2 + 1/10$, $c = -0.014\ldots$.

realizable by the two turning points) exist "stochastic turbulence" regions (where the map has a.c.i.m.).

Considering the set of pairs of admissible kneading sequences with the lexicographic order given above, we can enumerate the different topological types of the maps $f_{m_0,m_1,R,Z}$. The combinatorial complexity and the order of occurrence of the bifurcations are codified and completely characterized by the lexicographic order, depending on the M-parity, of the pairs of kneading sequences. The enumeration of its topological types, because it is a bimodal map, is completely characterized by two topological invariants, one of them is the topological entropy, which can be calculated by the zero of the kneading determinant (see [4]), and we will study the other invariant in a later work. It will be interesting to study how these invariants depend on the parameters R and Z.

Now we present the main result of this paper, which is a coding of the levels of maxima and minima of the solutions using the symbolic dynamics. In the theorem we suppose that the orbits of turning points are periodic, but the theorem can be extended to aperiodic orbits.

Theorem 3.1. *Let $\{\lambda_i\}_{i=1,\ldots,p}$ be the orbits of the turning points c_1 and c_2 of the map $f_{m_0,m_1,R,Z}$ and let $\{\gamma_j\}_{j=1,\ldots,q}$ be the orbit of the image by φ_c of the turning point c_3 of the initial function φ_c. Then the levels of maxima and minima of the solutions of the problem (8), (11) (and of the boundary value problem (1)–(4)) is the set $\Omega = \{\lambda_i, \gamma_j\}_{i=1,\ldots,p;j=1,\ldots,q}$.*

Sketch of the proof: Note that the map $\beta(\tau)$ restricted to the interval $[k, k+1]$ corresponds to the graph of the map $f_{m_0,m_1,R,Z}(\varphi_c)$. Then it follows from the derivative of composite functions that the images of turning points determine the local maxima and minima. □

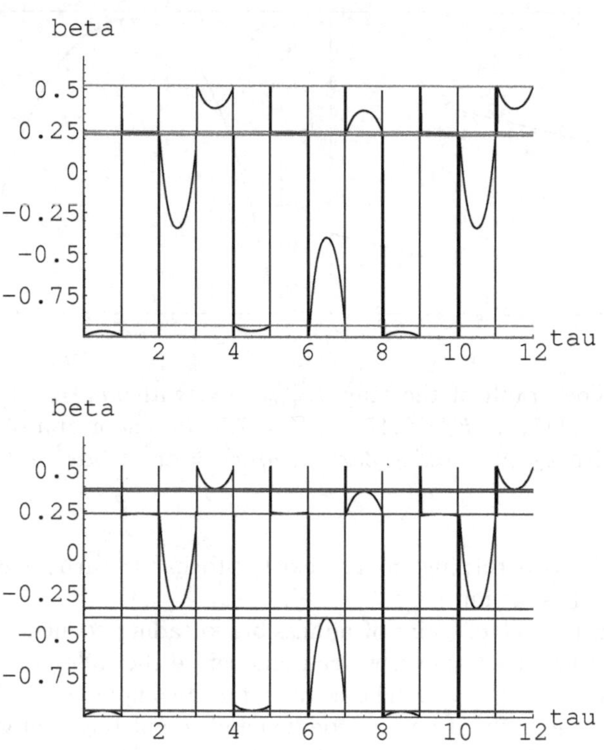

Figure 5: The graph of the solution of the problem (8), (11), with $v_0(x) = cx^3 + 3cx^2/2$, $i_0(x) = cx^3 + 3cx^2/2 + 1/10$, $c = -0.014\ldots$, $m_0 = 1/10$, $m_1 = 0.439\ldots$, $R = 2.295\ldots$, $E = 0.47\ldots$, $Z = 7.5$, $l = 1$, *a*) the λ_i levels, and *b*) the γ_j levels.

Example 3.2. *To illustrate the previous results, we consider* $v_0(x) = cx^3 + 3cx^2/2$, $i_0(x) = cx^3 + 3cx^2/2 + 1/10$ *and we fix* $c = -0.014\ldots$, $m_0 = -1/10$, $m_1 = 0.439\ldots$, $R = 2.295\ldots$, $Z = 7.5$, $E = 0.47\ldots$ *and* $l = 1$. *For these values the map* $f_{m_0,m_1,R,Z}$ *has a period* 6 *orbit given by* $\lambda_1 = 0.225\ldots$, $\lambda_2 = 0.525\ldots$, $\lambda_3 = -0.928\ldots$, $\lambda_4 = 0.245\ldots$, $\lambda_5 = -0.995\ldots$ *and* $\lambda_6 = 0.229\ldots$, *whose kneading sequence is* $(RLBLMA, LMARLB)^\infty$; *and the orbit of the image by* φ_c *of the turning point of the initial function* φ_c *is given by* $\gamma_1 = -0.964\ldots$, $\gamma_2 = 0.237\ldots$, $\gamma_3 = -0.341\ldots$, $\gamma_4 = 0.388\ldots$, $\gamma_5 = -0.961\ldots$, $\gamma_6 = 0.237\ldots$, $\gamma_7 = -0.397\ldots$, $\gamma_8 = 0.374\ldots$, $\gamma_9 = -0.964\ldots$, $\gamma_{10} = 0.237\ldots$, $\gamma_{11} = -0.337\ldots$ *and* $\gamma_{12} = 0.389\ldots$ *(see Figures 4, 5).*

Acknowledgment: Special thanks to Professor Alexander Sharkovsky for his comments and suggestions.

References

[1] Chen, G., Hsu, S.B. and Zhou, J., Nonistropic spatiotemporal chaotic vibration of the wave equation due to mixing energy transport and van der Pol boundary condition, *Int. J. Bifurcation and Chaos* **12** (2002), 535–559.

[2] Guckenheimer, J., Sensitive dependence on initial conditions for one-dimensional maps, *Comun. Math. Phys.* **70** (1979), 133–160.

[3] Maistrenko, Yu.L., Maistrenko, V.L., Vikul, S.I., and Chua, L.O., Bifurcations of attracting cycles from time-delayed Chua's circuit, *Int. J. Bifurcation and Chaos* **5** (1995), no. 3, 653–671

[4] Milnor, J. and Thurston, W., On iterated maps of the interval, in: J.C. Alexander (ed.) *Proceedings Univ. Maryland 1986-1987.* Lect. Notes in Math. **1342**, 465–563, Springer-Verlag, Berlin, New York, 1988.

[5] Romanenko, E.Yu. and Sharkovsky, A.N., From boundary value problems to difference equations: a method of investigation of chaotic vibrations, *Int. J. Bifurcation and Chaos* **9** (1999), 1285–1306.

[6] Severino, R., Sharkovsky, A.N., Sousa Ramos, J. and Vinagre, S., Symbolic dynamics in boundary value problems, *Grazer Math. Ber.* **346** (2004), 393–402.

[7] Sharkovsky, A.N., Ideal turbulence in an idealized time-delayed Chua's circuit, *Int. J. Bifurcation and Chaos* **4** (1994), 303–309.

[8] Sharkovsky, A.N., Universal phenomena in some infinite-dimensional dynamical systems, *Int. J. Bifurcation and Chaos* **5** (1995), 1419–1425.

[9] Sharkovsky, A.N., Iteration of continuous functions and dynamics of solutions for some boundary value problems, *Annales Mathematicae Silesianae* **13** (1999), 243–255.

[10] Sharkovsky, A.N., Difference Equations and Boundary Value Problems, in: *New Progress in Difference Equations*, Proceedings of the ICDEA 2001, Taylor & Francis, (2003), 3–22.

[11] Sharkovsky, A.N., Deregel, Ph. and Chua, L.O., Dry turbulence and period-adding phenomena from 1-D map with time delayed, *Int. J. Bifurcation and Chaos* **5** (1995), 1283–1302.

[12] Sharkovsky, A.N., Maistrenko, Yu.L., Deregel, Ph. and Chua, L.O., Dry turbulence from a time delayed Chua's circuit, *J. Circuits, Systems and Computers* **3** (1993), 645–668.

[13] Sharkovsky, A.N., Maistrenko, Yu.L. and Romanenko, E.Yu., *Difference Equations and Their Applications*, Kluwer Academic Publishers, 1993.

Author Index

Milton Keynes UK
Ingram Content Group UK Ltd.
UKHW040445071024
449327UK00020B/996

9 780367 392932